Tomas Cipra

Time Series in Economics and Finance

Springer

Tomas Cipra
Faculty of Mathematics and Physics
Charles University
Prague, Czech Republic

ISBN 978-3-030-46346-5 ISBN 978-3-030-46347-2 (eBook)
https://doi.org/10.1007/978-3-030-46347-2

Mathematics Subject Classification: 62M10, 91B84, 62M20, 62P20, 91B25, 91B30

This Springer imprint is published by the registered company Springer Nature Switzerland AG.
The registered company address is: Gewerbestrasse 11, 6330 Cham, Switzerland

Contents

Chapter 1
Introduction

Most data in economics and finance are observed in time (sometimes even online in real time) so that they have the character of time series. This monograph presents methods currently used for analysis of data in this context. Such methods are available not only in many monographs, textbooks, or papers but also in various journals or working papers, case studies, or guides to the corresponding software systems. This text tries to bring together as many methods as possible to cover the most recommended instruments for analysis and prediction of dynamic data in economics and finance.

The objective of this book is the practical applicability. Therefore, it centers on the description of methods used in practice (both simple and complex ones from the computational point of view). Their derivation is often concise (if any, particularly in more complicated cases), but one always refers to easily available sources. In any case, a lot of numerical examples illustrate the theory by means of real data which are usually chosen to be characteristic for the presented methodology.

Selected parts of the text are suitable for university programs (undergraduate, graduate, or doctoral) concerning econometrics or calculation finance as study, training, or reference materials. Moreover, due to the complete survey of actual methods and approaches the book can serve as a reference text in research work. On the other hand, it can also be recommended for people dealing with analysis of data in economics and finance (banks, exchanges, energetic planning, currency and commodity markets, insurance, statistical offices, demography, and others).

The presented material requires mostly the application of suitable software. Fortunately, the corresponding programs are easily available since they can be found in libraries of common statistical or financial software systems (R Statistical Software, MATLAB, EViews, and others can be recommended). There are several reasons supporting ready-made software: (1) calculations (e.g., in Excel) are usually troublesome (particularly for users with superficial knowledge of programming); (2) software manuals are usually helpful in various individual situations, and, moreover, the parameters of programs are preset as default values suitable for the immediate (routine) application; and (3) when browsing through the offer of

T. Cipra, *Time Series in Economics and Finance*,
https://doi.org/10.1007/978-3-030-46347-2_1

software systems, one discovers other methods or modifications which can be useful for the solved problem. On the other hand, the potential user should not be only a software consumer sharing all drawbacks of the given software product. Moreover, the qualified users should be capable of interpreting the computer outputs in a proper way since they understand principles of the chosen methods.

The monograph consists of several parts divided into particular chapters:

Part I (*Subject of time series*, Chap. 2) deals with the subject of time series which are looked upon as trajectories of random processes.

Part II (*Decomposition of economic time series*, Chaps. 3–5) is devoted to the classical approach decomposing economic time series to trend, periodic (seasonal and cyclical), and residual components. Some of more advanced methods are also addressed, e.g., tests of periodicity or randomness.

Part III (*Autocorrelation methods for univariate time series*, Chaps. 6 and 7) summarizes so-called Box–Jenkins methodology based on (linear) ARMA models and their modifications (ARIMA, seasonal ARMA, long memory processes) for univariate time series. Some more actual topics are also mentioned in this context (e.g., information criteria or tests of unit root). Finally, dynamic regression models are presented in Part III including distributed lag models.

Part IV (*Financial time series*, Chaps. 8–11) confines itself to financial time series which require special (namely nonlinear) models and instruments due to the typical volatility of financial data. Models nonlinear in mean and in variance are distinguished including tests of linearity and duration modeling. Further, Part IV addresses the modeling of financial assets by means of diffusion processes including Black–Scholes formula and modeling of the term structure of interest rates. Chapter 11 presents a very actual topic of risk measures (value at risk and others). Extreme value theory is also mentioned in this context, namely block maxima and threshold excesses.

Part V (*State space models of time series*, Chaps. 12–14) concludes the monograph considering the multivariate time series. At first, the popular vector autoregression (VAR) model is presented including tests of causality, impulse response, variance decomposition, cointegration, and EC models. The multivariate volatility modeling is also described including multivariate EWMA and GARCH models with a practical application for conditional value at risk. Finally, the (multivariate) state space models as the background of Kalman filtering are discussed including the state space model approach to exponential smoothing.

Some parts of this monograph serve as lecture notes for courses of time series analysis and econometrics at the Faculty of Mathematics and Physics of Charles University in Prague (it is also the reason why some real data used in practical examples are taken from the Czech economics and finance).

Acknowledgment The author thanks for various forms of help to Dr. Radek Hendrych. The research work contained in the monograph was supported by the grant 19-28231X provided by the Grant Agency of the Czech Republic.

Part I
Subject of Time Series

Chapter 2
Random Processes

2.1 Random Processes as Models for Time Series

Data typical for economic and financial practice are *time data*, i.e., values of an economic variable (or variables in multivariate case) observed in a time interval with a given frequency of records (each trading day, in moments of transactions, monthly, etc.). The *frequency of records* is understood either as the lengths of intervals between particular observations (e.g., calendar months) or the *regularity* of observations (e.g., each trading day). As to the regularity, financial data are often irregularly observed (*irregularly spaced data*), e.g., the stock prices in stock exchanges are quoted usually in moments of transactions from the opening to closing time of trading day, the frequency of transactions being usually lower in the morning after opening, during the lunch time, and later in the afternoon before closing (a possible approach in such a situation assigns the closing or prevailing price to this day). The important property of time data is the fact that they are ordered chronologically in time.

The term *time series* denotes any sequence of data $y_1, \ldots, y_n$ ordered chronologically in time. It could justify a simplifying view looking on a time series as a set of numbers ordered in time (historically it was the case, e.g., for astronomic observations). However, a very important aspect of time series is not only their dynamics but also their randomness. In order to be adequate, the analysis of time series should apply such models that are based on stochastic principles (i.e., on probability theory) and are capable of generating time sequences similar from the stochastic point of view to the trajectory that we just observe. Such models are denoted as *random processes* and can be looked on as specific algorithms based on random number generators. The knowledge of the algorithm that generated the observed time series as an output among many realizations may be highly useful for examining our specific data.

Random process (or *stochastic process*) $\{Y_t, t \in T\}$ is a set (or family) of random variables in the same probability space $(\Omega, \mathfrak{I}, \mathrm{P})$ indexed by means of values t from

T. Cipra, *Time Series in Economics and Finance*,
https://doi.org/10.1007/978-3-030-46347-2_2

T ($T \subset \mathrm{R}$), where t is interpreted as time. According to the form of the index set T, which is a subset of real line R, one distinguishes:

- *Random process in continuous time*: T is an interval in real line, e.g., $T = \langle 0, \infty)$, i.e., $\{Y_t, t \geq 0\}$.
- *Random process in discrete time*: T is formed by discrete real values, e.g., $T = \mathrm{N}_0$, i.e., $\{Y_0, Y_1, Y_2, \ldots\}$.

According to the states of random variables Y_t (i.e., according to the state space S) one also distinguishes:

- *Random process with discrete states*: e.g., *counting process* $Y_t \in \mathrm{N}_0$ for all $t \in T$ that registers the number of specified events in time.
- *Random process with continuous states*: e.g., *real process* $Y_t \in \mathrm{R}$ or *nonnegative process* $Y_t \in \langle 0, \infty)$ for all $t \in T$.
- *Multivariate random process*: Y_t is an m-variate random vector $\mathbf{Y}_t$ for all $t \in T$.

In any case, one can observe only *trajectories* (*realizations*) of random processes. Such a trajectory arises by a choice of an elementary event $\omega \in \Omega$ and is a deterministic function of time $\{Y_t(\omega), t \in T\}$ observable due to this specific choice. One denotes trajectories simply as $\{y_t, t \in T\}$ in discrete time and $\{y(t), t \in T\}$ in continuous time.

Remark 2.1 Unfortunately in the literature (and also in this text), it is common that the term *time series* is interpreted sometimes as the trajectory and sometimes as the random process. The real meaning follows from the context.

◇

2.2 Specific Problems of Time Series Analysis

The general objective of time series analysis including applications in economics and finance is to construct an adequate model of the underlying random process. Such a model usually enables:

- *to understand mechanism* (or algorithm) that has generated the observed time series
- *to test hypotheses* of a priori expectations and conjectures in a statistically credible way (e.g., whether the stock markets show a long-term growth)
- *to predict* (or to forecast or to extrapolate) the future development of the system (e.g., which interest or currency rates one can expect the next month with a given confidence)
- *to control and optimize* the dynamics of a system including adjustment of parameters, initial conditions, self-regulations, and the like (e.g., how to set up parameters of pension reform)

In any case, the data in the form of time series have a lot of specific features. It can help to analyze such data files, but on the other hand, it can cause complications that must be overcome by suitable procedures and adjustments. The next section will present examples of specific problems that are typical for time series analysis.

2.2.1 *Problems of Economic and Financial Data Observed in Time*

Economic and financial data observed in time typically feature problems implied just by their time character:

2.2.1.1 Problems Due to Choice of Observation Time Points

Time series in discrete time that prevail in economics and finance usually arise by the following ways:

- They are *discrete by nature* (e.g., daily interbank LIBOR rates).
- One *discretizes time series in continuous time* (e.g., closing quotations on stock exchanges assigned to the ends of particular trading days in the context of continuous trading).
- One *accumulates* (*aggregates*) values over given time intervals (e.g., accumulated sums of insurance benefits paid out in particular quarters); often one produces *averages* instead of aggregates.

In some cases, one is not allowed to select time points of observations oneself. However, if such a possibility exists, one should pay careful attention to it. It often means that one must find a trade-off among contradictory requirements: for instance, on one side due to numerical complexity not to use too high density of records of a continuous process (see, e.g., *ultra-high-frequency data UHFD* in finance) and on the other side not to apply so scarce data that one is not capable of identifying some characteristic features of the given process (e.g., if we are interested in seasonal fluctuations we must dispose of several observations during each year at least). As the distance between neighboring observations is concerned, it is common to observe data regularly in equidistant time points. On the contrary, in finance there are not unusual *irregularly spaced data* due to irregularities in market trading (see Sect. 2.1).

2.2.1.2 Problems Due to Calendar

The nature is responsible for a minor part of problems caused by the calendar (e.g., the number of days of one solar year is not integer, various geographic zones require time shifts). However, the major part of calendar problems is due to human conventions due to which we have, e.g.,

- Different lengths of calendar months
- Four or five weekends monthly
- Different numbers of working or trading days monthly
- Moving holidays (e.g., Easter once in the first quarter and next time in the second one)
- Wintertime or summertime

Such irregularities must be taken into account in an adequate way, e.g., differences in security trading or in quality of produced cars at the beginning, middle, and end of particular weeks, different times to maturities quoted on some security exchanges as the third Friday of particular months, and others. In practice, one usually applies simple methods eliminating these undesirable phenomena. Several examples follow:

- *Calendar conventions* are common in the framework of simple interest and discount models (e.g., the calendar Euro-30/360 introduces months with 30 days and years with 360 days).
- If comparing monthly productions of some products (cars), the volumes are adjusted using so-called *standard month* with 30 days: in such a case, one should multiply the January production by coefficient 30/31, the February production in the common year by 30/28, and in the leap year by 30/29, etc. Similarly if comparing securities traded monthly, one should multiply the January volume by (21/real number of January trading day), the February volume by (21/real number of February trading day), etc., as the average annual number of trading days is 252, i.e., 21 monthly.
- Some short-term calendar irregularities can be eliminated by means of accumulation. For instance, if it suffices to analyze data accumulated annually instead of original quarterly data, then some calendar problems (e.g., seasonal fluctuations, moving Easter, and others) can be reduced in this natural way.

In addition to calendar problems, one must frequently face such irregularities in time series that are consequences of operation risk (blackouts, breakdowns of web, failures of human factor including frauds, etc.). The irregularities of this type are classified as *outliers*, and statistical methods for time series with outliers should be *robustified* to become insensitive to such outlying values. Another type or irregularities are jumps in consequence of *interventions* (it can be, e.g., successful advertising campaign, decision of bank council on decrease of key interest rates, new legislative, and so on).

2.2.1.3 Problems Due to Length of Time Series

The length of time series is the number n of observations of the given time series (not the time range between the beginning and the end of time series). Therefore, e.g., the monthly time series over 10 years has the length of 120. It is logic that the volume of information available for analysis increases with the increasing length of time series. However, the length of time series is not a unique measure of information contained in the time series (e.g., the doubling of time series length by halving the original time intervals between neighboring points of observations does not mean usually the doubling of information on this time series): one must consider also the inner structure of given time series.

As the length of time series is concerned, usually a reasonable trade-off is necessary in practice. On one side, some time series methods require a sufficient length of series (e.g., the routine application of Box–Jenkins methodology is not recommended for time series shorter than 50). On the other hand, characteristic features of long time series usually change in time so that the construction of adequate model becomes more complex with increasing length of time series. Similarly, the typical problem in longer time series originates due to the fact that the measurements in the beginning of the given time series need not be comparable with the ones in its end, e.g., due to inflation, price growth, technical development, and the like. In such a case, one should adjust data by means of a suitable index (in practice, it can be not only the inflation rate but also the salary growth for time series used in formulas of pay-as-you-go pension systems and the like).

2.2.2 *Methodological Problems*

The choice of suitable time series method depends on various factors, e.g.,

- *Objective of analysis*, mainly the identification of generating model, the hypothesis testing, the prediction, the control and optimization (see the introduction to Sect. 2.2); in this context, it is also relevant how the analysis results will be exploited in practice, which will be the costs of analysis, which is the volume of analyzed data and the like.
- *Type of time series*, since some methods are not suitable universally for all time series (e.g., it has no sense to apply a Box–Jenkins model for an economic time series of ten annual observations that show an apparent linear growth).
- *Experience of analyst*, who is responsible for the analysis, and *software*, which will be exploited for the analysis.

The most popular methods and procedures of time series analysis are the following ones:

2.2.2.1 Decomposition of Time Series

Reality shows that time series of economic character can be usually decomposed to several specific components, namely

- Trend component Tr_t (see Chap. 3)
- Seasonal component I_t (see Sect. 4.1)
- Cyclical component C_t (see Sect. 4.2)
- Residual (random, irregular) component E_t (see Chap. 5)

This decomposition is motivated by the expectancy that particular components will show some regular features more distinctly than the original (compound) time series. The classical decomposition regards the trend, seasonal, and cyclical components as deterministic functions of time, while the residual component as a stochastic function of time (i.e., as a random process). These unobservable functions have distinctive features:

Trend presents long-term changes in the level of time series (e.g., a long-term increase or decrease). One can imagine that the trend component originates as a consequence of forces acting in the same direction. For instance, the interrelated forces causing the growing mortgage volumes are higher demands of some segments of population, salary movements, higher market rents, changes in real estate market, and the like. The trend component has a relative character: the climate changes that economists perceive as long-term movements are from the point of view of climatologists only short-term deviations.

Seasonal component describes periodic changes in time series that pass off during one calendar year and repeat themselves each year. These changes are caused by the rotation of seasons, and they affect significantly most economic activities (typical seasonal phenomena are, e.g., agricultural production, unemployment, accident rate of cars, sale volumes, deposit withdrawals, and the like). Mainly monthly and quarterly data are typical for seasonal analysis of economic time series. The semi-annual observations present the lowest frequency (denoted as *Nyquist frequency* in spectral analysis of time series) that enables the statistical identification of seasonality. The seasonal structure varies in time, e.g., the global warming reduces the winter drops in building industry. From the practical point of view, the *seasonal elimination* is usually necessary for obtaining inferences from economic time series; the government statistical offices (in the EU, USA, and elsewhere) must publish time series important for national economics both before the seasonal elimination and after it. Special software products deal professionally with seasonality (e.g., the software systems X-12-ARIMA or X-13ARIMA-SEATS used by the U.S. Census Bureau).

Cyclical component is the most controversial component of time series. Some authors avoid denoting this component as cyclical (or even periodic), and they speak rather on fluctuations around the trend, where increase phases (*booms*) alternate with decrease phases (*recessions*). The length of particular cycles, i.e., the distance between the neighboring upper turning points (i.e., local maxima) or between the neighboring lower turning points (i.e., local minima), is usually variable, and also the intensity of particular phases of each cycle can vary in time. The cyclical behavior can be caused by evident external effects, but sometimes its causes are difficult to find. The typical representative of this component is so-called *business cycle* which is a (regular) alternation of booms and recessions (see above)—the length of business cycles generally ranges from 5 to 7 years. The elimination of cyclical component is usually complex both due to factual reasons (it is not easy to find out causes of its origin) and due to calculation reasons (its character can vary in time similarly as in the case of seasonal component). Sometimes the seasonal and cyclical components are denoted collectively *periodic components* of time series.

Residual component (called also *random* or *irregular* component) remains in time series after eliminating trend and periodic components. It is formed by random movements (or fluctuations) of time series which have no recognizable systematic character. Therefore, it is not included among the *systematic components* described above. The residual component also covers the measurement and rounding errors and the errors made when modeling the given time series. In order to justify some statistical procedures used for the classical decomposition, one usually assumes that the residual component is so-called white noise (or even the normally distributed white noise). Here the term *white noise* denotes a sequence $\{\varepsilon_t\}$ of uncorrelated random variables with zero mean value and constant (finite) variance $\sigma^2 > 0$:

$$\mathrm{E}(\varepsilon_t) = 0, \mathrm{var}(\varepsilon_t) = \sigma^2 > 0, \mathrm{cov}(\varepsilon_s, \varepsilon_t) = 0 \text{ for } s \neq t \tag{2.1}$$

(some authors demand for the white noise even stronger assumptions written usually as $\varepsilon_t \sim iid(0, \sigma^2)$, where ε_t are independent and identically distributed random variables with zero mean value and constant variance). The name "white noise" derives from the spectral analysis and refers to the property of a constant spectrum with equal magnitude at all frequencies (or wavelengths) similarly as in the white light in optics. The values ε_t are also called *innovations* as they correspond to unpredictable movements (shocks) in time series.

Obviously, one can look upon the given economic time series as a trend linked with periodic components (i.e., seasonal and cyclical ones) and white noise. Moreover, the decomposition can be either additive or multiplicative:

Additive decomposition has the form

$$y_t = Tr_t + C_t + I_t + E_t. \tag{2.2}$$

In the additive decomposition, all components are measured in the units of the time series y_t, i.e., all components are absolute ones (not relative ones measured, e.g., as percent of the trend).

Multiplicative decomposition has the form

$$y_t = Tr_t C_t I_t E_t. \tag{2.3}$$

In the multiplicative decomposition, only the trend component is usually measured in the units of the time series y_t, i.e., it is the absolute one. The others are then considered relatively to the trend. For example, $I_1 = 1.15$ means that the value of time series explained by the trend and seasonal components is $1.15Tr_1$ in time $t = 1$. Obviously, the logarithmic transformation converts the multiplicative

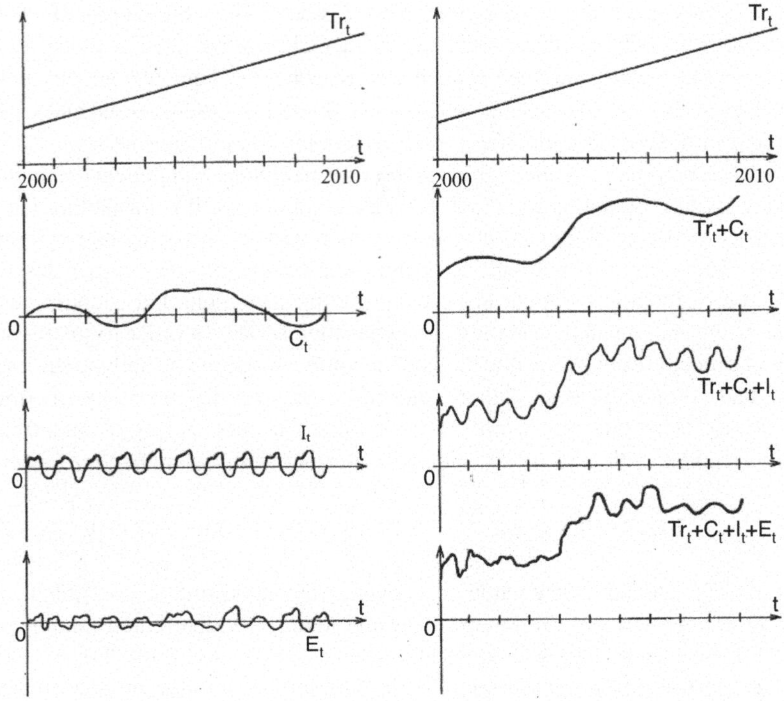

Fig. 2.1 Additive decomposition of time series

decomposition to the additive one, and vice versa, by means of the exponential transformation (one must pay attention to the changes of statistical properties of the transferred residual component in such a case).

If the observations are $y_1, \ldots, y_n$, then

$$\widehat{y}_t = \widehat{Tr}_t + \widehat{C}_t + \widehat{I}_t \quad \text{or} \quad \widehat{y}_t = \widehat{Tr}_t \cdot \widehat{C}_t \cdot \widehat{I}_t \tag{2.4}$$

is the *smoothed* time series (for $t \leq n$), or the *prediction* of time series (for $t > n$), based on the calculated values of systematic components, or on the extrapolated ones, respectively. Obviously, some systematic components can be missing in the decomposition of various economic time series, e.g., the series $y_t = Tr_t + E_t$ does not contain any periodic components at all. Figure 2.1 shows the scheme of additive decomposition in a graphical way.

2.2.2.2 Box–Jenkins Methodology

The decomposition methods are based mainly on the analysis of systematic components of time series (i.e., the trend, seasonal, and cyclical components), and they regard the particular observations as uncorrelated. The typical statistical instrument is here the regression analysis. On the contrary, Box–Jenkins methodology takes as the basis for construction of time series models the residual component (i.e., the component of random character). The key statistical instruments consist here in the *correlation analysis*, and therefore, this methodology can deal successfully with mutually correlated observations (see Chap. 6). Its formal principles were formulated by Box and Jenkins (1970).

For instance, one of the simplest model of Box–Jenkins methodology is so-called moving average process of the first order denoted as MA(1) [see also (6.10)]. It is suitable for such time series where all observations are mutually uncorrelated except for direct neighbors. This model can have the following concrete form:

$$y_t = \varepsilon_t + 0.7\varepsilon_{t-1}, \tag{2.5}$$

where y_t is the modeled time series and ε_t is the white noise (2.1). Other types of models applied in the framework of Box–Jenkins methodology are so-called autoregressive processes AR [see (6.31)] and processes ARMA [see (6.45)].

At first sight it could seem that the attention devoted by this methodology to the random component is excessive and that one loses the possibility to model nonstationary time series with evident trend or seasonal character (the so-called stationarity of a time series means that the behavior of this series is stable in a specific way; see Sect. 6.1). However, Box–Jenkins methodology is capable of managing also these cases by means of so-called integrated processes ARIMA and seasonal processes SARIMA, where the trend or seasonal components are modeled in a stochastic way (in contrast to the deterministic modeling when using the classical decomposition approach). For instance in a very simple model ARIMA (0, 1, 0)

$$y_t = y_{t-1} + \varepsilon_t, \tag{2.6}$$

the stochastic trend can be characterized in such a way that its increments over particular observation intervals are random in the form of white noise (hence it is logic why the process (2.6) is called the *random walk*). Due to this stochastic approach, Box–Jenkins methodology is very flexible modeling in a satisfactory way also non-standard time series that are unmanageable by the classical decomposition approach.

2.2.2.3 Analysis of Multivariate Time Series

In analysis of multivariate time series, one models several time series simultaneously including relations and correlations among them (see Chap. 12). Then the causality relations among various economic variables modeled dynamically in time can be addressed in this context. Another important phenomenon is here so-called *cointegration* when particular (univariate) time series from multivariate model have a common stochastic trend which can be eliminated completely combining particular time series in a suitable way (see Sect. 12.5). The popular instrument for modeling multivariate time series is the process VAR (vector autoregression; see Sect. 12.2).

2.2.2.4 Spectral Analysis of Time Series

Three approaches presented above can be summarized as the time series analysis in *time domain*. A distinct approach that regards the examined time series as an (infinite) mixture of sinusoids and cosinusoids with different amplitudes and frequencies (according to the Wiener–Khinchin theorem for stationary time series) is the time series analysis in *spectral domain* called briefly the spectral analysis of time series (sometimes one also speaks more generally on Fourier analysis). Applying special statistical instruments, e.g., *periodogram* or *spectral density*, one can obtain in this context the image which is the distribution of intensities of particular frequencies in the examined time series (so-called *spectrum* of time series), which of its frequencies are the most intensive ones including the estimation of the corresponding periodic components, etc.

The spectral analysis is important for applications in engineering (vibrograms, technical diagnostics, seismograms) and biology (electrocardiograms). On the other hand, it is not usual for economic time series [except for the tests of periodicity (see Sect. 4.2) or the investigation of cycles in economics (see Hatanaka 1996)]. In any case, a deeper study of theoretical backgrounds of this approach to time series demands special references, e.g., monographs Koopmans (1995) or Priestley (2001).

2.2.2.5 Special Methods of Time Series Analysis

There exist plenty of methods concerning special types or aspects of time series, e.g.:

- *Nonlinear models of time series*: for instance, *threshold models* are suitable for time series that change their character after exceeding particular threshold levels; *asymmetric models* are applied for time series whose momentary development is revised according to their previous development and, moreover, such a revision is asymmetric in dependence on the previous growth or decline.
- *Models of financial time series*: these time series have various typical features, e.g., so-called leptokurtic or heavy-tailed distribution, extreme values appearing in clusters, high frequency of records, and others; therefore, very specific nonlinear models are necessary for time series used in finance (e.g., models ARCH or GARCH with *conditional heteroscedasticity* whose variance called usually *volatility* depends in the given time on the previous behavior of time series; see Chap. 8).
- *Recursive methods in time series*: these methods provide results for a new time step (estimates, smoothed values, predictions, and others) using results from previous time periods and adjusting them by means of new observations; in

particular, one can make use of *Kalman filter* here as a formal recursive methodology; see Chap. 14).
- *Methods for time series with missing or irregular observations*: in such series, some observations are either missing (e.g., they are unobservable or false or outlied or secret) or are observed in irregular time intervals (e.g., due to time irregularities in trading on markets, one must also model so-called durations between neighboring values; see Sect. 9.4).
- *Robust analysis of time series*: here one identifies and eliminates the influence of *outliers* that contaminate analyzed records and distort results of classical methods (a very simple example how to robustify a classical statistical method in order to be insensitive to outliers is to replace the arithmetic average by the median when estimating the average level of a time series).
- *Intervention analysis of time series*: this analysis examines one-off impacts from outside that can influence significantly the course of time series (e.g., intervention of central bank, useful advertising campaign, and others; see Sect. 7.4);
- Plenty of other special methods.

2.2.3 Problems with Construction of Predictions

The construction of predictions is one of the important objectives of time series analysis. Some analysts compare particular models entirely according to the accuracy of generated predictions (other criteria need not be satisfactory for preferred models in such a case).

Particularly, the predictions in finance are extremely important. The financial management often deals with long-term liabilities and investments whose results are known in the far future, and therefore acceptable predictions play the key role in this context, e.g.,

- Prediction of profitability and risk of given investment portfolio in next year
- Prediction of volatility of bond yields during future 5 years
- Prediction of stock prices next trading day on stock exchanges
- Short-term prediction of correlations among American and European stock markets
- Long-term prediction of volumes of credit defaults in commercial banks
- Prediction of prices of real estates according to their characteristics

In this section, we mention only some general aspects concerning predictions in time series. Specific prediction methods will be described later after introducing particular time series models.

2.2.3.1 Point Prediction and Interval Prediction

This classification of predictions holds not only for time series, but it is also common, e.g., in the econometric regression analysis:

Point prediction is the quantity that presents a numerical estimate of future value of time series which is optimal in a certain sense (i.e., the estimate of time series value in so far unobserved future time point). For instance, the point prediction of exchange rate EUR/USD in three future months predicted just now is 1.0635. Obviously, the point prediction is always burdened by error so it must be taken with discretion.

Interval prediction is the prediction interval which is quite analogous to the confidence interval used in mathematical statistics; the only difference consists in the fact that one estimates an unknown (future) value of time series instead of an unknown parameter in this case. For instance, the 95% prediction interval presents the lower and upper bounds for the range in which the corresponding future value of time series will lie with the probability of 0.95. Let us consider again the previous example with the exchange rate EUR/USD: if the corresponding 95% prediction interval is (1.0605; 1.0665), then, e.g., a European company can expect with high confidence that it obtains for each euro at least 1.0605 dollars. From the practical point of view, the interval predictions seem to be more useful for users than the point predictions.

2.2.3.2 Quantitative Prediction and Qualitative Prediction

It has nothing in common with the classification of variables (the *quantitative variable* can be measured quantitatively on a numeric or quantitative scale, e.g., stock quotes, while the *qualitative variable* has no natural or logical order, e.g., the variable with values spring, summer, autumn, and winter denoting seasons of year). Roughly speaking, the classification of quantitative and qualitative predictions depends on the objectivity of their construction:

Quantitative prediction methods provide predictions based on the statistical analysis of observed data, i.e., such predictions are usually based on objective mathematical methods of statistics. Of course it does not guarantee that one obtains the best prediction results in this way: although the quantitative predictions are constructed applying objective methods, their worth depends significantly on the assumption that in the (future) prediction horizon the character of the given time series remains unchanged so that the model constructed using current and past data remains valid. One produces here only a technical *extrapolation* (autoprojection or mathematical extension) of past and current observations to the future. This fact must be reminded in the following chapters where we confine ourselves entirely to the quantitative predictions (so that the term extrapolation could be more appropriate even if it is not common in economic applications).

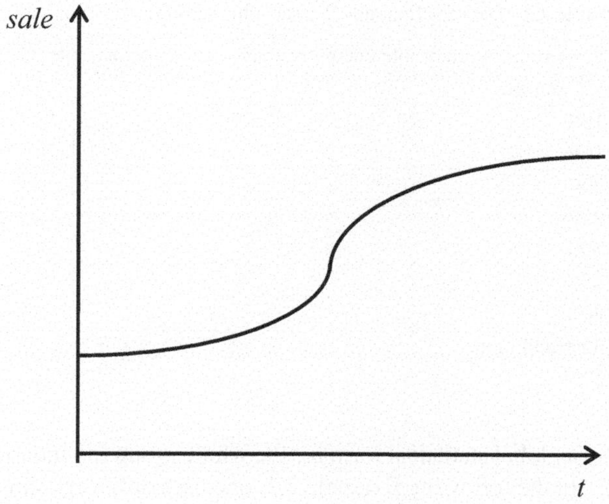

Fig. 2.2 *S*-curve plotting the sale of a new product

Qualitative prediction methods are based usually on opinions of experts (one calls them sometimes the "expert predictions"), and therefore in practice they have rather subjective character. Sometimes one is forced to apply these methods when historical data are missing, e.g., when one introduces a new bank product. Since we avoid these methods in the following chapters with regard to the character of this text, some simple examples of qualitative predictions will be given below to have an idea of this approach to predicting (sometimes the qualitative predictions are even better than the purely mathematical ones):

- *Subjective fitting by curve* is a (graphical) method when experts strive to estimate the future behavior of particular time series using their experience with time series of similar type. For instance, the graphical plot describing the sale of a new product (e.g., a new car make) has frequently the form of so-called *S-curve* shown in Fig. 2.2: after the starting stage, the sale accrues during the growth stage in dependence on the intensity of advertising campaign till the stable stage is achieved (later usually the drop of sale follows only). Therefore, the experts can suggest the prediction just according to a specific S-curve applying their subjective opinion on its form (e.g., on the length of particular stages).
- *Delphi method* is the prediction method based on the enquiry in an expert group and the gradual mediation of consensus for given prediction problem. This methodology has been developed by large-scale multinational corporations to forecast development in science, engineering, production, consumption, and the like. In particular, its application consists in several stages of anonymous enquiring where each of addressed experts presents his or her opinion on the given prediction. In each stage, one adds to the enquiry form the statistical results of previous stages so that experts can adjust their previous opinions and to converge gradually to the group opinion which is declared as the final prediction. It should be stressed that the results of particular stages are communicated only in

Table 2.1 Data for Example 2.1 (Delphi method)

Proportion of renewable energy resources 25 years forwards	Numbers of positive answers
5% and less	0
10%	1
15%	7
20%	11
25%	12
30%	10
35%	5
40%	4
45% and more	0
Σ	50

a global statistical form for the whole group (no individual answers are provided). In the following Example 2.1, one uses only very simple statistical instruments in this context, namely the mean values, standard deviations, and lower and upper quartile.

Example 2.1 (*Delphi method*). In Table 2.1, one summarizes enquiries by 50 experts on the proportion of renewable energy resources predicted for a given region 25 years forward in the first stage of Delphi method.

From the statistical sample in Table 2.1, one calculates the mean value

$$\frac{1 \cdot 10 + \ldots + 4 \cdot 40}{50} = 25.4\ \%$$

and the standard deviation

$$\sqrt{\frac{1 \cdot (10 - 25.4)^2 + \ldots + 4 \cdot (40 - 25.4)^2}{50}} \cong 7.5\%.$$

Further one finds the lower and upper quartile. The lower (or upper) quartile is the bound separating one-quarter of the lowest (or highest) observations, respectively. In our case, one-quarter of number of observations is 50/4 = 12.5, and therefore, the lower quartile is 20% and the upper quartile is 30%. In the next stage of this prediction method, one informs particular participants of the expert group on the statistical results obtained in the first stage (but not on the answers of particular experts) and so on.

◇

2.2.3.3 Prediction in Structural Model and Prediction in Time Series Model

Prediction in structural (econometric) model relates the future value of explained variable (which is the prediction task in this case) to future values of relevant explanatory variables. Such models may be successful even for long-term predictions since the long-term relations are frequent in economic and financial practice, e.g., due to efficiency of financial markets and arbitrage-free principle. However, in such a case one must construct predictions of future values for all participating explanatory variables in the model.

Prediction in time series model is constructed as the autoprojection of past and current values of the time series to future. Such a prediction often makes use explicitly of calculated values of error terms in the model. Sometimes the difference between the prediction based on structural model and the prediction based on time series model is not quite clear (e.g., in the vector autoregression VAR; see Sect. 12.2).

2.2.3.4 In-Sample Prediction and Out-of-Sample Prediction

In-sample prediction is that generated for the same set of data that was used to estimate the model's parameters. Obviously, one can expect good prediction results since one only recalculates selected values of the original sample by means of the constructed model so that all model assumptions remain valid in the "prediction horizon." Nevertheless, this procedure can serve a very simple test of in-sample fit of the model.

Out-of-sample prediction is on the contrary the prediction of time series values which have not participated in the construction of prediction model at all: either they were not available at that time (i.e., they were future values from the point of view of that time), or they were deleted from the sample on purpose (it is common in the situation when one tries to evaluate the prediction ability of a time series model: then the data deleted artificially are denoted as the *hold-out sample*). Obviously, predictions of hold-out sample represent a better evaluation of the prediction model than an examination of its in-sample fit (see above).

As a simple example let us consider the time series of 120 monthly observations in the period 2008M1–2017M12. The objective is to construct a model for this time series and to assess its quality (in particular, its prediction abilities). Two solutions are possible in this case: either (1) to construct the model using the whole time series 2008M1–2017M12 (and possibly to generate in-sample predictions) or (2) to construct the model using only the shorter time series 2008M1–2016M12 and to generate the out-of-sample predictions for 2017M1–2017M12 (which is the hold-out sample here; see Fig. 2.3) and to compare them with the real values from the hold-out sample. The second approach is more correct (of course, a suitable length of

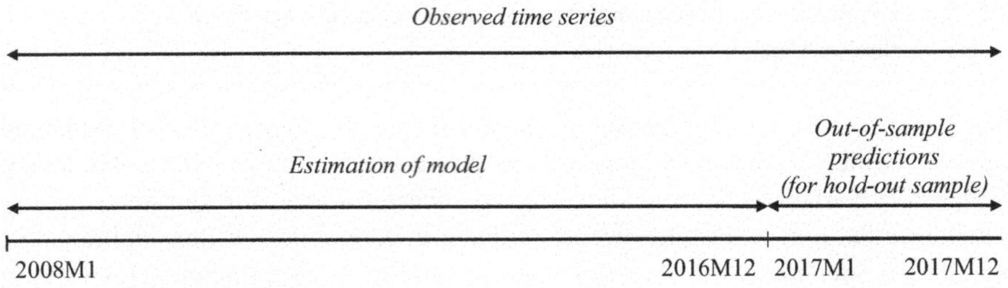

Fig. 2.3 Example of out-of-sample predictions

hold-out sample must be chosen appropriately) since here the data information on 2017M1–2017M12 is not used for the construction of prediction model.

2.2.3.5 Single-Prediction and Multi-prediction

Single-prediction is the prediction constructed for a single time (usually for the next one), e.g., at time n for time $n + 1$ denoted as $\widehat{y}_{n+1}(n)$, but also, e.g., at time n for time $n + 5$ denoted as $\widehat{y}_{n+5}(n)$.

Multi-prediction is the prediction simultaneously for more (future) times (e.g., at time n for times $n + 1, \ldots, n + h$). Obviously, one obtains a vector of several single predictions which are constructed at the same time (on the other hand, the sequence of one-step-ahead predictions constructed at times $n + 1, \ldots, n + h$ always after receiving particular observations $y_n, \ldots, y_{n+h-1}$ cannot be called multi-prediction).

The example in Fig. 2.3 may show some problems connected with multi-predictions, e.g., with assessment of their quality. Let an examined prediction technique applied at the end of 2006 for the hold-out sample 2007 provide good result only for the first month 2007M1 (i.e., the short-term prediction result) and bad results for the remaining months 2007M2–2007M12 (i.e., the long-term prediction results). The multi-prediction from time 2006M12 for times 2007M1–2007M12 is not enough to assess the quality of applied prediction methodology: one should generate a set of multi-predictions, and it is possible to do it in a systematic way using so-called *prediction windows*. Moreover, two types of prediction windows can be used (see Table 2.2 if predicting only three closest future values for simplicity):

- *Rolling windows*: the samples used for prediction (observable in the rolling windows) have a fixed length (e.g., 108 in Table 2.2), but their beginning is shifted.
- *Recursive windows*: the samples used for prediction (observable in the recursive windows) have a fixed beginning (e.g., 2008M1 in Table 2.2), but their length increases.

In both cases, ten multi-predictions are obtained (see Table 2.2) so that conclusions on the prediction methodology may be reliable.

Table 2.2 Rolling and recursive windows to assess the quality of multi-predictions (see Fig. 2.3)

Multi-prediction	Multi-predictions based on samples provided by	
Constructed for times	Rolling windows	Recursive windows
2017M1, M2, M3	2008M1–2016M12	2008M1–2016M12
2017M2, M3, M4	2008M2–2017M1	2008M1–2017M1
2017M3, M4, M5	2008M3–2017M2	2008M1–2017M2
2017M4, M5, M6	2008M4–2017M3	2008M1–2017M3
2017M5, M6, M7	2008M5–2017M4	2008M1–2017M4
2017M6, M7, M8	2008M6–2017M5	2008M1–2017M5
2017M7, M8, M9	2008M7–2017M6	2008M1–2017M6
2017M8, M9, M10	2008M8–2017M7	2008M1–2017M7
2017M9, M10, M11	2008M9–2017M8	2008M1–2017M8
2017M10, M11, M12	2008M10–2017M9	2008M1–2017M9

2.2.3.6 Static Prediction and Dynamic Prediction

This classification of predictions is used in cases when we explain the predicted variable by means of explanatory variables, and among these explanatory variables there are lagged (i.e., delayed) values of the predicted variable (this situation is common, e.g., in autoregressive models):

Static prediction makes use of the observed lagged values of predicted variable if they are available for prediction. For instance, predicting for time t +2 from time t by means of the estimated model $y_t = 0.64y_{t-1} + \varepsilon_t$ (so-called autoregression of the first order; see Sect. 6.2) with value y_{t+1} known at time t of prediction, the corresponding static prediction will be

$$\widehat{y}_{t+2}(t) = 0.64y_{t+1}. \tag{2.7}$$

Dynamic prediction does not exploit values of predicted variable lying in the prediction horizon (even if these values are known), but replaces them by corresponding predictions. Therefore, in the situation described above, the dynamic prediction will be

$$\widehat{y}_{t+2}(t) = 0,64\widehat{y}_{t+1}(t), \tag{2.8}$$

i.e., the value y_{t+1} in (2.7) is replaced by the one-step-ahead prediction $\widehat{y}_{t+1}(t)$ ignoring the possibility that the value y_{t+1} may be known at time t. Obviously, the dynamic predictions are not so accurate as the static predictions (if the static predictions are feasible).

2.2.3.7 Measures of Prediction Accuracy

The important aspect of prediction consists in the measuring of its accuracy based on the *error of prediction*. The error e_t of prediction $\widehat{y}_t$ (when predicting value y_t) is defined as

$$e_t = y_t - \widehat{y}_t. \tag{2.9}$$

The error of prediction cannot be calculated until the time when we know the actual value y_t (this value has been unknown at the time of prediction). However in practice, when assessing the quality of prediction, one sometimes "predicts" known values of time series to compare these predictions with the known actual values (see the hold-out sample described above).

The main source of prediction errors consists in the residual component of time series since it represents unpredictable (unsystematic) fluctuations in data. If the participation of this component in time series is significant, then the possibility of construction of reliable predictions is limited. On the other hand, the size of prediction error depends also on the quality of predictions for systematic components of time series. Therefore, significant prediction errors may indicate either an extraordinary participation of residual component or inappropriateness of prediction methodology.

In any case, the examination of error of prediction is useful. If the prediction technique masters predictions of systematic components, then the prediction errors reflect the influence of residual component only (see Fig. 2.4a). On the contrary, Fig. 2.4b–d shows the cases when the prediction technique failed due to inappropriate prediction of trend, seasonal, and cyclical components, respectively.

The measures of prediction accuracy assess the development of predictions in time. We will give below the usual measures of this type for a simple situation when one assesses in total the accuracy of predictions $\widehat{y}_{n+1}, \ldots, \widehat{y}_{n+h}$ of values $y_{n+1}, \ldots, y_{n+h}$ (here it does not matter if we assess a multi-stage prediction or a sequence of one-step-ahead predictions, static or dynamic predictions, or other types of predictions):

1. *Sum of squared errors SSE* (*sum of squared errors*):

$$SSE = \sum_{t=n+1}^{n+h} (y_t - \widehat{y}_t)^2 = \sum_{t=n+1}^{n+h} e_t^2. \tag{2.10}$$

SSE is the analogy of the criterion of least squares in regression models.

2. *Mean squared error MSE*:

$$MSE = \frac{1}{h}\sum_{t=n+1}^{n+h} (y_t - \widehat{y}_t)^2 = \frac{1}{h}\sum_{t=n+1}^{n+h} e_t^2. \tag{2.11}$$

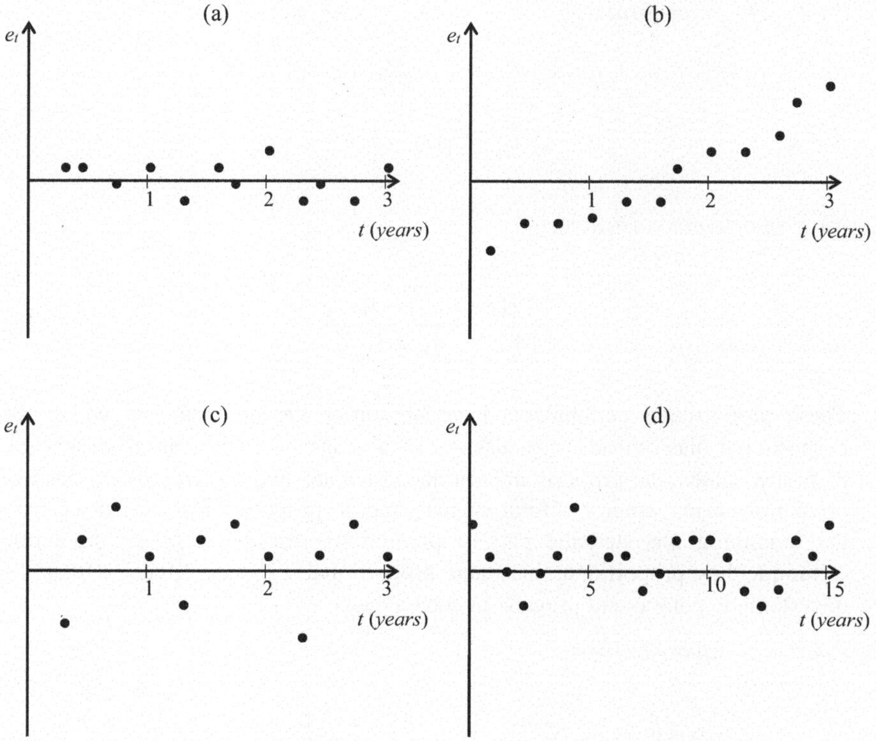

Fig. 2.4 Graphical plots of errors of prediction

MSE is a popular quadratic loss function. Some software decompose *MSE* to three components:

$$\frac{1}{h}\sum_{t=n+1}^{n+h}(y_t-\widehat{y}_t)^2=\left(\overline{\widehat{y}}-\overline{y}\right)^2+\left(s_{\widehat{y}}-s_y\right)^2+2\left(1-r_{\widehat{y}y}\right)s_{\widehat{y}}s_y \tag{2.12}$$

($\overline{\widehat{y}}$, $\overline{y}$, $s_{\widehat{y}}$, s_y are the corresponding sample means and (biased) sample standard deviations of values $\widehat{y}$ and y; $r_{\widehat{y}y}$ is the sample correlation coefficient between $\widehat{y}$ and y). Usually, one uses relative values of these components, namely

(a) *Proportional bias*:

$$\frac{\left(\overline{\widehat{y}}-\overline{y}\right)^2}{\frac{1}{h}\sum_{t=n+1}^{n+h}(y_t-\widehat{y}_t)^2} \tag{2.13}$$

(b) *Proportional variance*:

$$\frac{\left(s_{\hat{y}} - s_y\right)^2}{\frac{1}{h}\sum_{t=n+1}^{n+h}\left(y_t - \widehat{y}_t\right)^2} \tag{2.14}$$

(c) *Proportional covariance*:

$$\frac{2\left(1 - r_{\hat{y}y}\right)s_{\hat{y}}\, s_y}{\frac{1}{h}\sum_{t=n+1}^{n+h}\left(y_t - \widehat{y}_t\right)^2} \tag{2.15}$$

These proportional components have obviously the unit sum, in which the proportional bias indicates the distance of average of predictions from average of future values, the proportional variance indicates the distance of variance of predictions from variance of future values, and the proportional covariance covers the remaining unsystematic part of prediction error (any "good" prediction technique has proportional bias and proportional variance small so that the unsystematic component prevails in such a case).

3. *Root mean squared error*:

$$RMSE = \sqrt{\frac{1}{h}\sum_{t=n+1}^{n+h}\left(y_t - \widehat{y}_t\right)^2} = \sqrt{\frac{1}{h}\sum_{t=n+1}^{n+h} e_t^2}. \tag{2.16}$$

RMSE modifies *MSE* in order to be measured in the same units as the given time series.

4. *Mean absolute error MAE*:

$$MAE = \frac{1}{h}\sum_{t=n+1}^{n+h}\left|y_t - \widehat{y}_t\right| = \frac{1}{h}\sum_{t=n+1}^{n+h}\left| e_t\right|. \tag{2.17}$$

MAE penalizes large prediction errors not so strictly as *MSE*. Therefore, it is recommended for assessment of prediction accuracy in time series with outliers (see Sect. 2.2.1).

The measures of prediction accuracy given above depend on the scale of the given time series. Therefore, they are applicable only if one mutually compares various prediction techniques in similar time series. Now we will present further measures which do not depend on the time series scale:

5. *Mean absolute percentage error MAPE*:

$$MAPE = \frac{100}{h}\sum_{t=n+1}^{n+h}\left|\frac{y_t - \widehat{y}_t}{y_t}\right|. \tag{2.18}$$

MAPE usually ranges from 0 to 100%. This measure is preferred in practice by some authors [see, e.g., Makridakis (1993)]. The result less than 100% means that the given prediction model is better than the random walk model (the random walk model predicts the zero level permanently), i.e., $MAPE = 100\%$ constantly. *MAPE* is not reliable for time series ranging in vicinity of zero.

6. *Adjusted mean absolute percentage error AMAPE*:

$$AMAPE = \frac{100}{h}\sum_{t=n+1}^{n+h}\left|\frac{y_t - \widehat{y}_t}{(y_t + \widehat{y}_t)/2}\right|. \tag{2.19}$$

AMAPE rectifies the asymmetry of the criterion MAPE in (2.18), namely that it provides the same result even if one swaps the real value and its prediction (e.g., real value 0.7 and prediction 0.9 give the same value in (2.19) as real value 0.9 and prediction 0.7).

7. *Theil's U-statistic* [see Theil (1966)]

$$U = \frac{\sqrt{\sum_{t=n+1}^{n+h}(y_t - \widehat{y}_t)^2}}{\sqrt{\sum_{t=n+1}^{n+h}\widehat{y}_t^{\,2}} + \sqrt{\sum_{t=n+1}^{n+h}y_t^2}}. \tag{2.20}$$

U lies always between 0 and 1 (e.g., $U=0$ means the perfect coincidence of prediction with reality).

Further group of measures of prediction accuracy only indicates whether the model predicts correct signs of future values (i.e., whether these values will be positive or negative) or predicts correct direction changes (i.e., whether an increase changes to a decrease and the like). From the strategic point of view, such predictions are often more important than numerical predictions:

8. *Percentage of correct sign predictions*:

$$\frac{100}{h}\sum_{t=n+1}^{n+h} z_t, \text{ where } z_t = \begin{cases} 1 & \text{for } y_t \cdot \widehat{y}_t > 0, \\ 0 & \text{otherwise.} \end{cases} \tag{2.21}$$

9. *Percentage of correct direction change predictions*:

$$\frac{100}{h}\sum_{t=n+1}^{n+h} z_t, \text{ where } z_t = \begin{cases} 1 & \text{for } (y_t - y_{t-1}) \cdot (\widehat{y}_t - y_{t-1}) > 0, \\ 0 & \text{otherwise.} \end{cases} \tag{2.22}$$

Remark 2.2 One should stress once more that the given measures concern the statistical accuracy of predictions only. In any way they do not justify an economic or financial adequacy of predictions. For instance, small values of *MSE* do not mean that we dispose of a successful outline how to predict future market strategies (e.g., sometimes it can be desirable from the strategic point of view to underestimate or overestimate the future development and the like).

◇

2.2.3.8 Prediction Combinations

As the accuracy of predictions is concerned, a surprisingly positive effect brings combining several predictions constructed in different ways (i.e., by different approaches, by employing different information sources, and the like) to a final summary prediction [see, e.g., Clements and Harvey (2011), Timmermann (2006), and others]. Moreover, the predictions based on different modeling approaches may have similar predictive accuracy so that it is difficult to identify a single best forecast. This philosophy based on mixing various predictions has been successfully applied in practice, e.g., to forecast interest rates, currency market volatility and exchange rates, inflation, money supply, stock returns, meteorological data, city populations, outcomes of football games, and many others.

Let us consider a simple case when we combine two predictions in the form

$$\widehat{y}_t = w \cdot \widehat{y}_{1t} + (1 - w) \cdot \widehat{y}_{2t}. \tag{2.23}$$

The weights w and $1 - w$ ($w \geq 0$) should be chosen to minimize the mean squared error

$$\begin{aligned} \text{var}(e) &= \text{var}(y_t - \widehat{y}_t) = \text{var}(w \cdot (y_t - \widehat{y}_{1t}) + (1 - w) \cdot (y_t - \widehat{y}_{2t})) \\ &= \text{var}(w \cdot e_1 + (1 - w) \cdot e_2). \end{aligned} \tag{2.24}$$

If both the combined predictions are unbiased (i.e., $\text{E}(e_1) = 0$ and $\text{E}(e_2) = 0$) with finite mean squared errors denoted as ${\sigma_1}^2 = \text{E}({e_1}^2)$ and ${\sigma_2}^2 = \text{E}({e_2}^2)$ and covariance denoted as $\sigma_{12} = \text{cov}(e_1, e_2) = \text{E}(e_1 \cdot e_2)$, then the corresponding optimal weights are

$$w = \frac{\sigma_2^2 - \sigma_{12}}{\sigma_1^2 + \sigma_2^2 - 2\sigma_{12}}, \qquad 1 - w = \frac{\sigma_1^2 - \sigma_{12}}{\sigma_1^2 + \sigma_2^2 - 2\sigma_{12}}. \tag{2.25}$$

According to (2.25), greater weights are assigned to more precise models with smaller ${\sigma_1}^2$ or ${\sigma_2}^2$. Moreover, the weights can be negative if $\sigma_{12} > {\sigma_2}^2$ or $\sigma_{12} > {\sigma_1}^2$ (the negative weight assigned to a prediction component means that this component is replaced in the prediction combination by other prediction components with lower prediction errors). The weakly correlated prediction errors enable us to rewrite the weights as functions of the relative variance ${\sigma_2}^2/{\sigma_1}^2$:

$$w \approx \frac{\sigma_2^2/\sigma_1^2}{1 + \sigma_2^2/\sigma_1^2}, \qquad 1 - w \approx \frac{1}{1 + \sigma_2^2/\sigma_1^2}.$$

The result (2.25) can be generalized easily if one combines m predictions. In the literature and in software systems (*R* packages, EViews, and others), there are suggested many strategies how to combine predictions, e.g.:

- *Equal-weighted predictions*:

$$\widehat{y}_t = \frac{1}{m} \sum_{i=1}^{m} \widehat{y}_{it}. \tag{2.26}$$

- *Median prediction*:

$$\widehat{y}_t = \text{med}(\widehat{y}_{1t}, \ \ldots, \widehat{y}_{mt}). \tag{2.27}$$

- *Trimmed mean of predictions*: One orders (increasingly) the predictions $\widehat{y}_{1t} \leq \ldots \leq \widehat{y}_{mt}$ and then trims $\lambda{\cdot}100\%$ predictions at the top and $\lambda{\cdot}100\%$ predictions at the bottom:

$$\widehat{y}_t = \frac{1}{m(1 - 2\lambda)} \sum_{i=\lfloor \lambda m + 1 \rfloor}^{\lfloor (1-\lambda)m \rfloor} \widehat{y}_{it} \tag{2.28}$$

  (the symbol $\lfloor \ldots \rfloor$ denotes the integer part).

- *Predictions weighted in inverse proportional way to MSE*:

$$\widehat{y}_t = \frac{1}{\sum_{j=1}^{m} MSE_j^{-1}} \sum_{i=1}^{m} MSE_i^{-1} \widehat{y}_{it}, \tag{2.29}$$

  where $MSE_i = \mathrm{E}(e_i^2)$ is the mean squared error of prediction $\widehat{y}_{it}$.

- *Predictions weighted in inverse proportional way to ranking*:

$$\widehat{y}_t = \frac{1}{\sum_{j=1}^{m} R_j^{-1}} \sum_{i=1}^{m} R_i^{-1} \widehat{y}_{it}, \tag{2.30}$$

where R_i is the rank of prediction $\widehat{y}_{it}$ (e.g., the smallest prediction has the rank 1). This weighting scheme which weights predictions inversely to their rank seems to be surprisingly robust [see Timmermann (2006)].

2.3 Random Processes with Discrete States in Discrete Time

The majority of methods of time series analysis in this publication concerns time series that are modeled as random processes with continuous states in discrete time (see Sect. 2.1). Therefore, the term "time series" means here usually the trajectory of values $y_1, \ldots, y_n$ from a continuous interval on the real line which are observed in (regular) discrete moments. For the sake of completeness, the remaining Sects. 2.3–2.5 of this chapter are devoted to examples of time series with different character (e.g., to time series which are modeled as random processes with discrete states in continuous time; see Sect. 2.4) in order to get an idea on further possibilities in this modeling framework.

Let us start with several *examples of random processes with discrete states in discrete time*:

2.3.1 *Binary Process*

Binary process is the two-valued random process in discrete time

$$\{Y_t, t = 1, 2, \ldots\}, \text{where } Y_t \sim iid, P(Y_t = 1) = P(Y_t = -1) = 1/2 \tag{2.31}$$

(the definition can be more general with asymmetric probabilities p and q). Its trajectory may be interpreted as a record of results when tossing an ideal coin (e.g., {1, −1, −1, 1, 1, 1, −1, ...}; see Fig. 2.5).

2.3.2 *Random Walk*

Random walk (*RW*) on line is the integer-valued random process in continuous time

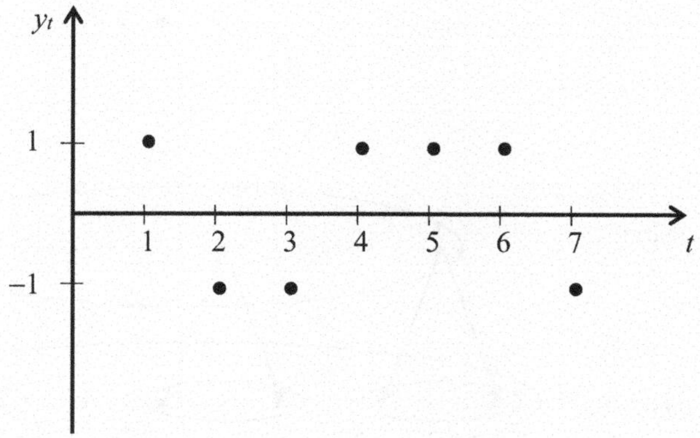

Fig. 2.5 Trajectory of binary process

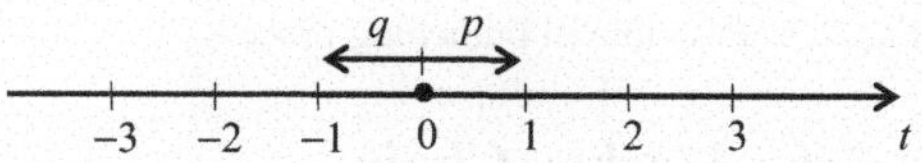

Fig. 2.6 Principle of random walk on real line

$$\{Y_t, t = 0, 1, \ldots\}, \text{ where } Y_0 = 0; \quad Y_t = \sum\nolimits_{i=1}^{t} X_i, \quad t = 1, 2, \ldots;$$
$$X_t \sim iid; P(X_t = 1) = p, P(X_t = -1) = q \; (p + q = 1). \tag{2.32}$$

Its trajectory may be interpreted as a record of movement of particle which moves across integer values on real line: the particle is at time 0 in the origin and during each time unit moves to the right with probability p and to the left with probability q (see Fig. 2.6). In the case with $p = q = ½$, one calls it the *symmetric* random walk.

2.3.3 Branching Process

Branching process (or *Galton–Watson process*) is the integer-valued nonnegative random process in discrete time

$$\{Y_t, t = 0, 1, \ldots\}, \text{where } Y_0 = 1; \quad Y_{t+1} = \sum\nolimits_{j=1}^{Y_t} Z_{tj}, \quad t = 0, 1, \ \ldots \ ;$$
$$Z_{tj} \sim iid \text{ random variables with values } 0, 1, \ldots. \tag{2.33}$$

Here Y_t describes the random number of members of tth generation: the initial 0th generation has only one member, and the jth member of tth generation gives rise to a random number Z_{tj} of members of $(t+1)$th generation (see Fig. 2.7). For example, if in a pyramid game each player finds further three participants, then the corresponding trajectory is $\{1, 3, 9, 27, \ldots\}$.

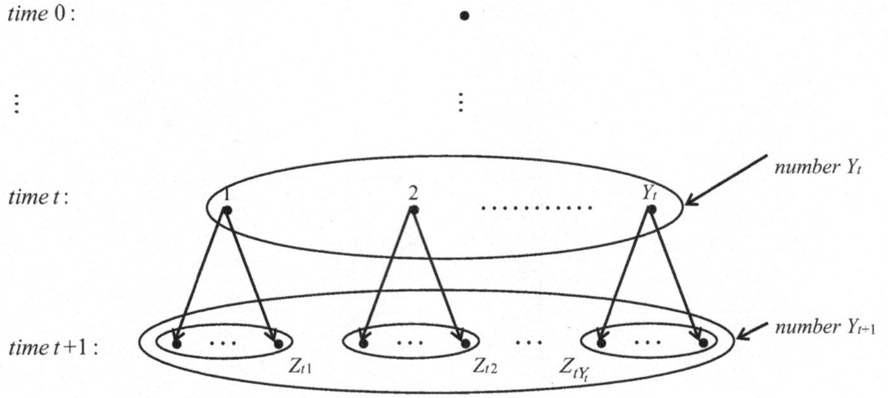

Fig. 2.7 Principle of branching process

2.3.4 Markov Chain

Markov chain is a general scheme frequently used in practice to model random processes with discrete states in discrete time (the binary process, the random walk, and the branching process are its special cases). Let for simplicity the possible states of this process be integer numbers $i = \ldots, -1, 0, 1, \ldots$. The process can move across these states in particular discrete times $t = 0, 1, \ldots$ with given transition probabilities. Moreover, so-called *Markov property* is here fundamental

$$P(Y_{t+1} = j \mid Y_t = i,\ Y_{t-1} = i_{t-1}, \ldots,\ Y_0 = i_0) = P(Y_{t+1} = j \mid Y_t = i) \tag{2.34}$$

for all $t = 0, 1, \ldots$ and $i, j, i_0, \ldots, i_{t-1} = \ldots, -1, 0, 1, \ldots$ (in other words, the probability of moving to the next state depends only on the present state and not on the previous states). The probability on the right-hand side of (2.34) is so-called *transition probability* from state i at time t to state j at time $t + 1$. The important special case is the *homogenous* Markov chain whose transition probabilities do not depend on time, i.e.,

$$p_{ij} = P(Y_{t+1} = j \mid Y_t = i) \tag{2.35}$$

for all t. For example, the symmetric random walk (see above) is the homogenous Markov chain with starting value $Y_0 = 0$ and with transition probabilities

$$p_{ij} = \begin{cases} 1/2 & \text{for } j = i \pm 1 \\ 0 & \text{otherwise} \end{cases} \quad \text{for all } i,j = \ldots, -1, 0, 1, \ldots. \tag{2.36}$$

The maximum likelihood estimate (MLE) of transition probabilities (2.35) has a simple form

$$\widehat{p}_{ij} = \frac{n_{ij}}{\sum_{k=-\infty}^{\infty} n_{ik}}, \tag{2.37}$$

where n_{ij} is the number of transitions from state i to state j during time unit using a sample of observed trajectories.

Further one introduces in this context the *n-step* transition probabilities defined (for simplicity, we constrain ourselves to homogenous Markov chains) as

$$p_{ij}(n) = P(Y_{t+n} = j \mid Y_t = i), \qquad n = 1,\ 2,\ \ldots \tag{2.38}$$

and the probabilities representing the *probability distribution* of Markov chain at time t

$$p_i(t) = P(Y_t = i). \tag{2.39}$$

The following matrix symbols are usually applied in this context:

$$\mathbf{P} = (p_{ij}); \qquad \mathbf{P}(n) = (p_{ij}(n)); \qquad \mathbf{p}(n) = (\ldots,\ p_{-1}(n), p_0(n), p_1(n),\ \ldots)' \tag{2.40}$$

for so-called *transition matrices* ($\mathbf{P}(0) = \mathbf{I}$, $\mathbf{P}(1) = \mathbf{P}$) and *distribution vectors* of Markov chain. It holds when multiplying (infinite) transition matrices

$$\mathbf{P}(n) = \mathbf{P}^n; \mathbf{p}(n)' = \mathbf{p}(0)' \cdot \mathbf{P}^n. \tag{2.41}$$

Some homogenous Markov chains can have so-called *stationary distribution* $\boldsymbol{\pi}$

$$\lim_{n \to \infty} p_i(n) = \pi_i > 0, \tag{2.42}$$

which corresponds to a stable limit behavior of Markov chain. The vector $\boldsymbol{\pi}$ fulfills

$$\boldsymbol{\pi}' = \boldsymbol{\pi}' \cdot \mathbf{P}. \tag{2.43}$$

Example 2.2 (*Markov chain*). A bonus system in motor car (Casco) insurance has three bonus levels denoted as 0, 1, 2 (presenting, e.g., 100 %, 80 %, and 60 % of basic insurance premiums): if the clients report no claims in the given year, their bonus improves next year by one level or they remain at the best level 2; if reporting one or more claims they grow worse next year by one level or they remain at the worst level 0 (i.e., no malus level is introduced). The insurance company disposes of stable insurance portfolio with 10,000 clients: 5000 are "good" drivers with

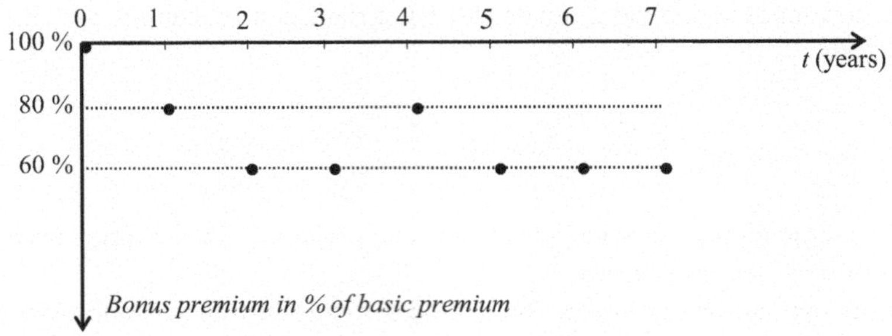

Fig. 2.8 Development of bonus system of a client in Example 2.2

estimated probability of loss-free year about 0.9 and 5000 are "bad" drivers with estimated probability of loss-free year about 0.8. The objective is to estimate the stabilized numbers of clients in particular bonus levels.

The behavior of clients can be described by homogenous Markov chain with annual time units and with three possible states (then the trajectory for an individual client is an annual time series jumping across particular levels 100 %, 80 %, and 60 %; see Fig. 2.8).

Obviously, the transition matrices of good or bad drivers are

$$\begin{pmatrix} 0.1 & 0.9 & 0 \\ 0.1 & 0 & 0.9 \\ 0 & 0.1 & 0.9 \end{pmatrix} \quad \text{or} \quad \begin{pmatrix} 0.2 & 0.8 & 0 \\ 0.2 & 0 & 0.8 \\ 0 & 0.2 & 0.8 \end{pmatrix}. \tag{2.44}$$

The stationary distribution of good drivers must fulfill according to (2.43) the system of linear equations

$$(\pi_0 \ \pi_1 \ \pi_2) = (\pi_0 \quad \pi_1 \quad \pi_2) \begin{pmatrix} 0.1 & 0.9 & 0 \\ 0.1 & 0 & 0.9 \\ 0 & 0.1 & 0.9 \end{pmatrix}. \tag{2.45}$$

Its solution is $\pi_0 = 0.010\,989$, $\pi_1 = 0.098\,901$, and $\pi_2 = 0.890\,109$ so that $5000 \times 0.010\,989 = 55.0$ clients have the bonus level 0; similarly 494.5 clients have the bonus level 1 and 4450.5 clients have the bonus level 2 among 5000 good drivers in the portfolio.

Quite analogously we get that 238.1 clients have the bonus level 0, 952.4 clients have the bonus level 1, and 3809.5 clients have the bonus level 2 among 5000 bad drivers in the portfolio. Hence in limit, the majority of good and also bad drivers will achieve the best bonus level 2 (although this number is significantly lower among bad drivers than among good drivers). In any case, the given bonus system is very favorable for insured.

◇

2.4 Random Processes with Discrete States in Continuous Time

Let us present the following well-known *examples of random processes with discrete states in continuous time*:

2.4.1 Poisson Process

Poisson process (with intensity λ) is the integer-valued nonnegative random process in continuous time $\{N_t, t \geq 0\}$, where

$$\begin{cases} (i) \ \ N_0 = 0; \\ (ii) \ \ N_{t_2} - N_{t_1}, \ \ \ldots, \ \ N_{t_n} - N_{t_{n-1}} \ \text{are independent for arbitrary } 0 \leq t_1 < \ldots < t_n; \\ (iii) \ \ N_t - N_s \sim \text{Poisson distribution with intensity } \lambda \cdot (t-s) \text{ for arbitrary } 0 \leq s < t. \end{cases} \tag{2.46}$$

Here N_t at time $t \geq 0$ describes the number of occurrences of an observed event during the time interval $\langle 0, t\rangle$. In particular, Poisson process is suitable for modeling the occurrences of *rare events* in economy and finance, e.g., insurance claims, credit defaults, turbulences in stock prices, and the like). For instance, Fig. 2.9 plots the number n_t of mortgage defaults in a bank credit portfolio from the beginning of accounting year to time t modeled as Poisson process.

In particular, the (random) number N_t of occurrences of given event to time t has Poisson distribution

$$P(N_t = i) = e^{-\lambda \cdot t} \frac{(\lambda \cdot t)^i}{i!} \quad \text{for } i = 0, 1, \ldots \tag{2.47}$$

with mean value $\lambda \cdot t$. Hence it follows (without any additional assumptions) that the periods $T_1, T_2, \ldots$ between particular occurrences of events are *iid* random variables with exponential distribution and mean value $1/\lambda$ (this conclusion has a logic interpretation: the mean number of occurrences per time unit is $\lambda \cdot 1$ so that the mean period between two occurrences must be $1/\lambda$). The efficient estimate of the intensity λ is $\widehat{\lambda} = n/T$, where n is the observed number of occurrences during period T.

2.4.2 Markov Process

Markov process (similarly as Markov chain in Sect. 2.3) is a general scheme for random processes with discrete states in continuous time. Let for simplicity the

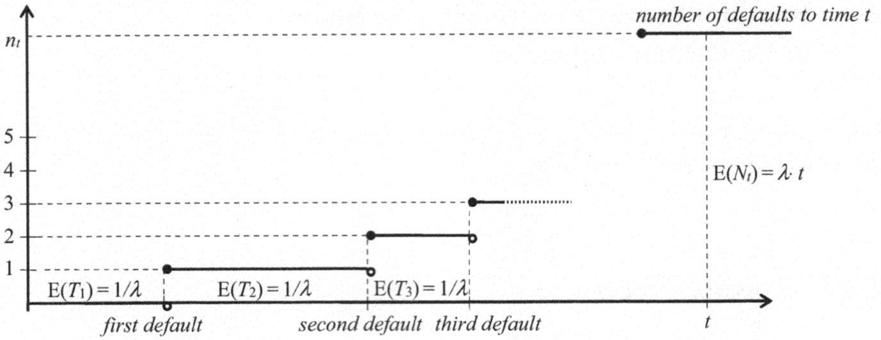

Fig. 2.9 Numbers of mortgage defaults in a bank credit portfolio from the beginning of year to time t modeled as Poisson process (T_1, T_2, ... are random periods between defaults)

possible states of this process be again integer numbers $i = \ldots, -1, 0, 1, \ldots$, and the process can move across these states in any positive times with given transition probabilities. Again the *Markov property* must hold

$$P(Y_{s+t} = j \mid Y_s = i, \ Y_{t_n} = i_n, \ldots, Y_{t_1} = i_1) = P(Y_{s+t} = j \mid Y_s = i) \tag{2.48}$$

for all times $0 \le t_1 < \ldots < t_n < s \le s + t$ and $i, j, i_1, \ldots, i_n = \ldots, -1, 0, 1, \ldots$. The probability on the right-hand side of (2.48) is the *transition probability* from state i at time s to state j at time $s + t$. In the case of homogenous Markov process, it depends only on time t

$$p_{ij}(t) = P(Y_{s+t} = j \mid Y_s = i) \quad \text{for all } s \ge 0. \tag{2.49}$$

Similarly the probabilities representing the *probability distribution* of Markov process at time t are

$$p_i(t) = P(Y_t = i). \tag{2.50}$$

Other important concepts here are so-called *transition intensities*

$$q_{ii} = \lim_{h \to 0+} \frac{p_{ii}(h) - 1}{h}; \ q_{ij} = \lim_{h \to 0+} \frac{p_{ij}(h)}{h}, \quad \text{where } i \ne j, \tag{2.51}$$

i.e.,

$$p_{ii}(h) = 1 + q_{ii} \cdot h + o(h); \ p_{ij}(h) = q_{ij} \cdot h + o(h), \quad \text{where } i \ne j. \tag{2.52}$$

Markov process can be defined directly by transition intensities: in such a case the transition probabilities and the probability distribution of Markov process are obtained by solving so-called *Kolmogorov differential equations*.

Poisson process with intensity $\lambda > 0$ (see above) is the special case of homogenous Markov process with

$$p_{ij}(h) = \begin{cases} \lambda \cdot h + o(h) & \text{for } j = i + 1; \\ 1 - \lambda \cdot h + o(h) & \text{for } j = i; \\ o(h) & \text{for } j > i + 1; \\ 0 & \text{for } j < i, \end{cases} \tag{2.53}$$

i.e., in the interval of small length h the given event occurs just once with probability $\lambda \cdot h + o(h)$ (which is proportional approximately to the length of this interval) and more than once with probability $o(h)$.

Analogously one can define *continuous Markov process* [i.e., the Markov property holds for continuous states in continuous time; see, e.g., Malliaris and Brock (1982)].

2.5 Random Processes with Continuous States in Continuous Time

Finally, two *examples of random processes with continuous states in continuous time* will be presented:

2.5.1 Goniometric Function with Random Amplitude and Phase

For instance, the *sinusoid with random amplitude and phase* is the random process with continuous states in continuous time $\{Y_t, t \geq 0\}$ defined as

$$Y_t(\omega) = A(\omega) \cdot \sin(\nu \cdot t + \Phi(\omega)), \tag{2.54}$$

where A is nonnegative random variable and Φ is random variable with uniform distribution on $\langle 0, 2\pi)$, both A and Φ being mutually independent.

The realization ω_0 with denotation $y_t = Y_t(\omega_0)$, $a = A(\omega_0)$, $\varphi = \Phi(\omega_0)$ gives the trajectory in the form of (deterministic) sinusoid

$$y_t = a \cdot \sin(\nu \cdot t + \phi). \tag{2.55}$$

2.5.2 Wiener Process

Wiener process (or also *Brownian motion*) is the random process with continuous states in continuous time $\{W_t, t \geq 0\}$, where

$$\begin{cases} (i) & W_0 = 0; \\ (ii) & \text{particular trajectories are continuous in time;} \\ (iii) & W_{t_2} - W_{t_1}, \ \ldots, \ W_{t_n} - W_{t_{n-1}} \ \text{are independent for arbitrary } 0 \leq t_1 < \ldots < t_n; \\ (iv) & W_t - W_s \sim N(0, t-s) \ \text{for arbitrary } 0 \leq s < t. \end{cases} \tag{2.56}$$

In particular, the increments $W_{t+h} - W_t$ have the normal distribution $N(0, h)$, and the correlation structure of this process fulfills

$$\text{cov}(W_s, W_t) = \min(s, t), \quad \text{var}(W_t) = t. \tag{2.57}$$

Further, more sophisticated properties of Wiener process (valid with probability one) are, e.g.,

- The particular trajectories are continuous but not differentiable functions of time (i.e., the derivations do not have to exist in any time point).
- The particular trajectories attain any real value infinitely times.
- The particular trajectories have the fractal form (i.e., they "look similarly in any zoom").

Wiener process is the basic concept of majority of financial models. After transforming (to achieve a necessary trend, volatility, and the like), one can apply it to model continuous movements of interest rates or asset prices (when jumps can occur, one must combine Wiener process with Poisson process from Sect. 2.4). Important modifications in practice are the following processes (they are described in more details later in Chap. 10):

1. Wiener process with *drift* μ and *volatility* (or *diffusion coefficient*) σ:

$$\{Y_t = \mu \cdot t + \sigma \cdot W_t, \ t \geq 0\}, \tag{2.58}$$

where $\text{E}(Y_t) = \mu \cdot t$ and $\text{var}(Y_t) = \sigma^2 \cdot t$.

2. *Exponential* Wiener process (or also *geometric* Brownian motion):

$$\{Y_t = \text{e}^{X_t} = \text{e}^{\mu \cdot t + \sigma \cdot W_t}, \ t \geq 0\}, \tag{2.59}$$

where $\text{E}(Y_t) = \exp\{(\mu + \sigma^2/2) \cdot t\}$ and $\text{var}(Y_t) = \exp[(2\mu + \sigma^2) \cdot t] \cdot [\exp(\sigma^2 \cdot t) - 1]$.

Remark 2.3 One can refer to further examples of random processes with discrete states or in continuous time which are more complex so that a specialized literature should be consulted, e.g.:

- *Binary process* originating by *clipping* a stationary process (with continuous states) where simply the values of this stationary process higher or equal to zero are replaced by the value "1" and the values lower than zero are replaced by "0" (a general threshold can be used instead of zero; see Kedem (1980)).
- *Counting process* of nonnegative integer random variables usually correlated over time that modifies the Box–Jenkins methodology for integer-valued processes:

  – *DARMA process* (i.e., discrete mixed process) models a general stationary series of counts with a given marginal distribution (binomial, geometric, Poisson); see, e.g., Jacobs and Lewis (1983), McKenzie (1988). Sometimes Markov chains present a suitable model scheme for such processes; see MacDonald and Zucchini (1997).
  – *INAR process* (i.e., integer autoregressive process) generates integer-valued time series in a manner similar to the autoregressive recursive scheme for continuous random variables; see, e.g., Al-Osh and Alzaid (1987), Kedem and Fokianos (2002), Weiss (2018). In this context, one makes use of the so-called *thinning operator*

$$p \circ X = \sum_{i=1}^{X} Y_i, \tag{2.60}$$

  where $\{Y_t, t = 1, 2, \ldots\}$ are *iid* Bernoulli (i.e., zero-one) random variables with the probability of success equal to p

$$P(Y_t = 1) = 1 - P(Y_t = 0) = p. \tag{2.61}$$

- *CARMA process* (i.e., continuous-time ARMA process) extends the classical Box–Jenkins methodology over continuous time. It has various applications in financial time series, e.g., for the pricing of options; see Brockwell (2009).

◇

2.6 Exercises

Exercise 2.1 Realize practically (e.g., in a group of students) the Delphi method to predict some actual economic or financial themes.

Exercise 2.2 Repeat the calculation from Example 2.2 (the bonus system in motor car insurance), but for five bonus levels (e.g., 100%, 90%, 80%, 70%, and 60% of basic insurance premiums). Moreover, apply for this bonus system the modified rule: if reporting one or more claims the clients grow worse next year by two levels or they remain at the worst level 0 (with 100% of basic insurance premiums).

Part II
Decomposition of Economic Time Series

Chapter 3
Trend

This chapter and Chaps. 4 and 5 describe various methods of additive and multiplicative decomposition which result in the elimination of particular components of time series. In practice, it can have various motivations:

1. First and foremost, the analysis of separated (eliminated) components of time series is useful from the practical point of view since one can detect in such a way various patterns in behavior of time series, identify particular external effects influencing records, and compare several time series and the like (e.g., using the trend, securities dealers can compare the growth rate of various stocks, or using the seasonal component, banks can assess the demand for commercial credits during particular years).
2. Important objectives of decomposition are also predictions of future development of particular components (e.g., which will be the growth rate of contracted mortgages) or predictions of the (non-decomposed) time series constructed by compounding predictions of particular components (which are relatively simple and accurate predictions).
3. Sometimes due to the character of solved problems it is convenient to reveal the behavior of given time series adjusted by removing some components. For example, the economic and financial time series are frequently *seasonally adjusted* (this seasonal adjustment is even demanded for economic time series reported officially by government statistical offices; see also Sect. 2.2.2).

The methods of elimination of particular components of time series differ by various levels of objectivity, accuracy, and computational complexity. The choice of relevant method depends on the motivation for decomposition and on the type of analyzed time series. The methods based on the regression approach are often very popular in this context mostly under the assumption that the residual component is uncorrelated and homoscedastic in time [see the concept of white noise in (2.1)]. Moreover, the normal distribution of the residual component is sometimes assumed (and justified by the Central Limit Theorem since the residuals are resultants of many random effects). If the time series is contaminated by outliers, then one should use

T. Cipra, *Time Series in Economics and Finance*,
https://doi.org/10.1007/978-3-030-46347-2_3

robust decomposition methods which are insensitive to outliers (e.g., one should apply the median instead of the sample average when estimating the constant level of time series) .

3.1 Trend in Time Series

This chapter is devoted to methods suggested to eliminate the trend component from time series and to extrapolate this component to the future. In this context, one speaks of *smoothing* of time series since the seasonal (sometimes even periodic) and random fluctuations of time series are damped down simultaneously. While in Sect. 3.1 we will deal with the classical methods of elimination of trend, in Sects. 3.2 and 3.3 we will present the adaptive approaches that take into account local changes in the character of trend (e.g., the changes in the slope of linear trend).

3.1.1 Subjective Methods of Elimination of Trend

These methods include simple approaches to the elimination of trend based on (computer) graphics (in general, the graphical methods used in time series analysis are mostly supported by special software available for this purpose).

A simple method of this type is based on *upper and lower turning points* (see Fig. 3.1). Here one connects by polygonal lines firstly the upper turning points (i.e., the local maxima of given trajectory) and then the lower turning points (i.e., the local minima of trajectory), and finally it suffices to plot the middles between the upper and lower lines as the smoothing result for each time point. The method is subjective since one can do suitable corrections subjectively (e.g., one can ignore outliers or

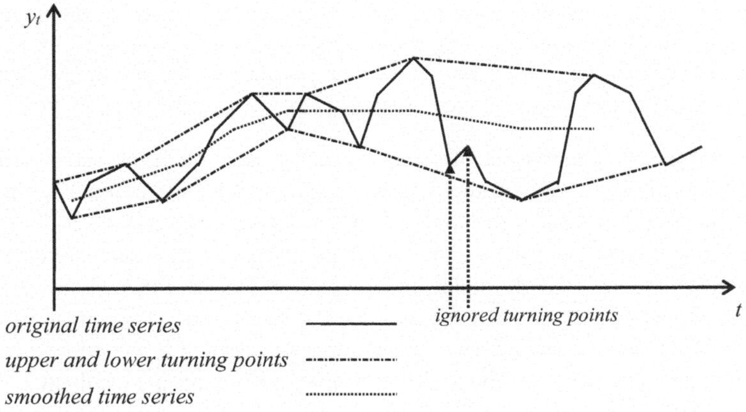

Fig. 3.1 Method based on upper and lower turning points

irrelevant local turning points which are not characteristic for cycle identification; see Fig. 3.1).

3.1.2 Trend Modeling by Mathematical Curves

Section 3.1.2 describes the methods that express the trend analytically by simple curves used in mathematics (e.g., by the line or logarithmic curve). Such estimated curves enable to calculate in a natural way their future values, i.e., as a matter of fact, to construct predictions of the trend component (under the assumption that its future character will sustain in time).

Using this philosophy, one usually assumes that the analyzed time series can be modeled as

$$y_t = Tr_t + E_t \tag{3.1}$$

(or one has transformed the time series to this form by methods described in the following chapters, e.g., by means of the seasonal adjustment). Moreover, the residual component in (3.1) has the properties of white noise. These assumptions permit to apply suitable regression methods when estimating the parameters of trend curves, and then to take directly the corresponding regression extrapolations for Tr_t as the predictions for y_t.

The choice of type of the most appropriate mathematical curve for particular time series is based on a preliminary analysis, usually by means of graphical records of time series or by using expected properties of the trend component following, e.g., from the economic theory (however, it is obvious that one cannot suppress completely subjective impacts here). Several reference tests for the choice of the most appropriate mathematical curves for given trajectory $y_1, \ldots, y_n$ are shown in Table 3.6. There also exist systematic typologies where the controlled movement along particular knots of the typological tree offers the most appropriate curve according to answers to selecting questions (e.g., "the analyzed trajectory is/isn't symmetric around the point of inflection ?").

Now we will survey favorite trend curves including the formulas for estimation of their parameters and for construction of their (point and interval) predictions in the time series models of the type (3.1):

3.1.2.1 Linear Trend

It is a simple trend in the form of straight line

$$Tr_t = \beta_0 + \beta_1 \; t, \qquad t = 1, \ldots, n. \tag{3.2}$$

The OLS estimates b_0 and b_1 of parameters β_0 and β_1 fulfill the system of normal equations

$$\begin{aligned} b_0 n &+ b_1 \sum_{t=1}^{n} t &= \sum_{t=1}^{n} y_t, \\ b_0 \sum_{t=1}^{n} t &+ b_1 \sum_{t=1}^{n} t^2 &= \sum_{t=1}^{n} t y_t \end{aligned} \tag{3.3}$$

with solution given by the formulas

$$\begin{aligned} b_1 &= \frac{\sum_{t=1}^{n} t y_t - \bar{t} \sum_{t=1}^{n} y_t}{\sum_{t=1}^{n} t^2 - n\, \bar{t}^2} = \frac{\sum_{t=1}^{n} t y_t - \frac{n+1}{2} \sum_{t=1}^{n} y_t}{\frac{n(n^2-1)}{12}}, \\ b_0 &= \bar{y} - \bar{t}\, b_1 = \bar{y} - \frac{n+1}{2} b_1. \end{aligned} \tag{3.4}$$

The prediction $\widehat{y}_T$ of future value y_T has the form

$$\widehat{y}_T = b_0 + b_1 T \tag{3.5}$$

and $(1 - p)\cdot 100\%$ prediction interval for this value (e.g., 95% interval if $p = 0.05$; see Sect. 2.2.3) under the normality assumption (or normality achieved asymptotically) is

$$\left(b_0 + b_1 T - t_{1-\ p/2}(n-2) \cdot s \cdot f_T,\quad b_0 + b_1 T + t_{1-\ p/2}(n-2) \cdot s \cdot f_T\right), \tag{3.6}$$

where

$$s = \sqrt{\frac{\sum_{t=1}^{n} (y_t - \widehat{y}_t)^2}{n-2}} = \sqrt{\frac{\sum_{t=1}^{n} y_t^2 - \sum_{t=1}^{n} \widehat{y}_t^2}{n-2}},$$

$$f_T = \sqrt{1 + \frac{1}{n} + \frac{\left(T - \frac{n+1}{2}\right)^2}{\frac{n(n^2-1)}{12}}}. \tag{3.7}$$

Example 3.1 (*linear trend*). Table 3.1 and Fig. 3.2 present the elimination of linear trend in the time series y_t of the Swiss gross national income (at current prices in billions of CHF) for particular years 1980–2015 ($t = 1, ..., 36$). The spreadsheet of EViews 7 in Table 3.2 and the predictions for particular years 2016–2025 in Table 3.1 coincide with the calculations according to the formulas (3.4) and (3.5)

Table 3.1 Annual data 1980–2015, eliminated linear trend, and predictions for years 2016–2025 in Example 3.1 (*Swiss gross national income in bn CHF*)

t	Year	y_t (bn CHF)	$\hat{y}_t$ (bn CHF)	t	Year	y_t (bn CHF)	$\hat{y}_t$ (bn CHF)
1	1980	203.9	211.7	24	2003	505.9	517.5
2	1981	220.7	225.0	25	2004	520.5	530.8
3	1982	231.4	238.3	26	2005	550.8	544.1
4	1983	239.4	251.6	27	2006	579.2	557.4
5	1984	257.7	264.9	28	2007	577.4	570.7
6	1985	273.2	278.2	29	2008	559.0	584.0
7	1986	284.6	291.5	30	2009	599.1	597.3
8	1987	294.9	304.8	31	2010	642.8	610.6
9	1988	315.3	318.1	32	2011	624.3	623.9
10	1989	339.4	331.4	33	2012	637.6	637.2
11	1990	365.8	344.7	34	2013	649.6	650.5
12	1991	382.4	358.0	35	2014	649.8	663.8
13	1992	389.2	371.3	36	2015	660.3	677.1
14	1993	399.6	384.6	37	2016		690.4
15	1994	406.1	397.9	38	2017		703.7
16	1995	414.5	411.2	39	2018		717.0
17	1996	419.0	424.5	40	2019		730.3
18	1997	435.0	437.8	41	2020		743.6
19	1998	448.9	451.0	42	2021		756.9
20	1999	460.4	464.3	43	2022		770.2
21	2000	489.3	477.6	44	2023		783.5
22	2001	488.9	490.9	45	2024		796.7
23	2002	482.4	504.2	46	2025		810.0

Source: AMECO (European Commission Annual Macro-Economic Database). (https://ec.europa.eu/economy_finance/ameco/user/serie/SelectSerie.cfm)

$$b_1 = \frac{\sum_{t=1}^{36} t y_t - \frac{36+1}{2} \sum_{t=1}^{36} y_t}{\frac{36\left(36^2-1\right)}{12}} = \frac{51\ 655.23}{3\ 885} = 13.296\ 07,$$

$$b_0 = \bar{y} - \frac{36+1}{2} b_1 = 444.400\ 9 - 18.5 \cdot 13.296\ 07 = 198.423\ 6,$$

$$\hat{y}_{37} = b_0 + b_1 \cdot 37 = 198.423\ 6 + 13.296\ 07 \cdot 37 = 690.4,$$

etc. Further according to (3.6) and (3.7), we calculated the 95% prediction intervals, e.g.:

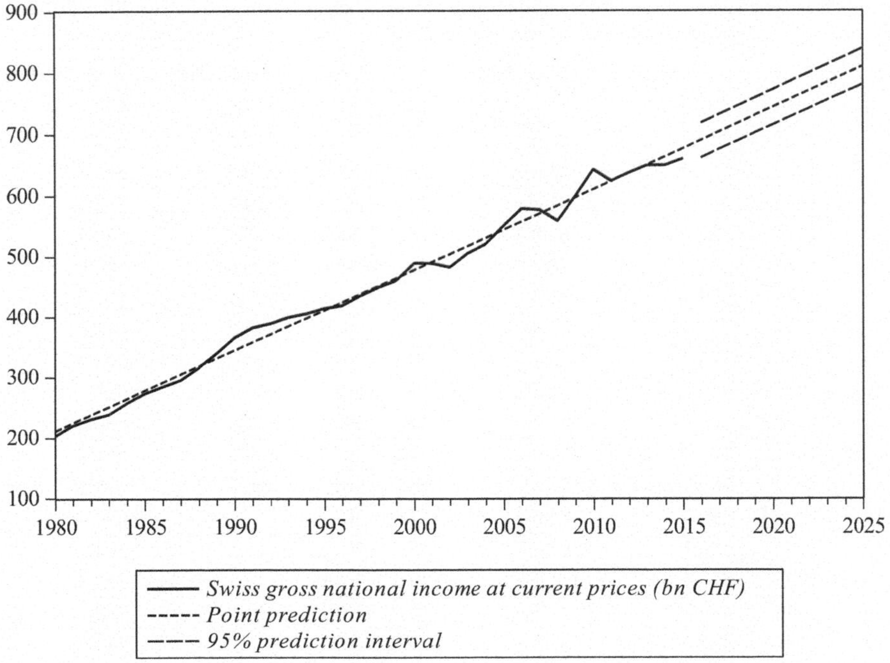

Fig. 3.2 Annual data 1980–2015, eliminated linear trend, and predictions for years 2016–2025 in Example 3.1 (*Swiss gross national income in bn CHF*)

Table 3.2 Spreadsheet of EViews 7 for the Swiss gross national income from Example 3.1 (*elimination of linear trend*)

Dependent variable: M1				
Method: Least squares				
Sample: 1980–2015				
Included observations: 36				
Variable	Coefficient	Std. Error	*t*-Statistic	Prob.
C	198.4236	4.469807	44.39198	0.0000
T	13.29607	0.210670	63.11334	0.0000
R-squared	0.991537	S.E. of regression		13.13100

$$s = \sqrt{\frac{\sum_{t=1}^{36} y_t^2 - \sum_{t=1}^{36} \widehat{y}_t^2}{34}} = 13.131\ 00,$$

$$f_{37} = \sqrt{1 + \frac{1}{36} + \frac{\left(37 - \frac{36+1}{2}\right)^2}{\frac{36(36^2-1)}{12}}} = 1.056,$$

$$\begin{aligned} \widehat{y}_{37} \pm t_{0.975}(36-2) \cdot s \cdot f_{37} &= 690.4 \pm 2.032\ 2 \cdot 13.131\ 00 \cdot 1.056 \\ &= 690.4 \pm 28.2, \end{aligned}$$

i.e.,

$$(662.2; 718.6),$$

etc. (see Fig. 3.2).

◇

Remark 3.1 As the polynomial trends of higher orders are concerned, the *quadratic trend* can be also found in economic and financial applications

$$Tr_t = \beta_0 + \beta_1 \; t + \beta_2 \; t^2, \qquad t = 1, \; \ldots, \; n. \tag{3.8}$$

The prediction $\widehat{y}_T$ of future value y_T has the form

$$\widehat{y}_T = b_0 + b_1 T + b_2 T^2, \tag{3.9}$$

and the corresponding (1-p)·100% prediction interval is

$$\left(\widehat{y}_T - t_{1-p/2}(n-3) \cdot s \cdot f_T, \quad \widehat{y}_T + t_{1-p/2}(n-3) \cdot s \cdot f_T\right), \tag{3.10}$$

where

$$s = \sqrt{\frac{\sum_{t=1}^{n} (y_t - \widehat{y}_t)^2}{n-3}} = \sqrt{\frac{\sum_{t=1}^{n} y_t^2 - \sum_{t=1}^{n} \widehat{y}_t^2}{n-3}},$$

$$f_T = \sqrt{1 + (1, T, T^2)(\mathbf{X}'\mathbf{X})^{-1}(1, T, T^2)'},$$

$$\mathbf{X} = \begin{pmatrix} 1 & 1 & 1 \\ 1 & 2 & 4 \\ \vdots & \vdots & \vdots \\ 1 & n & n^2 \end{pmatrix}. \tag{3.11}$$

◇

3.1.2.2 Exponential Trend

It is the two-parametric trend in the form of exponential

$$Tr_t = \alpha \beta^t, \qquad t = 1, \; \ldots, \; n \qquad (\beta > 0) \tag{3.12}$$

(the parameters are denoted as α and β). This trend has two typical characteristics, namely that both its *coefficient of growth* (i.e., the ratio of neighboring values Tr_{t+1}/Tr_t) and the ratio of neighboring differences

$$\frac{Tr_{t+2} - Tr_{t+1}}{Tr_{t+1} - Tr_t} \tag{3.13}$$

are constant with value β in time. If $\alpha > 0$, then the exponential trend is increasing for $\beta > 1$ and decreasing for $0 < \beta < 1$. The both parameters of exponential trend can be estimated by taking its logarithm which transfers this trend to the linear one

$$\ln Tr_t = \ln \alpha + t \ln \beta. \tag{3.14}$$

Then it is sufficient to find the antilogarithm for the estimated parameters ln α and ln β (in any case, if one conjectures that an analyzed trend could be exponential, one should plot the corresponding time series using the logarithmic scale). Moreover, by taking the antilogarithm of the prediction intervals in the linear model (3.14), one can also construct the prediction intervals in the original exponential model. On the other hand, practical experiences with the exponential trend (3.12) (and also with other models of nonlinear regression which can be transferred to the linear regression using a suitable transformation) show that more consistent estimation results can be obtained using the *weighted least squares method* (WLS) with weights which are obtained by a suitable transformation of the original weights since it is not possible to assume the multiplicative form and the normal-logarithmic distribution of residual components in the original model (3.12) before transformation. In particular for the exponential trend, WLS method consists in minimizing the expression

$$\sum_{t=1}^{n} v_t (y_t - \alpha\beta^t)^2, \tag{3.15}$$

where the weights v_t are chosen in advance. However instead of the expression (3.15), one minimizes the sum of weighted least squares of the form

$$\sum_{t=1}^{n} w_t (\ln y_t - \ln \alpha - t \ln \beta)^2, \tag{3.16}$$

where the weights w_t are constructed in dependence on the original weights v_t in such a way that the minimization of (3.15) and (3.16) provides nearly identical estimates α and β. It can be shown that in our case of logarithmic transformation one can put

$$w_t = y_t^2 v_t, \qquad t = 1, \ \ldots, \ n. \tag{3.17}$$

Since the most usual choice of original weights is $v_t = 1$, $t = 1, ..., n$ (if there is a priori no reason to prefer some of given observations), the transformed weights are simply $w_t = y_t^2$, $t = 1, \ldots, n$. Minimizing the expression (3.16) with the weights (3.17), one obtains the following system of normal equations:

$$\begin{aligned}\left(\sum y_t^2\right)\ln\alpha &+ \left(\sum ty_t^2\right)\ln\beta = \sum y_t^2 \ln y_t,\\ \left(\sum ty_t^2\right)\ln\alpha &+ \left(\sum t^2y_t^2\right)\ln\beta = \sum ty_t^2 \ln y_t\end{aligned}$$

with explicit solution

$$\begin{aligned}\ln a &= \frac{\sum t^2y_t^2 \cdot \sum y_t^2 \ln y_t - \sum ty_t^2 \cdot \sum ty_t^2 \ln y_t}{\sum y_t^2 \cdot \sum t^2y_t^2 - \left(\sum ty_t^2\right)^2},\\ \ln b &= \frac{\sum y_t^2 \cdot \sum ty_t^2 \ln y_t - \sum ty_t^2 \cdot \sum y_t^2 \ln y_t}{\sum y_t^2 \cdot \sum t^2y_t^2 - \left(\sum ty_t^2\right)^2}.\end{aligned} \tag{3.18}$$

Example 3.2 (*exponential trend*). Table 3.3 and Fig. 3.3 present the elimination of exponential trend in the time series y_t of the US gross national income (at current prices in billions of USD) for particular years 1960–2016 ($t = 1, \ldots, 57$). The auxiliary results for formulas (3.18) given in Table 3.4 enable to calculate the estimated parameters:

$$\ln a = 7.203\,85, \quad \text{i.e.} \quad a = 1\,344.60,$$
$$\ln b = 0.047\,816, \quad \text{i.e.} \quad b = 1.048\,98$$

so that the eliminated exponential trend (regarded as the smoothed time series in practice) can be calculated in Table 3.3 as

$$\widehat{y}_t = 1\,344.60 \cdot 1.048\,98^t.$$

Figure 3.3 also plots the modified exponential trend (see (3.19) below) that fits obviously better the given time series than the exponential trend.

◇

3.1.2.3 Modified Exponential Trend

This trend of the type

$$Tr_t = \gamma + \alpha\beta^t, \quad t = 1, \ldots, n \quad (\beta > 0) \tag{3.19}$$

is the three-parametric generalization of the exponential trend (the parameters are denoted as α, β ,and γ). It is suitable to model trends with constant ratio of

Table 3.3 Annual data 1960–2016 and eliminated exponential trend in Example 3.2 (*US gross national income in bn USD*)

Year	t	y_t (bn USD)	$\widehat{y}_t$ (bn USD)	Year	t	y_t (bn USD)	$\widehat{y}_t$ (bn USD)	Year	t	y_t (bn USD)	$\widehat{y}_t$ (bn USD)
1960	1	542.7	1410.5	1979	20	2619.3	3498.8	1998	39	9167.6	8679.1
1961	2	562.0	1479.5	1980	21	2852.8	3670.1	1999	40	9725.3	9104.2
1962	3	604.7	1552.0	1981	22	3207.1	3849.9	2000	41	10,421.2	9550.1
1963	4	638.1	1628.0	1982	23	3374.7	4038.5	2001	42	10,788.6	10,017.8
1964	5	686.4	1707.7	1983	24	3621.0	4236.3	2002	43	11,098.9	10,508.5
1965	6	744.6	1791.4	1984	25	4038.3	4443.7	2003	44	11,591.4	11,023.2
1966	7	815.9	1879.1	1985	26	4320.9	4661.4	2004	45	12,372.6	11,563.0
1967	8	862.4	1971.2	1986	27	4530.4	4889.7	2005	46	13,221.8	12,129.4
1968	9	943.3	2067.7	1987	28	4847.2	5129.2	2006	47	14,140.8	12,723.4
1969	10	1020.7	2169.0	1988	29	5275.8	5380.4	2007	48	14,585.8	13,346.6
1970	11	1076.9	2275.2	1989	30	5618.3	5643.9	2008	49	14,791.2	14,000.3
1971	12	1165.9	2386.6	1990	31	5922.9	5920.3	2009	50	14,494.5	14,686.0
1972	13	1283.9	2503.5	1991	32	6117.2	6210.3	2010	51	15,121.1	15,405.3
1973	14	1435.1	2626.1	1992	33	6459.4	6514.4	2011	52	15,802.9	16,159.8
1974	15	1557.0	2754.8	1993	34	6758.4	6833.5	2012	53	16,596.1	16,951.3
1975	16	1688.6	2889.7	1994	35	7195.8	7168.2	2013	54	17,073.7	17,781.5
1976	17	1874.0	3031.2	1995	36	7602.3	7519.3	2014	55	17,899.1	18,652.4
1977	18	2087.0	3179.7	1996	37	8075.4	7887.6	2015	56	18,496.1	19,565.9
1978	19	2355.0	3335.4	1997	38	8620.4	8273.9	2016	57	19,041.6	20,524.2
Σ		21,944.2		Σ		101,057.6		Σ		266,430.3	

Source: AMECO (European Commission Annual Macro-Economic *Database)* (https://ec.europa.eu/economy_finance/ameco/user/serie/SelectSerie.cfm)

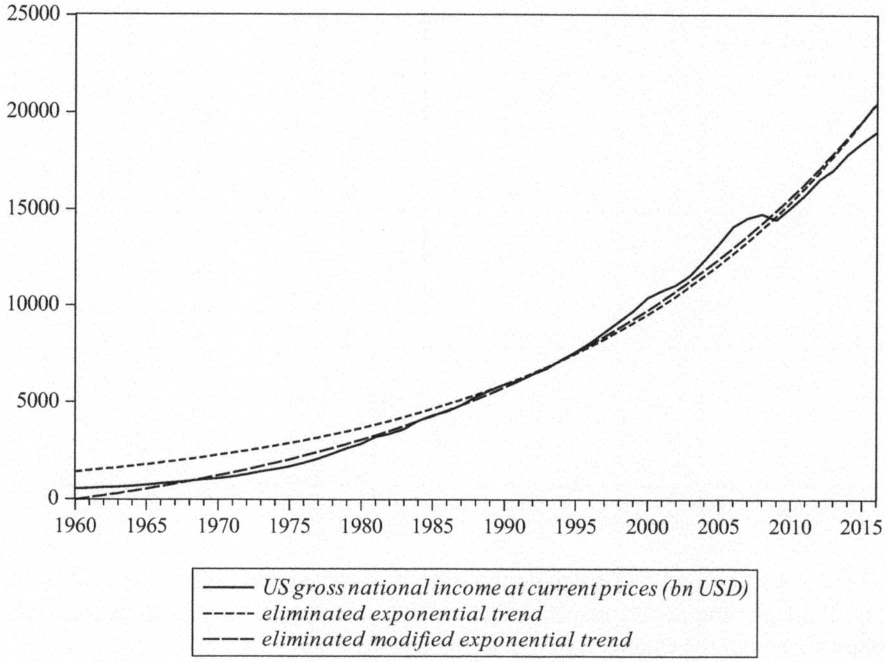

Fig. 3.3 Annual data 1960–2016 and eliminated exponential trend in Example 3.2 (*US gross national income in bn USD*)

Table 3.4 Auxiliary results in Example 3.2 (*exponential trend*)

$\sum y_t^2$	$\sum t y_t^2$	$\sum t^2 y_t^2$	$\sum \ln y$	$\sum y_t^2 \ln y_t$	$\sum t y_t^2 \ln y_t$
4 528 876 399	215 566 247 678	10 576 070 915 906	475	42 932 861 468	2 058 612 296 639

neighboring differences, but in addition to this property of exponential trend, the modified trend can be bounded asymptotically (sometimes one formulates it practically as a movement to saturation level; see Fig. 3.4).

The approximative parameter estimation for the modified exponential trend (optionally before using more accurate iterative procedures of nonlinear optimization) consists in the following approach. We split the observed time series into three parts of the same length m (if $n \neq 3m$, then we omit one or two observations preferably in the beginning of time series) and add up the observations in particular thirds so that one can write

$$\sum\nolimits_1 y_t \sim \sum\nolimits_1 Tr_t = m\gamma + \frac{\alpha\beta(\beta^m - 1)}{\beta - 1},$$

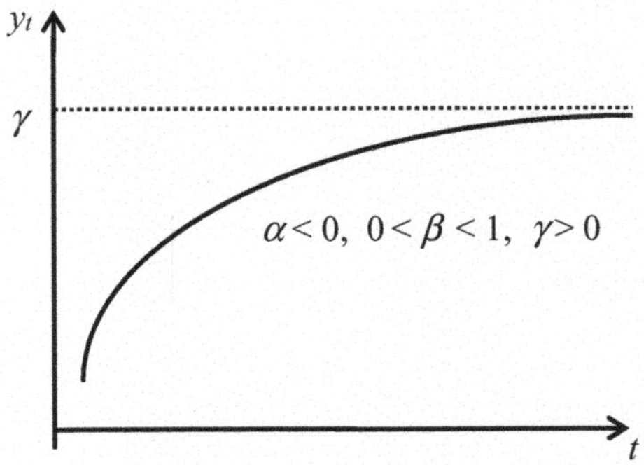

Fig. 3.4 Modified exponential trend

$$\sum\nolimits_2 y_t \sim \sum\nolimits_2 Tr_t = m\gamma + \frac{\alpha\beta^{m+1}(\beta^m - 1)}{\beta - 1},$$

$$\sum\nolimits_3 y_t \sim \sum\nolimits_3 Tr_t = m\gamma + \frac{\alpha\beta^{2m+1}(\beta^m - 1)}{\beta - 1},$$

where, e.g., $\sum_1 y_t$ and $\sum_1 Tr_t$ denote the sum of observed and trend values from the first third of time series, respectively. Solving this system of equations, one can stepwise obtain the estimates *b*, *a*, *c* of parameters β , α, γ as

$$b = \left(\frac{\sum_3 y_t - \sum_2 y_t}{\sum_2 y_t - \sum_1 y_t}\right)^{1/m}, \tag{3.20}$$

$$a = \frac{b-1}{b(b^m - 1)^2}\left(\sum\nolimits_2 y_t - \sum\nolimits_1 y_t\right), \tag{3.21}$$

$$c = \frac{1}{m}\left(\sum\nolimits_1 y_t - \frac{ab(b^m - 1)}{b - 1}\right). \tag{3.22}$$

Another approach is also possible: If fixing the value of parameter β, the model (3.19) will become obviously the linear model in which one estimates simply the parameters α and γ for various fixed values β and chooses finally the variant minimizing *SSE*.

Example 3.3 (*modified exponential trend*). Table 3.5 and Fig. 3.5 present the elimination of the modified exponential trend in the time series y_t of the Japan gross national income (at current prices in billions of JPY) for particular years 1960–2016 ($t = 1, ..., 57$).

The data are divided into three groups ($m = 19$), for which particular sums are calculated (see Table 3.5). Then the formulas (3.20)–(9.22) provide stepwise the following results:

Table 3.5 Annual data 1960–2016 and eliminated modified exponential trend in Example 3.3 (*Japan gross national income in bn JPY*)

Year	t	y_t	Year	t	y_t	Year	t	y_t
1960	1	16,421	1979	20	227,692	1998	39	519,390
1961	2	19,817	1980	21	246,449	1999	40	511,280
1962	3	22,480	1981	22	264,544	2000	41	516,340
1963	4	25,717	1982	23	278,328	2001	42	513,933
1964	5	30,225	1983	24	289,723	2002	43	507,189
1965	6	33,640	1984	25	308,153	2003	44	507,117
1966	7	39,080	1985	26	331,538	2004	45	513,112
1967	8	45,807	1986	27	346,885	2005	46	515,652
1968	9	54,222	1987	28	361,520	2006	47	521,152
1969	10	63,708	1988	29	388,725	2007	48	530,313
1970	11	75,124	1989	30	419,067	2008	49	518,002
1971	12	82,724	1990	31	452,267	2009	50	484,216
1972	13	94,845	1991	32	479,613	2010	51	495,651
1973	14	115,496	1992	33	492,078	2011	52	486,254
1974	15	137,541	1993	34	495,227	2012	53	490,386
1975	16	152,089	1994	35	499,681	2013	54	496,725
1976	17	170,819	1995	36	505,821	2014	55	506,607
1977	18	190,438	1996	37	517,710	2015	56	522,127
1978	19	209,883	1997	38	530,218	2016	57	525,062
Σ		1,580,076	Σ		7,435,239	Σ		9,680,508

Source: AMECO (European Commission Annual Macro-Economic Database) (https://ec.europa.eu/economy_finance/ameco/user/serie/SelectSerie.cfm)

$$b = \left(\frac{\sum_3 y_t - \sum_2 y_t}{\sum_2 y_t - \sum_1 y_t}\right)^{1/m} = \left(\frac{9\ 680\ 508 - 7\ 435\ 239}{7\ 435\ 239 - 1\ 580\ 076}\right)^{1/19} = 0.950\ 804,$$

$$a = \frac{b-1}{b(b^m - 1)^2}\left(\sum_2 y_t - \sum_1 y_t\right)$$

$$= \frac{0.950\ 804 - 1}{0.950\ 804\left(0.950\ 804^{19} - 1\right)^2}(7\ 435\ 239 - 1\ 580\ 076) = -797\ 015,$$

$$c = \frac{1}{m}\left(\sum_1 y_t - \frac{ab(b^m - 1)}{b-1}\right)$$

$$= \frac{1}{19}\left(1\ 580\ 076 - \frac{(-797\ 015)\cdot 0.950\ 804\cdot\left(0.950\ 804^{19} - 1\right)}{0.950\ 804 - 1}\right) = 583\ 001.$$

The eliminated modified exponential trend (regarded again as the smoothed time series) can be calculated as

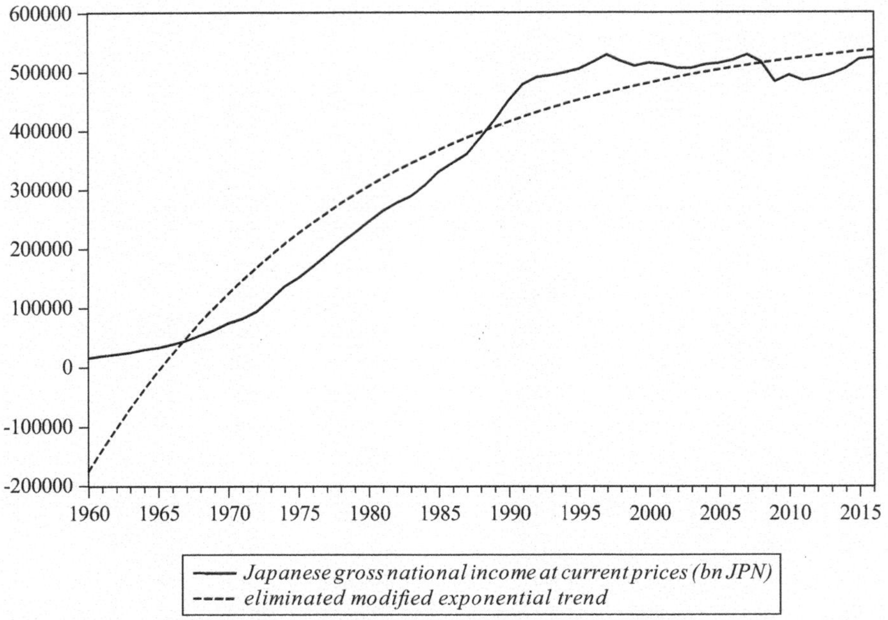

Fig. 3.5 Annual data 1960–2016 and eliminated modified exponential trend in Example 3.3 (*Japan gross national income in bn JPY*)

$$\hat{y}_t = 583\ 001 - 797\ 015 \cdot 0.950\ 804^t$$

and it is plotted in Fig. 3.5 (moreover, the saturation "insurmountable" level for the Japan gross national income according to this model should be approximately 583 000 bn JPY).

◇

3.1.2.4 Logistic Trend

This three-parametric trend has the form

$$Tr_t = \frac{\gamma}{1 + \alpha\beta^t}, \qquad t = 1, \quad \ldots, \quad n \qquad (\beta > 0, \ \gamma > 0). \tag{3.23}$$

It is plotted schematically in Fig. 3.6.

The logistic trend has the inflection (i.e., the change of convex course to concave one and vice versa) in the point $t = -\ln\alpha/\ln\beta$. It can be also bounded asymptotically (see again the saturation level γ in Fig. 3.6). If deriving this trend with respect to time t (time is regarded in this context as a continuous variable), one obtains

$$\frac{\mathrm{d}Tr_t}{\mathrm{d}t} = -\frac{\ln \beta}{\gamma} Tr_t(\gamma - Tr_t). \tag{3.24}$$

It is another important indicator of the growth of trend curves (in general, the first derivative of a trend curve is usually called the *growth function*). According to (3.24), the velocity of growth of logistic trend is directly proportional to the achieved level Tr_t and to the distance of the achieved level from the saturation level, i.e., $\gamma - Tr_t$; see Fig. 3.6. Moreover, the first derivative (3.24) is symmetric around inflection point $-\ln\alpha/\ln\beta$. Hence the logistic trend can be classified as so-called *S-curve symmetric around inflection point* (S-curves have been discussed in Sect. 2.2.3, e.g., as a suitable instrument for modeling sales of new products; see Fig. 2.2).

As the estimation of logistic trend is concerned, its parameters can be estimated by means of various methods. For example, the logistic trend can be regarded as the reciprocal value of modified exponential trend so that one can apply the formulas (3.20)–(3.22) for the time series with values $1/y_t$. Another approach consists in so-called *difference parametric estimation* which is based on the time series of the first differences $y_{t+1} - y_t$. Here we approximate the trend component Tr_t in (3.24) by the real observations y_t so that one can write

$$\frac{\mathrm{d}y_t}{\mathrm{d}t} \sim -\frac{\ln \beta}{\gamma} y_t(\gamma - y_t). \tag{3.25}$$

If we approximate further

$$\frac{\mathrm{d}y_t}{\mathrm{d}t} \sim \frac{y_{t+1} - y_t}{(t+1) - t} = y_{t+1} - y_t = d_t, \tag{3.26}$$

where d_t denotes the time series of the first differences, then it follows from (3.25)

$$\frac{d_t}{y_t} \sim -\ln \beta + \frac{\ln \beta}{\gamma} y_t. \tag{3.27}$$

Using the classical least squares method in the linear regression model

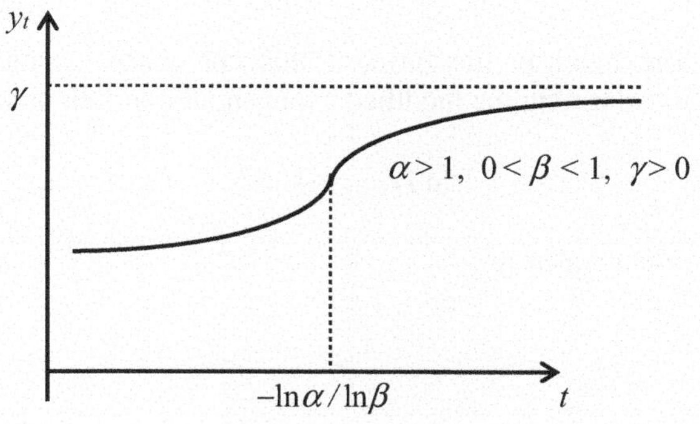

Fig. 3.6 Logistic trend

$$\frac{d_t}{y_t} = -\ln\beta + \frac{\ln\beta}{\gamma} y_t + \varepsilon_t \tag{3.28}$$

one obtains the OLS estimates of $-\ln\beta$ and $\ln\beta/\gamma$ and hence the estimates of parameters β and γ. In order to obtain the estimate of α, we finally approximate Tr_t by y_t in (3.23)

$$\alpha\beta^t \sim \frac{\gamma}{y_t} - 1. \tag{3.29}$$

After taking the logarithm and making the sum over $t = 1, \ldots, n$ one gets so-called *Rhodes formula*

$$\ln\alpha = -\frac{(n+1)\ln\beta}{2} + \frac{1}{n}\sum_{t=1}^{n} \ln\left(\frac{\gamma}{y_t} - 1\right), \tag{3.30}$$

which enables to estimate the parameter α.

Example 3.4 (*logistic trend*). Fig. 3.7 presents the elimination of logistic trend for the data from Example 3.3 (Japan gross national income at current prices in billions of JPY) estimated as

$$\widehat{y}_t = \frac{518\ 158}{1 + 37.143\ 7 \cdot 0.845\ 748^t}.$$

Obviously the logistic trend fits the given time series better than the modified exponential trend in Fig. 3.5 (e.g., the estimated values of modified exponential trend have turned out negative at the beginning of the given time series). The saturation level for the Japan gross national income is approximately 518 160 bn JPY in this case.

◇

3.1.2.5 Gompertz Trend

The trend in the form of this curve arises similarly to the logistic trend by transforming the modified exponential trend. In this case, one puts

$$\ln Tr_t = \gamma + \alpha\beta^t, \qquad t = 1, \ \ldots, \ n \qquad (\beta > 0) \tag{3.31}$$

or equivalently

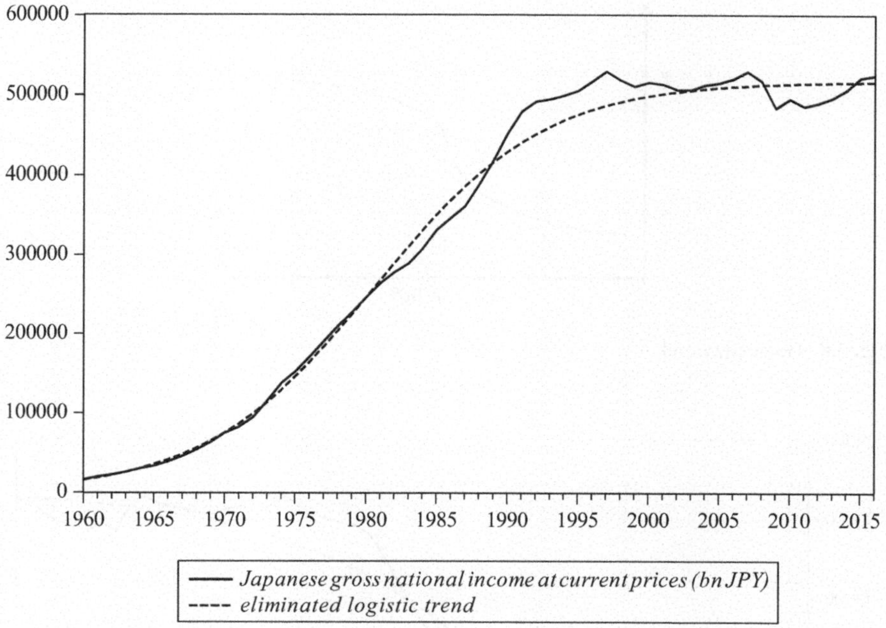

Fig. 3.7 Annual data 1960–2016 and eliminated logistic trend in Example 3.4 (*Japan gross national income in bn JPY*)

$$Tr_t = \exp(\gamma + \alpha\beta^t), \qquad t = 1, \ \ldots, \ n \qquad (\beta > 0). \tag{3.32}$$

If applying the parameter values from Fig. 3.8, then Gompertz trend has the inflection in the point $t = -\ln(-\alpha)/\ln\beta$ and is bounded asymptotically. However, the first derivative of this curve (i.e., the growth function) is not symmetric around inflection point, but it is skewed to the right. Hence Gompertz trend is classified as the *S-curve asymmetric around inflection point*. The estimation procedure is similar to that for the modified exponential trend using the time series with values ln y_t.

Example 3.5 (*Gompertz trend*). Fig. 3.9 presents the elimination of Gompertz trend for the data from Example 3.3 (Japan gross national income at current prices in billions of JPY) estimated as

$$\hat{y}_t = \exp(13.2004 - 489.071 \cdot 0.909\ 765^t).$$

The model implies the saturation level for the Japan gross national income approximately 540 580 bn JPY.

Examples 3.1–3.5 demonstrate that time series of the same type (in our case the gross national incomes) can be modeled using different trend curves. The choice of the appropriate curve may depend on the economic or financial hypotheses: the national income will not be saturated in the future (then e.g. the linear or even

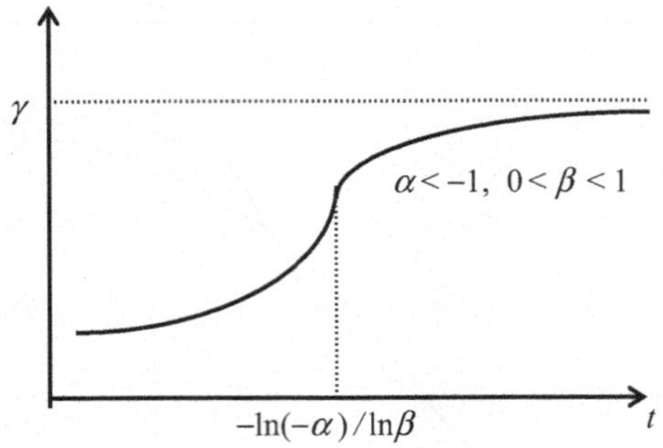

Fig. 3.8 Gompertz trend

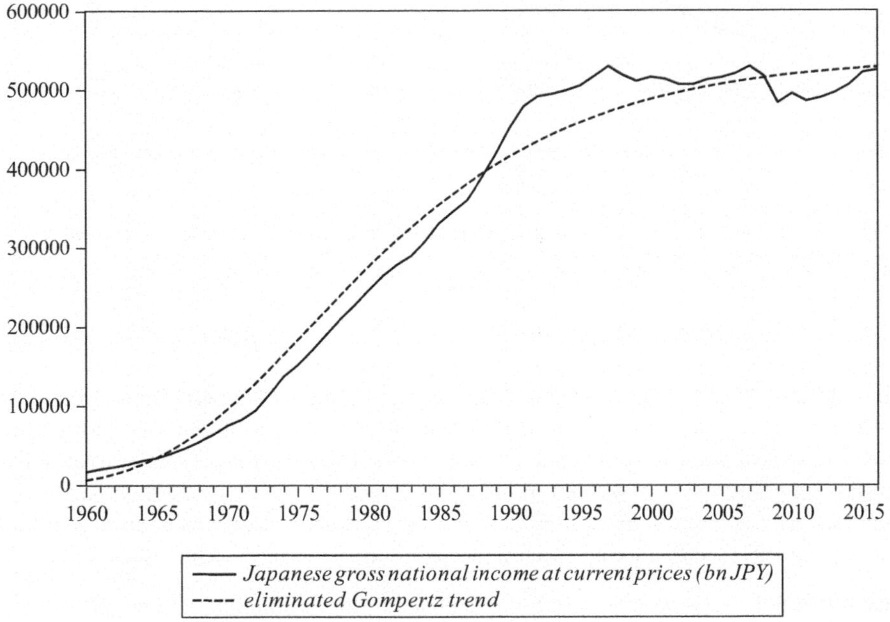

Fig. 3.9 Annual data 1960–2016 and eliminated Gompertz trend in Example 3.5 (*Japan gross national income in bn JPY*)

exponential trend) or there will be different levels of saturation (in the Japanese case the lowest level 518 160 bn JPY applying the logistic trend or the highest one 583 000 bn JPY applying the modified exponential trend).

◇

Remark 3.2 The examples of trend curves given in this section can be parametrized in other ways, e.g., *logistic trend* (3.23) as

$$Tr_t = \frac{\beta_0}{1 + \exp(\beta_1 + \beta_2 t)}, \qquad t = 1, \quad \ldots, \quad n, \tag{3.33}$$

or *Gompertz trend* (3.32) as

$$Tr_t = \frac{\beta_0}{\exp(\beta_1 \cdot \beta_2^t)}, \qquad t = 1, \quad \ldots, \quad n. \tag{3.34}$$

Moreover, there is plenty of other trend curves, e.g., *logarithmic trend*

$$Tr_t = \beta_0 + \beta_1 \ln t, \qquad t = 1, \quad \ldots, \quad n, \tag{3.35}$$

which seems to resemble the modified exponential trend in Fig. 3.4 except for the fact that it grows indefinitely (not to a saturation level), or *Johnson trend*

$$Tr_t = \frac{\beta_0}{\exp\left(\frac{\beta_1}{\beta_2 + t}\right)}, \qquad t = 1, \quad \ldots, \quad n. \tag{3.36}$$

◇

3.1.2.6 Splines

Sometimes the trend changes its character in time and cannot be modeled by means of a single mathematical curve over the whole range of observations (or only in a complicated way). In such a case, one can use the technique of so-called *spline functions*. Here instead of applying sophisticated mathematical functions, one splits the given time series to several segments and estimates the trends in particular segments by simpler functions linked to each other. Moreover, the joint curve must be sufficiently smooth which can be guaranteed, e.g., by means of conditions for the existence of two-sided derivatives of appropriate orders in joint points. Splines consist frequently of piecewise polynomials with pieces defined by a sequence of knots where the pieces join smoothly.

As an example, Fuller (1976) used for the time series of average wheat yields (in the USA in years 1908–1971) the trend which is compounded from the following curves ($t = 1$ corresponds to the year 1908):

$$\begin{aligned} Tr_t &= 13.97, & t &= 1, \ldots, 25, \\ Tr_t &= 13.97 + 0.0123(t - 25)^2, & t &= 25, \ldots, 54, \\ Tr_t &= 24.314 + 0.664(t - 54), & t &= 54, \ldots, 64. \end{aligned}$$

In the first joint point $t = 25$, there exists the two-sided derivative of the first order, while in the second joint point $t = 54$, the corresponding one-sided first derivatives obviously differ from each other.

The simplest case of the spline function is a *piecewise linear function* that is linear in all segments but with different slopes. However, such a function is not flexible enough and, moreover, is not smooth in knots. In practice it is common to use *cubic splines* with such cubic polynomials in particular segments that the two-sided derivatives of the second order exist in the corresponding knots (higher order polynomials can have erratic behavior at the boundaries of the domain).

Penalized splines present a different approach to this issue [see, e.g., Eilers and Marx (2010), Durbin and Koopman (2012)]. Suppose that we wish to approximate a time series $y_1, \ldots, y_T$ by a relatively smooth function Tr_t. The penalized spline method chooses Tr_t by minimizing

$$\sum_{t=1}^{T} (y_t - Tr_t)^2 + \lambda \sum_{t=3}^{T} \left(\Delta^2 Tr_t\right)^2 \tag{3.37}$$

with respect to Tr_t for given $\lambda > 0$. The penalty is based on the level of variation in Tr_t measured by the second difference $\Delta^2 Tr_t = Tr_t - 2Tr_{t\text{-}1} - Tr_{t\text{-}2}$ [see (3.61)]. If λ is small, the values of Tr_t will be close to the values of y_t but Tr_t may not be smooth enough. If λ is large, the Tr_t series will be smooth but the values of Tr_t may not be close enough to the values of y_t.

Remark 3.3 In order to choose the appropriate trend curve for the given time series, one can make use of simple reference tests based on characteristic features of particular curves. A survey of such tests is given in Table 3.6.

◊

Table 3.6 Reference tests for choice of appropriate trend curves

Trend	Reference test
Linear	First differences $y_{t+1} - y_t$ are approximately constant
Quadratic	Second differences $y_{t+2} - 2y_{t+1} + y_t$ are approximately constant
Exponential	Ratios of neighboring values y_{t+1}/y_t (or first differences of logarithmic values $\ln y_{t+1} - \ln y_t$) are approximately constant
Modified exponential	Ratios of neighboring first differences $(y_{t+2} - y_{t+1})/(y_{t+1} - y_t)$ are approximately constant
Logistic	(1) Histogram of first differences $y_{t+1} - y_t$ looks like density $N(0, 1)$ (2) Ratios of neighboring first differences of reciprocal values $(1/y_{t+2} - 1/y_{t+1})/(1/y_{t+1} - 1/y_t)$ are approximately constant
Gompertz	Ratios of neighboring first differences of logarithmic values $(\ln y_{t+2} - \ln y_{t+1})/(\ln y_{t+1} - \ln y_t)$ are approximately constant

3.2 Method of Moving Averages

The method of moving averages is a representative of so-called *adaptive approaches* to the trend component (as well as the exponential smoothing in Sect. 3.3). The convenient property of such approaches consists in the possibility to manage systematic components (e.g., trend) which vary their character in time (in particular, it is not possible to apply mathematical curves with steady parameters in such a case). On the other hand, one assumes that in local segments of time series such trend elimination by means of mathematical curves is possible when these curves are constructed using different parameters in different segments. In other words, only a local elimination of trend is acceptable. For instance, a time series cannot be smoothed by means of the linear trend

$$\beta_0 + \beta_1 \ \tau, \qquad \tau = 1, \ \ldots, \ n, \tag{3.38}$$

but for short segments with middles in particular times t one can apply local trends

$$\beta_0(t) + \beta_1 \ (t)\tau, \qquad \tau = \ \ldots, \ t-1, \ t, \ t+1, \ \ldots. \tag{3.39}$$

Obviously, the process of trend elimination according to (3.39) adapts itself to the actual local run of time series, and, moreover, the intensity of this adaptation can be controlled. Another advantage of adaptive methods is the numerical simplicity and the construction of predictions which respond flexibly to eventual changes in the character of time series.

As the *moving averages* are concerned, this term denotes linear combinations of time series values with the unit sum of weights, e.g.,

$$\frac{1}{8}\left(y_{t-2} + 2y_{t-1} + 2y_t + 2y_{t+1} + y_{t+2}\right). \tag{3.40}$$

The construction of such a linear combination is equivalent amazingly to the local elimination of trends by means of specific mathematical curves which is shown in Sect. 3.2.1 (we still assume the basic form (3.1) of time series models).

3.2.1 Construction of Moving Averages by Local Polynomial Fitting

This approach is based on the axiom that each "reasonable" function can be approximated in an acceptable way by a polynomial. Respecting the previous discussion, at first let us fit by a suitable polynomial the initial time series segment of length $2m + 1$ and take the value of this polynomial at time $t = m + 1$ (i.e., in the middle of this segment) as the smoothed value $\widehat{y}_{m+1}$ of given time series at this time.

In order to obtain the smoothed value at time $t = m + 2$, we use similarly the observations $y_2, \ldots, y_{2m+2}$, etc. It will be shown in this section that such an algorithm is formally equivalent to the construction of smoothed values as linear combinations of original observations with fixed weights, i.e., to the construction of moving averages.

For instance, let us fit by the polynomial of the third order (i.e., by the cubic parabola) gradually the time series segments of length $2m + 1 = 5$ denoted formally as

$$y_{t+\tau}, \qquad \tau = -2, -1, 0, 1, 2. \tag{3.41}$$

The parameters of this polynomial can be estimated by means of the least squares method (i.e., as OLS estimates) minimizing the expression

$$\sum_{\tau=-2}^{2} \left(y_{t+\tau} - \beta_0 - \beta_1\tau - \beta_2\tau^2 - \beta_3\tau^3\right)^2. \tag{3.42}$$

If deriving with respect to particular parameters, we obtain the system of four normal equations for the estimates b_0, b_1, b_2, b_3 of parameters $\beta_0, \beta_1, \beta_2, \beta_3$ written as

$$\sum_{\tau=-2}^{2} y_{t+\tau}\tau^j - b_0\sum_{\tau=-2}^{2}\tau^j - b_1\sum_{\tau=-2}^{2}\tau^{j+1} - b_2\sum_{\tau=-2}^{2}\tau^{j+2} - b_3\sum_{\tau=-2}^{2}\tau^{j+3} = 0, \ j = 0, 1, 2, 3. \tag{3.43}$$

Since it holds for each odd i that

$$\sum_{\tau=-2}^{2} \tau^i = 0 \tag{3.44}$$

(it is one of the reasons for the choice of time series segments with the odd number $2m + 1$ of observations), this system of equations simplifies to the form

$$\begin{aligned}
5b_0 \qquad\quad + 10b_2 \qquad\qquad &= \sum y_{t+\tau}, \\
10b_1 \qquad\quad + 34b_3 &= \sum \tau\, y_{t+\tau}, \\
10b_0 \qquad\quad + 34b_2 \qquad\qquad &= \sum \tau^2 y_{t+\tau}, \\
34b_1 \qquad\quad + 130b_3 &= \sum \tau^3 y_{t+\tau}.
\end{aligned} \tag{3.45}$$

However, we are interested only in the estimate b_0 since it is the value of the fitting polynomial $b_0 + b_1\tau + b_2\tau^2 + b_3\tau^3$ at the point $\tau = 0$. Therefore, b_0 is taken in our method as the smoothed value of time series in the middle of the investigated segment $y_{t-2}, \ldots, y_{t+2}$. Obviously, it is sufficient to use only the first and third equation of system (3.45) with solution

$$b_0 = \frac{1}{35}\left(17\sum y_{t+\tau} - 5\sum \tau^2 y_{t+\tau}\right) = \frac{1}{35}\left(-3y_{t-2} + 12y_{t-1} + 17y_t + 12y_{t+1} - 3y_{t+2}\right), \quad (3.46)$$

so that the fitted trend component, which presents simultaneously the smoothed value of time series at time t, is also equal to

$$\widehat{y}_t = \frac{1}{35}\left(-3y_{t-2} + 12y_{t-1} + 17y_t + 12y_{t+1} - 3y_{t+2}\right). \quad (3.47)$$

It can be written symbolically as

$$\widehat{y}_t = \frac{1}{35}(-3,\ 12,\ 17,\ 12,\ -3)y_t = \frac{1}{35}(-3,\ 12,\ \mathbf{17},\ \ldots)y_t. \quad (3.48)$$

Example 3.6 In this example one applies the formula (3.47) to smooth the time series given in Table 3.7. The smoothed value at time $t = 3$ is

$$\widehat{y}_3 = \frac{1}{35}(-3 \cdot 1 + 12 \cdot 8 + 17 \cdot 27 + 12 \cdot 64 - 3 \cdot 125) = 27 = y_3.$$

Analogously

$$\widehat{y}_4 = 64 = y_4,$$

etc. This result corresponds to the fact that one smoothes the cubic time series by the cubic polynomial in this example (one would obtain the same results for any polynomials with the order higher than three).

◊

Table 3.7 Cubic time series from Example 3.6

t	1	2	3	4	5	6	7	8	9	10
y_t	1	8	27	64	125	216	343	512	729	1000

In general, we can fit segments of length $2m + 1$ by polynomials of order r to obtain the *moving averages of length 2m + 1 and order r*. The smoothed value $\widehat{y}_t$ at time t is the linear combination of expressions

$$\sum_{\tau=-m}^{m} \tau^j y_{t+\tau} \tag{3.49}$$

with even j ($j \leq r$), which can be derived if generalizing the system of equations (3.45). After rearrangement it gives a linear combination of values $y_{t-m}, ..., y_{t+m}$ with fixed coefficients called *weights of moving average*. One can verify easily the following properties of moving averages:

1. The sum of weights of moving average is equal to one (if one applies the moving average to any series of constant values, then obviously the smoothed values must be again the original constant values).
2. The weights are symmetric around the middle value (since for even j the values $y_{t-\tau}$ and $y_{t+\tau}$ in the expressions of type (3.49) have symmetric coefficients).
3. If r is even, then the moving averages of orders r and r +1 with the same length $2m + 1$ are identical (looking, e.g., at the system of equations (3.45), then obviously its solution for the unknown b_0 does not depend on including or not including the unknown b_3 to this system).

Let us note that the described moving averages produce only the smoothed values $\widehat{y}_{m+1}, \ldots, \widehat{y}_{n-m}$ (i.e., m values at the beginning and m values at the end remain unsmoothed).

Another note concerns the case when it is desirable to smooth time series using segments with an even length $2m$; then the positions of smoothed values should be just in the middle of original unit time intervals which has no reasonable practical interpretation. We will solve both mentioned problems later (see, e.g., the centered moving averages in Sect. 3.2.2).

Table 3.8 Weights of moving averages

Order length	2 and 3	4 and 5
3	(0, 1, 0)	(0, 1, 0)
5	$\frac{1}{35}$(−3, 12, **17**, ...)	(0, 0, **1**, ...)
7	$\frac{1}{21}$(−2, 3, 6, **7**, ...)	$\frac{1}{231}$(5, −30, 75, **131**, ...)
9	$\frac{1}{231}$(−21, 14, 39, 54, **59**, ...)	$\frac{1}{429}$(15, −55, 30, 135, **179**, ...)
11	$\frac{1}{429}$(−36, 9, 44, 69, 84, **89**, ...)	$\frac{1}{429}$(18, −45, −10, 60, 120, **143**, ...)
13	$\frac{1}{143}$(−11, 0, 9, 16, 21, 24, **25**, ...)	$\frac{1}{2431}$(110, −198, −135, 110, 390, 600, **677**, ...)

Table 3.8 summarizes the weights of moving averages of various lengths and orders ($r = 2, ..., 5$). Since the moving averages are symmetric, one gives only the first half of weights (the middle one is bold-faced). The weights for the second and third order or for the fourth and fifth order are equal (see above). The moving averages of order zero and one are omitted since they have the form of arithmetic averages

$$\frac{y_{t-m} + \ldots + y_{t+m}}{2m+1}.$$

However for the sake of completeness, this table includes, e.g., the moving averages of length 3 and order 3 in spite of the fact that it holds $\widehat{y}_t = y_t$ in such a case.

We have stressed above that the application of moving averages of length $2m+1$ does not deliver the smoothed values for the first m and the last m observations and any predictions at all. Let us go back to fitting always five neighboring observations by the cubic parabola (see above), and let the fitted segment be the last one with values $y_{n-4}, ..., y_n$. In contrast to the previous construction, now we are interested in the values of the cubic parabola that fit the last segment for $\tau = 1$ and 2 (these values have been ignored before). Therefore in addition we also need the estimates of parameters β_1, β_2, and β_3 in the parabola model (before it has been sufficient to estimate only β_0). Solving the system of equations (3.45), one can easily find these estimates in the form

$$b_1 = \frac{1}{72}\left(65\sum_{\tau=-2}^{2}\tau y_{t+\tau} - 17\sum_{\tau=-2}^{2}\tau^3 y_{t+\tau}\right),$$

$$b_2 = \frac{1}{14}\left(\sum_{\tau=-2}^{2}\tau^2 y_{t+\tau} - 2\sum_{\tau=-2}^{2} y_{t+\tau}\right),$$

$$b_3 = \frac{1}{72}\left(5\sum_{\tau=-2}^{2}\tau^3 y_{t+\tau} - 17\sum_{\tau=-2}^{2}\tau y_{t+\tau}\right). \tag{3.50}$$

Together with the value (3.46) of b_0 one obtains for the last two observations y_{n-1} and y_n the following smoothed values:

$$\begin{aligned}\widehat{y}_{n-2+k} &= b_0 + b_1 k + b_2 k^2 + b_3 k^3 \\ &= \frac{1}{35}(-3,\ 12,\ 17,\ 12,\ -3)y_{n-2} + \frac{k}{12}(1,\ -8,\ 0,\ 8,\ -1)y_{n-2} + \\ &\frac{k^2}{14}(2,\ -1,\ -2,\ -1,\ 2)y_{n-2} + \frac{k^3}{12}(-1,2,0,-2,1)y_{n-2},\ k = 1,2. \end{aligned} \tag{3.51}$$

When substituting $k = 1$ and $k = 2$, one gets explicitly

$$\widehat{y}_{n-1} = \frac{1}{35}(2, -8, 12, 27, 2)y_{n-2}, \quad \widehat{y}_n = \frac{1}{70}(-1, 4, -6, 4, 69)y_{n-2}. \tag{3.52}$$

Due to the apparent symmetry we can also immediately rewrite it for the first and second value at the beginning of time series

$$\widehat{y}_1 = \frac{1}{70}(69, 4, -6, 4, -1)y_3, \qquad \widehat{y}_2 = \frac{1}{35}(2, 27, 12, -8, 2)y_3. \tag{3.53}$$

Moreover, this approach enables to construct predictions in the given time series: e.g., the prediction of value y_{n+1} can be constructed when substituting $k = 3$ to (3.51), i.e.,

$$\widehat{y}_{n+1}(n) = \frac{1}{5}(-4, 11, -4, -14, 16)y_{n-2}. \tag{3.54}$$

Apparently it can be recommended only for short-term predictions: longer prediction horizons worsen significantly the confidence of predictions.

The previous moving averages are called logically *beginning moving averages* [see, e.g., (3.53)], *end moving averages* (see, e.g., (3.52)), and *prediction moving averages* [see, e.g., (3.54)]. One should stress that these moving averages lose the advantageous properties in comparison with the ones in Table 3.8: their weights are not symmetric around the middle value and the weights, e.g., for the second and third order are not identical. On the other side, the weights for these moving averages also used to be tabulated in the literature [see, e.g., Kendall (1976)]. Table 3.9 shows the beginning moving averages of the second and third order with various lengths. Tables 3.10, 3.11, and 3.12 summarize the one-step-ahead moving averages of the first, second, and third order with various lengths (i.e., they predict value y_{n+1} at time n). For instance, according to the first row of Table 3.10, the one-step-ahead prediction using the moving averages of the first order with length 3 has the form

$$\widehat{y}_{n+1}(n) = \frac{1}{3}(-2y_{n-2} + y_{n-1} + 4y_n).$$

There is an important question concerning the moving average methods, namely which length and order of moving averages to choose for given observations. In practice, we usually make subjective decisions preferring simple averages of low orders. For example, the length of moving averages depends on the desired level of time series smoothing (obviously the longer the moving average is chosen, the higher level of smoothing will follow). These are practical problems which deserve some further comments.

As the *choice of length of moving averages* is concerned, one of the important principles declares that this length should correspond to the period of seasonal or cyclical fluctuations that are to be eliminated from the time series. Fig. 3.10 shows schematically applications of moving averages of non-adequate lengths. Here one has used the moving averages of lengths 3 and 5 to eliminate a biennial cycle. In the first case (see Fig. 3.10a), the smoothing causes the "inverse cycle" since in each

Table 3.9 Beginning moving averages of the second and third order

Order r = 2											
Length 5			*Length 7*				*Length 9*				
$\widehat{y}_1$	$\widehat{y}_2$	$\widehat{y}_3$	$\widehat{y}_1$	$\widehat{y}_2$	$\widehat{y}_3$	$\widehat{y}_4$	$\widehat{y}_1$	$\widehat{y}_2$	$\widehat{y}_3$	$\widehat{y}_4$	$\widehat{y}_5$
31	9	−3	32	5	1	−2	109	126	378	14	−21
9	13	12	15	4	3	3	63	92	441	273	14
−3	12	**17**	3	3	4	6	27	63	464	447	39
−5	6	12	−4	2	4	**7**	1	39	447	536	54
3	−5	−3	−6	1	3	6	−15	20	390	540	**59**
35	35	35	−3	0	1	3	−21	6	293	459	54
			5	−1	−2	−2	−17	−3	156	293	39
			42	14	14	21	−3	−7	−21	42	14
							21	−6	−238	−294	−21
							165	330	2310	2310	231
Order r = 3											
Length 5			*Length 7*				*Length 9*				
$\widehat{y}_1$	$\widehat{y}_2$	$\widehat{y}_3$	$\widehat{y}_1$	$\widehat{y}_2$	$\widehat{y}_3$	$\widehat{y}_4$	$\widehat{y}_1$	$\widehat{y}_2$	$\widehat{y}_3$	$\widehat{y}_4$	$\widehat{y}_5$
69	2	−3	39	8	−4	−2	85	56	−28	−56	−21
4	27	12	8	19	16	3	28	65	392	84	14
−6	12	**17**	−4	16	19	6	−2	56	515	144	39
4	−8	12	−4	6	12	**7**	−12	36	432	145	54
−1	2	−3	1	−4	2	6	−9	12	234	108	**59**
70	35	35	4	−7	−4	3	0	−9	12	54	54
			−2	4	1	−2	8	−20	−143	4	39
			42	42	42	21	8	−14	−140	−21	14
							−7	16	112	0	−21
							99	198	1 386	462	231

group of three neighboring observations there are either two upper turning points and one lower turning point, or vice versa. In the second case (see Fig. 3.10b) just the opposite situation occurs: the smoothed time series follows the original time series upward to the upper turning points and downward to the lower turning points.

On the contrary, as the *choice of order of moving averages* is concerned, one can decide on it by means of the following objective criterion based on differencing time series (see also Remark 3.4). Let the given time series y_t fulfill the model (3.1), where Tr_t is a polynomial of the *r*th order (we denote it as $\beta_0 + \beta_1\, t + \ldots + \beta_r\, t^r$) and E_t is the

Table 3.10 One-step-ahead moving averages of the first order

Length	$\widehat{y}_{n+1}(n)$
3	$\frac{1}{3}(-2, 1, 4)$
5	$\frac{1}{10}(-4, -1, 2, 5, 8)$
7	$\frac{1}{7}(-2, -1, 0, 1, 2, 3, 4)$
9	$\frac{1}{36}(-8, -5, -2, 1, 4, 7, 10, 13, 16)$
11	$\frac{1}{55}(-10, -7, -4, -1, 2, 5, 8, 11, 14, 17, 20)$
13	$\frac{1}{26}(-4, -3, -2, -1, 0, 1, 2, 3, 4, 5, 6, 7, 8)$

Table 3.11 One-step-ahead moving averages of the second order

Length	$\widehat{y}_{n+1}(n)$
5	$\frac{1}{5}(3, -3, -4, 0, 9)$
7	$\frac{1}{7}(3, -1, -3, -3, -1, 3, 9)$
9	$\frac{1}{42}(14, 0, -9, -13, -12, -6, 5, 21, 42)$
11	$\frac{1}{165}(45, 9, -17, -33, -39, -35, -21, 3, 37, 81, 135)$
13	$\frac{1}{143}(33, 11, -6, -18, -25, -27, -24, -16, -3, 15, 38, 66, 99)$

Table 3.12 One-step-ahead moving averages of the third order

Length	$\widehat{y}_{n+1}(n)$
5	$\frac{1}{5}(-4, 11, -4, -14, 16)$
7	$\frac{1}{7}(-4, 6, 4, -3, -8, -4, 16)$
9	$\frac{1}{126}(-56, 49, 64, 24, -36, -81, -76, 14, 224)$
11	$\frac{1}{66}(-24, 12, 24, 19, 4, -14, -28, -31 -16, 4, 96)$
13	$\frac{1}{143}(-44, 11, 36, 38, 24, 1, -24, -44, -52, -41, -4, 66, 176)$

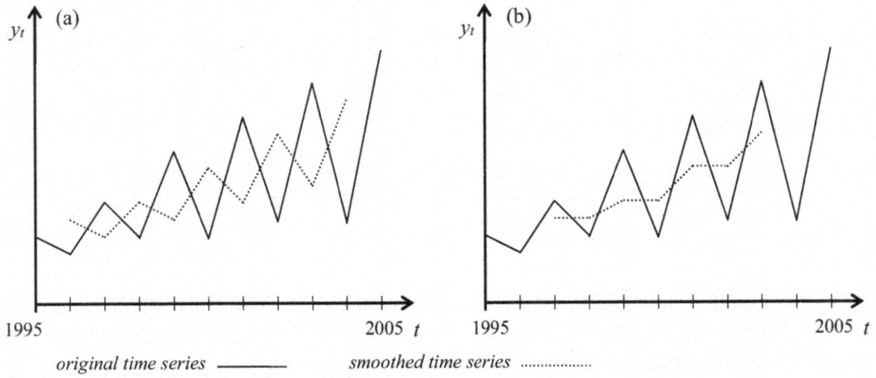

Fig. 3.10 Applications of moving averages of non-adequate lengths

white noise (we denote it for simplicity as ε_t and its variance as σ^2). The corresponding criterion, which should find r as the order of moving averages in question, consists in differencing gradually the analyzed time series. When differencing y_t, we decrease the order of its polynomial trend by one in each step: e.g., the order of the polynomial

$$(\beta_0 + \beta_1 t + \ldots + \beta_r t^r) - (\beta_0 + \beta_1 (t-1) + \ldots + \beta_r (t-1)^r)$$

is $r - 1$, etc. It is important in our context that $\Delta^{r+1} Tr_t = 0$, i.e., the polynomial trend Tr_t can be eliminated completely after applying $r + 1$ gradual differences (only r differences are not sufficient since they produce a constant which may be nonzero). As differencing the white noise ε_t is concerned, then its kth difference

$$\Delta^k \varepsilon_t = \varepsilon_t - \binom{k}{1}\varepsilon_{t-1} + \binom{k}{2}\varepsilon_{t-2} - \ldots + (-1)^k \varepsilon_{t-k} \tag{3.55}$$

has zero mean value and variance fulfilling

$$\mathrm{var}\left(\Delta^k \varepsilon_t\right) = \left(1 + \binom{k}{1}^2 + \binom{k}{2}^2 + \ldots + 1\right)\sigma^2 = \binom{2k}{k}\sigma^2. \tag{3.56}$$

If we denote

$$V_k = \frac{\sum_{t=k+1}^{n} \left(\Delta^k y_t\right)^2}{\binom{2k}{k}(n-k)}, \tag{3.57}$$

then obviously V_k for $k \geq r+1$ is the estimate of variance σ^2 of ε_t which is consistent in routine situations. These conclusions hold even if we approximate the trend component by polynomials of order r only locally as it is the case in our approach to constructing moving averages. Let us calculate the values V_1, V_2, ... gradually till we note that these values start to converge to a constant. In such a case, if the values V_{r+1}, V_{r+2}, ... are close to this constant value, then our criterion speaks in favor of the order r of corresponding moving averages (moreover, the mentioned constant represents the estimate of variance of the residual component of given time series). However, applying this criterion one must be very cautious. Usually the values V_k are not mutually independent and they incline to grow or descend without converging manifestly to a constant. Sometimes the sequence of V_k is decreasing, or it decreases to a minimal value V_{r+1} and then starts to grow slowly to an asymptotic level (in the latter case, one can take r as the upper limit for the order of corresponding moving averages). Lately, numerically simple methods have been suggested that look for the order of moving averages by minimizing a suitable criterion function.

Remark 3.4 In general, a very important operation in time series analysis is construction of so-called *d*th differences. Theoretically, they can be based on the concept of operators which simplify various time series models (see, e.g., Sect. 6.4.2):

- *lag operator B* delays a variable defined in time by one time unit:

$$By_t = y_{t-1}; \tag{3.58}$$

- *j*th *power* of lag operator B delays a variable defined in time by j time units:

$$B^j y_t = B^{j-1}(By_t) = B^{j-1}y_{t-1} = \ldots = y_{t-j}; \tag{3.59}$$

- *difference operator* Δ produces the *first difference*:

$$\Delta y_t = y_t - y_{t-1} = (1 - B)y_t; \tag{3.60}$$

- *second power* Δ^2 of difference operator produces the *second difference*:

$$\begin{aligned}\Delta^2 y_t &= \Delta(\Delta y_t) = (y_t - y_{t-1}) - (y_{t-1} - y_{t-2}) = y_t - 2y_{t-1} + y_{t-2} \\ &= (1 - B)^2 y_t;\end{aligned} \tag{3.61}$$

- dth *power* Δ^d of difference operator produces the dth *difference*:

$$\begin{aligned}\Delta^d y_t &= \Delta^{d-1}(\Delta y_t) = y_t - \binom{d}{1} y_{t-1} + \binom{d}{2} y_{t-2} - \ldots + (-1)^d y_{t-d} \\ &= (1 - B)^d y_t;\end{aligned} \tag{3.62}$$

- *seasonal difference operator* Δ_s for the length of season s (e.g., $s = 12$ for monthly observations) produces the *seasonal difference*:

$$\Delta_s y_t = y_t - y_{t-s} = (1 - B^s)y_t. \tag{3.63}$$

◊

Example 3.7 Table 3.13 and Fig. 3.11 present smoothing and predicting by moving averages in the time series y_t of the US nominal short-term interest rates (in % p.a.) for particular years 1961–2015 ($t = 1, \ldots, 55$).

The values V_k calculated in Table 3.14 according to (3.57) indicate that the upper limit for the order of corresponding moving averages is probably equal to 3. Therefore, we will apply for this time series the moving averages with the order $r = 3$. Table 3.13 and Fig. 3.11 present the calculated moving averages of this order which have lengths 5 and 9 (i.e., $m = 2$ and $m = 4$). Figure 3.11 shows that the moving averages of length 5 follow very closely the original observations so that the eliminated trend includes some periodic and random fluctuations that should be left aside from the trend component. On the contrary, the moving averages of length 9 smooth such short-term fluctuations in a sufficient way and, therefore, they should be preferred for the trend elimination in our case. For smoothing one has applied the weights from Table 3.8 so that, e.g., the moving averages of length 5 give

Table 3.13 Annual data 1961–2015 and smoothing and predicting by moving averages in Example 3.7 (*US nominal short-term interest rates in % p.a.*)

Year	y_t	$\widehat{y}_t$ $r=3, m=2$	$\widehat{y}_t$ $r=3, m=4$	Year	y_t	$\widehat{y}_t$ $r=3, m=2$	$\widehat{y}_t$ $r=3, m=4$
1961	2.37	2.37	2.26	1989	9.28	8.95	7.98
1962	2.77	2.77	2.88	1990	8.28	8.24	7.26
1963	3.17	3.17	3.30	1991	5.98	5.98	6.05
1964	3.57	3.53	3.62	1992	3.83	3.93	5.05
1965	3.97	4.18	3.91	1993	3.30	3.51	4.35
1966	4.86	4.43	4.46	1994	4.75	4.71	4.43
1967	4.30	4.67	5.21	1995	6.04	5.68	5.00
1968	5.35	5.43	5.63	1996	5.51	5.83	5.49
1969	6.74	6.52	5.43	1997	5.74	5.60	6.03
1970	6.28	6.03	5.34	1998	5.56	5.49	6.06
1971	4.32	4.49	5.79	1999	5.41	5.96	5.60
1972	4.18	4.76	5.97	2000	6.53	5.69	4.83
1973	7.19	6.77	5.95	2001	3.77	4.12	3.62
1974	7.89	7.49	5.94	2002	1.79	1.88	2.57
1975	5.77	6.15	5.99	2003	1.22	1.13	2.08
1976	5.00	4.93	6.07	2004	1.62	1.83	2.40
1977	5.33	5.47	6.19	2005	3.56	3.51	3.48
1978	7.37	7.45	7.80	2006	5.20	5.18	4.09
1979	10.11	9.75	9.90	2007	5.30	4.99	3.94
1980	11.56	12.33	11.14	2008	2.91	2.99	3.13
1981	13.97	12.77	11.70	2009	0.69	0.97	1.88
1982	10.60	11.10	11.39	2010	0.34	0.23	0.74
1983	8.67	9.20	10.24	2011	0.34	0.35	0.11
1984	9.54	8.99	8.87	2012	0.43	0.37	0.22
1985	8.38	8.32	7.66	2013	0.27	0.30	0.42
1986	6.83	7.15	7.71	2014	0.23	0.21	0.48
1987	7.19	7.06	8.03	2015	0.32	0.32	0.16
1988	7.98	8.23	8.09	2016	–	0.83	−0.76

Source: AMECO (European Commission Annual Macro-Economic *Database*). (https://ec.europa.eu/economy_finance/ameco/user/serie/SelectSerie.cfm)

$$\widehat{y}_3 = \frac{1}{35}(-3 \cdot 2.37 + 12 \cdot 2.77 + 17 \cdot 3.17 + 12 \cdot 3.57 - 3 \cdot 3.97) = 3.17\%.$$

In the beginning and end of time series we have used the beginning and end moving averages, respectively: e.g., applying again the moving averages of length 5 one gets according to (3.53)

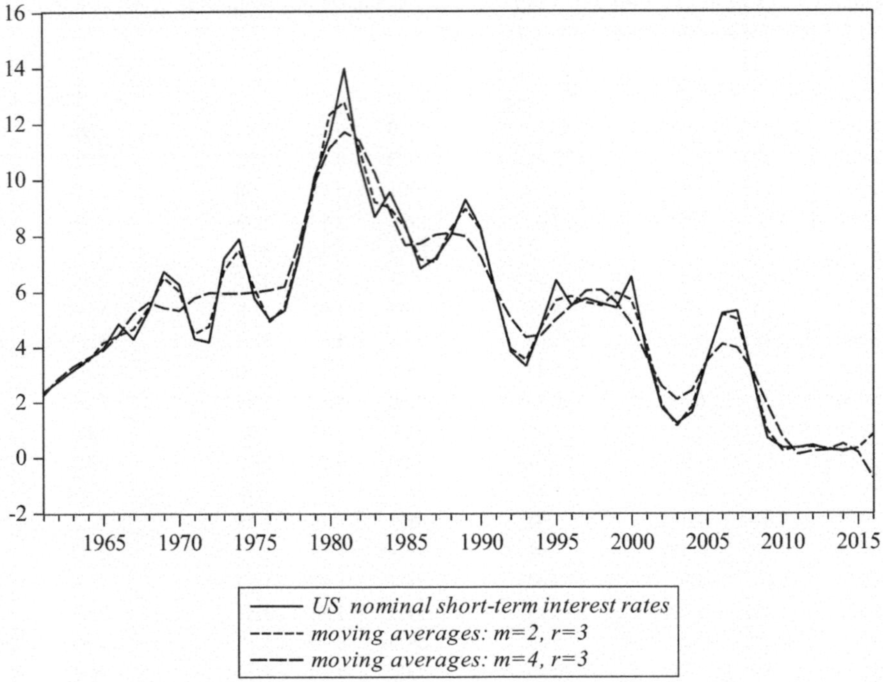

Fig. 3.11 Annual data 1961–2015 and smoothing and predicting by moving averages in Example 3.7 (*US nominal short-term interest rates in % p.a.*)

Table 3.14 Choice of order r of moving averages in Example 3.7

k	V_k
1	1.029
2	0.474
3	0.309
4	0.243
5	0.209
6	0.190
7	0.176
8	0.166
9	0.158
10	0.150

$$\hat{y}_1 = \frac{1}{70}(69 \cdot 2.37 + 4 \cdot 2.77 - 6 \cdot 3.17 + 4 \cdot 3.57 - 1 \cdot 3.97) = 2.37\%.$$

Table 3.13 contains also the one-step-ahead predictions of US nominal short-term interest rates for year 2016. For example, using the moving averages of order 3 and length 9 (see Table 3.12), one obtains the following negative interest rate:

$$\widehat{y}_{31}(30) = \frac{1}{126}(-56 \cdot 5.30 + 49 \cdot 2.91 + \ldots + 224 \cdot 0.32) = -0.76\%.$$

Obviously for short-term predictions, the "short" moving averages should be preferred: e.g., the moving averages of order 3 and length 5 give a quite different prediction 0.83% p.a. in this case. In general, the predictions based on moving averages cannot be regarded as highly credible.

◊

3.2.2 *Other Types of Moving Averages*

3.2.2.1 Arithmetic Moving Averages

In practice, simpler moving averages are popular. The simplest ones are the *arithmetic moving averages*. For instance, the arithmetic moving averages of length 5 are

$$\overline{y}_t^{(5)} = \frac{1}{5}(1,\ 1,\ 1,\ 1,\ 1)y_t = \frac{1}{5}(y_{t-2} + y_{t-1} + y_t + y_{t+1} + y_{t+2}). \tag{3.64}$$

In general, the arithmetic moving averages of length $2m + 1$ have the form

$$\overline{y}_t^{(2m+1)} = \frac{1}{2m+1}(y_{t-m} + y_{t-m+1} + \ldots + y_{t+m}). \tag{3.65}$$

They correspond to the moving averages from Sect. 3.2.1 with order 0 or 1 and the same length $2m + 1$ (i.e., the time series segments of length $2m + 1$ are fitted using constant or linear trend). Therefore, it holds, e.g., for the length 5 and order 0

$$\overline{y}_{n-1}^{(5)} = \overline{y}_n^{(5)} = \widehat{y}_{n+\tau}(n) = \frac{1}{5}(y_{n-4} + y_{n-3} + \ldots + y_n) \tag{3.66}$$

(for any $\tau > 0$) and for the length 5 and order 1

$$\overline{y}_{n-1}^{(5)} = \frac{1}{10}(y_{n-3} + 2y_{n-2} + 3y_{n-1} + 4y_n),$$

$$\overline{y}_n^{(5)} = \frac{1}{5}(-y_{n-4} + y_{n-2} + 2y_{n-1} + 3y_n),$$

$$\widehat{y}_{n+1}(n) = \frac{1}{10}(-4y_{n-4} - y_{n-3} + 2y_{n-2} + 5y_{n-1} + 8y_n), \ldots . \tag{3.67}$$

3.2.2.2 Centered Moving Averages

The centered moving averages modify the arithmetic moving averages in order to be applicable when one smoothes economic time series over particular seasons with an even number of observations (usually 4 for the quarterly data or 12 for the monthly data). In such a situation, the methodological problem appears, namely whereabouts to allocate the particular averages: e.g., the arithmetic average of the values over January till December belongs to the midpoint between time points for June and July values. However, when averaging such two neighboring moving averages (the first one corresponds to the center of interval "June–July" and the second one to the center of interval "July–August"), then the result can be undoubtedly allocated to the time point "July". In other words, we construct moving averages of the type

$$\begin{aligned}\overline{y}_t^{(12)} &= \frac{1}{2}\left(\frac{1}{12}\left(y_{t-6} + y_{t-5} + \ldots + y_t + \ldots + y_{t+5}\right)\right. \\ &\quad \left. + \frac{1}{12}\left(y_{t-5} + y_{t-4} + \ldots + y_t + \ldots + y_{t+6}\right)\right) \\ &= \frac{1}{24}\left(y_{t-6} + 2y_{t-5} + 2y_{t-4} + \ldots + 2y_{t+5} + y_{t+6}\right) \end{aligned} \tag{3.68}$$

(obviously with length 13). That is, when calculating, e.g., the July value, one exploits the February till December values of the given year (all with the weights 1/12) and the January values of the present and future year (both with weights 1/24). In general, one can write

$$\overline{y}_t^{(2m)} = \frac{1}{4m}\left(y_{t-m} + 2y_{t-m+1} + \ldots + 2y_{t+m-1} + y_{t+m}\right). \tag{3.69}$$

These values are denoted as the *centered moving averages* (quarterly ones $\overline{y}_t^{(4)}$ for $m = 2$ or monthly ones $\overline{y}_t^{(12)}$ for $m = 6$).

3.2.2.3 Robust Moving Averages

This method denotes the moving averages that are capable of restraining or completely filtering off the *outliers* (i.e., the outlying observations) in time series. A simple example is the *moving medians* of odd length $2m + 1$

$$\text{med}y_t^{(2m+1)} = \text{med}\left(y_{t-m},\ y_{t-m+1},\ \ldots,\ y_{t+m}\right). \tag{3.70}$$

Figure 3.12 compares the moving medians with the arithmetic moving averages in the case of one or two outliers (obviously the length 3 of moving averages is

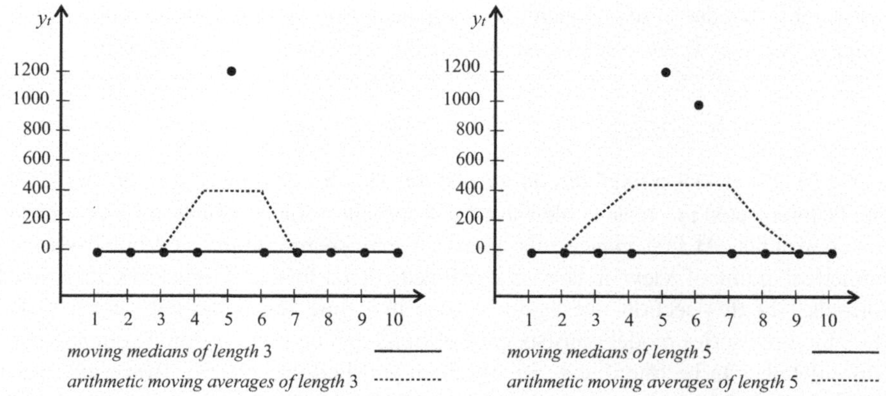

Fig. 3.12 Application of robust moving averages

insufficient in the second case with two outliers so that it has been necessary to prolong it to 5).

Remark 3.5 In practice, various software systems offer a plenty of other moving averages presented frequently as filters. For instance in macroeconomics, especially in real business cycle theory, so-called *Hodrick-Prescott filter* is popular to remove the cyclical component (see, e.g., EViews). This filter constructs the smoothed values $\widehat{y}_1, \ \ldots, \ \widehat{y}_n$ for given time series $y_1, \ldots, y_n$ by minimizing the expression

$$\sum_{t=1}^{n} (y_t - \widehat{y}_t)^2 + \lambda \sum_{t=2}^{n-1} \left((\widehat{y}_{t+1} - \widehat{y}_t) - (\widehat{y}_t - \widehat{y}_{t-1}) \right)^2. \tag{3.71}$$

The positive constant λ controls the intensity of smoothing of given time series (obviously if $\lambda \to \infty$, then the method eliminates the linear trend). Another example is the moving averages based on *OWA operators* (ordered weighted averaging) which calculate weighted averages of ordered values in particular segments of time series. They can be of interest when we want to over- or underestimate the results; see Merigó and Yager (2013).

◊

3.3 Exponential Smoothing

The exponential smoothing is another adaptive approach to the trend component that is frequently used in practice (see also the introduction to Sect. 3.2). It is a special case of the moving averages, in which the values observed up to the present period get weights that decrease exponentially with the age of particular observations. Such moving averages $\widehat{y}_t$ are constructed minimizing the expressions of the form

$$(y_t - \widehat{y}_t)^2 + (y_{t-1} - \widehat{y}_{t-1})^2\beta + (y_{t-2} - \widehat{y}_{t-2})^2\beta^2 + \ldots, \tag{3.72}$$

where $(0 < \beta < 1)$ is a fixed *discount constant*. The discounting of weights in (3.72) can be interpreted in a reasonable way: the observations more distant in the past have lower weights. At first glance this approach may seem complicated but from the numerical point of view it is easily realized, in particular if one uses recursive formulas. In this section, we again assume that time series have the form (3.1) (i.e., the time series model consists only of trend and additive residual component). More details can be found, e.g., in Abraham and Ledolter (1983), Bowerman and O'Connell (1987), Montgomery and Johnson (1976), and others.

3.3.1 Simple Exponential Smoothing

The simple exponential smoothing is recommended for time series in which the trend can be viewed as locally constant (i.e., constant in short segments of time series)

$$Tr_t = \beta_0. \tag{3.73}$$

A natural objective is to find an estimate of parameter β_0. However, since the exponential smoothing approach is declared as adaptive, this estimate will depend on the time moment of its construction. Let us denote $b_0(t)$ the estimate of parameter β_0 constructed at time t, i.e., using observations y_t, y_{t-1}, y_{t-2}, ... known at time t. Then the estimate $b_0(t)$ will represent both the fitted trend at time t and the smoothed value $\widehat{y}_t$ of given time series. Because of (3.72), we will find it minimizing the expression

$$\sum_{j=0}^{\infty} (y_{t-j} - \beta_0)^2 \beta^j, \tag{3.74}$$

where β $(0 < \beta < 1)$ is a fixed discount constant. It should be pointed out that the sum in the minimized expression (3.74) is infinite, although in real situations we always know only a finite number of values y_1, ..., y_t. However, the hypothetical extension of time series to the past simplifies significantly the corresponding formulas due to simpler limit results. In any case, the numerical calculations based on this abstraction exploit only the observed values y_1, ..., y_t of given time series (see below).

If we derive (3.74) with respect to β_0 and put this derivative equal to zero, then due to the convexity of minimized function we get the estimate $b_0(t)$ of parameter β_0 at time t as

$$\widehat{y}_t = (1-\beta)\sum_{j=0}^{\infty} \beta^j y_{t-j}. \tag{3.75}$$

Hence one can see that the smoothed value of time series at time t is the weighted average of values of this time series till time t with weights decreasing exponentially to the past

$$1-\beta,\ (1-\beta)\beta,\ (1-\beta)\beta^2,\ \ldots. \tag{3.76}$$

Since the formula (3.75) is not comfortable for practical calculations, it is transferred to the recursive form

$$\widehat{y}_t = \alpha y_t + (1-\alpha)\widehat{y}_{t-1}, \tag{3.77}$$

where $\alpha = 1-\beta\ (0 < \alpha < 1)$ is so-called *smoothing constant*. The recursive formula (3.77) also clearly demonstrates merits of exponential smoothing: (1) the calculations are simple; (2) the method is parsimonious since a low saving capacity is sufficient: it suffices to save only the previous smoothed value instead of the whole history of time series; (3) it is possible to control the intensity of smoothing: the smoothing with a higher constant α (e.g., $\alpha = 0.3$) responds quickly to changes in the character of data so that its smoothing ability is lower since the role of the first summand is dominant in (3.77), while the smoothing ability of (3.77) is enhanced for a lower constant α (e.g., $\alpha = 0.1$).

When we apply the simple exponential smoothing to construct predictions in time series, then due to (3.73) one can put

$$\widehat{y}_{t+\tau}(t) = \widehat{y}_t \tag{3.78}$$

for any $\tau > 0$. In particular, the most usual prediction in practice is

$$\widehat{y}_{n+\tau}(n) = \widehat{y}_n. \tag{3.79}$$

In addition to the formulas (3.75) and (3.77), there exists the third form of smoothing formula making use of (3.78), namely

$$\widehat{y}_t = \widehat{y}_{t-1} + \alpha(y_t - \widehat{y}_{t-1}) = \widehat{y}_{t-1} + \alpha(y_t - \widehat{y}_t(t-1)) = \widehat{y}_{t-1} + \alpha \cdot e_t, \tag{3.80}$$

The form (3.80) is sometimes denoted as the "error" formula: in order to correct the previous smoothed value $\widehat{y}_{t-1}$ one exploits the (reduced) one-step-ahead error e_t of prediction $\widehat{y}_t(t-1)$ constructed at time $t-1$ as soon as the value y_t is observed.

The practical realization of the recursive formula (3.77) of simple exponential smoothing assumes the choice of initial value $\widehat{y}_0$ and smoothing constant α :

1. The choice of $\widehat{y}_0$: One usually uses the arithmetic average of a short initial segment of time series (e.g., $y_1, ..., y_6$) or directly y_1 (then it is obviously $\widehat{y}_1 = y_1$).
2. The choice of α : First of all, the interval $0 < \alpha \leq 0.3$ is recommended in practice, and the following choices of α from this interval are possible in practice:

   (a) The fixed choice $\alpha = 0.1$ or $\alpha = 0.2$.
   (b) The choice

$$\alpha = \frac{1}{m+1}, \tag{3.81}$$

   where $2m + 1$ is the length of arithmetic moving averages which would be adequate for the given time series (this choice is derived comparing so-called *mean age* of arithmetic moving averages of this length, i.e.,

$$\sum_{k=0}^{2m} \frac{k}{2m+1}, \tag{3.82}$$

   and the mean age of weights of simple exponential smoothing, i.e.,

$$\sum_{k=0}^{\infty} k\alpha(1-\alpha)^k, \tag{3.83}$$

   where the mean ages (3.82) and (3.83) must coincide, and hence α expressed as (3.81) follows); however, this approach is handicapped by the fact that at first one must decide on an adequate length of moving averages.

   (c) The estimate of α : one admits the grid points 0.01; 0.02; ..., 0.30 as possible values of the smoothing constant and chooses such a value of α from this grid that gives the best predictions with minimum *SSE* [see (2.10)] in the given time series. This approach is included in many software systems (see, e.g., EViews).

Example 3.8 Figure 3.13 presents the simple exponential smoothing in the time series y_t of annual averages of exchange rates USD/EUR(ECU) for particular years 1960–2016 with $t = 1, ..., 57$ (the former basket currency ECU of the European Community was introduced as late as in 1979, but it was formally amended for the period since 1960). The smoothed values for $\alpha = 0.01$, 0.02, and 0.3 including the one-step-ahead prediction for year 2017 are plotted in the figure. As the possibility of estimation of α is concerned (see above), EViews has found the value close to one

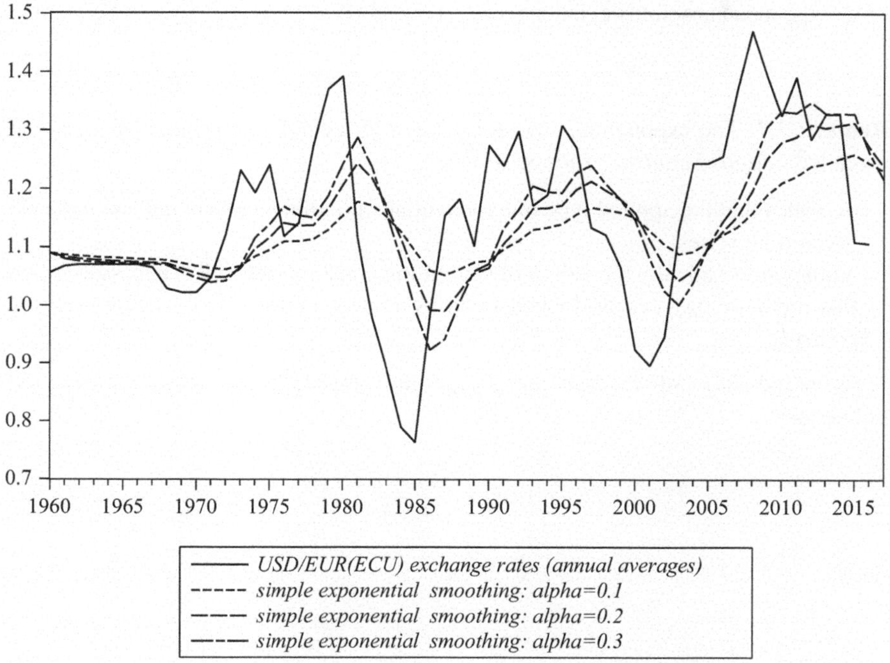

Fig. 3.13 Annual data 1960–2016 and single exponential smoothing in Example 3.8 (*annual averages of exchange rates USD/EUR(ECU)*). Source: AMECO (European Commission Annual Macro-Economic Database) (https://ec.europa.eu/economy_finance/ameco/user/serie/SelectSerie.cfm)

($\alpha = 0.999$), which means that the smoothed time series nearly coincides with the original time series.

◊

Remark 3.6 Assuming the normal distribution of residual components one can construct by means of exponential smoothing not only point predictions but also *interval predictions* [(3.84) is only approximative without this assumption]. For example, the $(1-p)\cdot 100\%$ prediction interval (i.e., 95% interval if $p = 0.05$) is recommended in the form

$$\left(\widehat{y}_{n+\tau}(n) - u_{1-p/2} \cdot d_\tau \cdot MAE, \quad \widehat{y}_{n+\tau}(n) + u_{1-p/2} \cdot d_\tau \cdot MAE\right), \tag{3.84}$$

where for any $\tau > 0$ one substitutes

$u_{1-p/2}$ the $(1-p/2)$-quantile of standard normal distribution $N(0, 1)$;
d_τ the constant equal to $\sqrt{(\pi/2)} \approx 1.25$ in order to transform *MSE* to *MAE*;

MAE the mean absolute deviation [see (2.17)]: $MAE = (1/n)\sum^{n}_{t=1}|y_t - \widehat{y}_t(t-1)|$.

◊

Remark 3.7 The exponential smoothing algorithm can be controlled by means of so-called *adaptive control process*, which

- indicates that the applied type of exponential smoothing stops being adequate for given time series;
- adjusts automatically the values of smoothing constants when it is necessary (then this methodology can be looked upon as a special case of so-called *stochastic control*).

Here the indicators of "default" are, e.g., significantly high values $I_t(\alpha)$ constructed online as

$$I_t(\alpha) = \frac{|Y_t(\alpha)|}{D_t(\alpha)}, \tag{3.85}$$

where

$$Y_t(\alpha) = \sum_{j=1}^{t} e_j(\alpha), \qquad D_t(\alpha) = \frac{1}{t}\sum_{j=1}^{t} |e_j(\alpha)|. \tag{3.86}$$

The symbol $e_j(\alpha)$ denotes the error of prediction of value y_j (the prediction is constructed at time $j-1$, i.e., one-step-ahead, applying the smoothing constant equal to α). The indicator $I_t(\alpha)$ is usually calculated online, and its increased values exceeding a given boundary K indicate that a change of α or even a change of the type of exponential smoothing is necessary (e.g., one should apply the double exponential smoothing from Sect. 3.3.2 instead of the simple exponential smoothing). The boundary K is usually fixed in the range from 4 to 6 (it can be constructed in a similar way to the critical value of statistical tests with fixed significance levels).

One of the methods to change automatically the smoothing constant α makes use of the approach that three procedures of exponential smoothing are realized in parallel with three different values of this constant: one can use, e.g., the values $\alpha - 0.05$, α, $\alpha + 0.05$. Here only the procedure using at time t the "middle" smoothing constant α delivers the output results for users, and (in each time point) the methodology compares the values $D_t(\alpha - 0.05)$, $D_t(\alpha)$, $D_t(\alpha + 0.05)$ calculated according to (3.86). If it holds

$$D_t(\alpha) \leq \min\left(D_t(\alpha - 0.05),\ D_t(\alpha + 0.05)\right), \tag{3.87}$$

then the algorithms goes on without changes. However, if it is, e.g.,

$$D_t(\alpha + 0.05) < D_t(\alpha), \tag{3.88}$$

then the "middle" process transfers α to $\alpha + 0.05$, and the algorithm goes on using the triplet of smoothing constants α, $\alpha + 0.05$, $\alpha + 0.10$, respectively.

◇

3.3.2 Double Exponential Smoothing

The double exponential smoothing (sometimes denoted as *Brown's method*) is suitable for time series in which the trend can be viewed as locally linear (i.e., linear in short segments)

$$Tr_{t-j} = \beta_0 + \beta_1 \cdot (-j). \tag{3.89}$$

The estimates $b_0(t)$ and $b_1(t)$ constructed at time t for the parameters β_0 and β_1 are obtained by minimizing the expression

$$\sum_{j=0}^{\infty} \left[y_{t-j} - (\beta_0 + \beta_1(-j))\right]^2 \beta^j, \tag{3.90}$$

where again β ($0 < \beta < 1$) is a fixed discount constant. If we put the partial derivatives of (3.90) with respect to β_0 and β_1 both equal to zero, we get the system of normal equations

$$\sum_{j=0}^{\infty} \beta^j y_{t-j} - \beta_0 \sum_{j=0}^{\infty} \beta^j + \beta_1 \sum_{j=0}^{\infty} j\beta^j = 0, \quad \sum_{j=0}^{\infty} j\beta^j y_{t-j} - \beta_0 \sum_{j=0}^{\infty} j\beta^j + \beta_1 \sum_{j=0}^{\infty} j^2\beta^j = 0, \tag{3.91}$$

which can be simplified by means of the formulas

$$\sum_{j=0}^{\infty} \beta^j = \frac{1}{1-\beta}, \qquad \sum_{j=0}^{\infty} j\beta^j = \frac{\beta}{(1-\beta)^2}, \qquad \sum_{j=0}^{\infty} j^2\beta^j = \frac{\beta(1+\beta)}{(1-\beta)^3} \tag{3.92}$$

to the form

$$\beta_0 - \frac{\beta}{1-\beta} \cdot \beta_1 = (1-\beta) \sum_{j=0}^{\infty} \beta^j y_{t-j}, \quad \beta \cdot \beta_0 - \frac{\beta(1+\beta)}{1-\beta} \cdot \beta_1 = (1-\beta)^2 \sum_{j=0}^{\infty} j\beta^j y_{t-j}. \tag{3.93}$$

In order to simplify the denotation, one introduces so-called *simple smoothing statistics*

$$S_t = (1-\beta)\sum_{j=0}^{\infty}\beta^j y_{t-j} \tag{3.94}$$

(due to (3.75), this S_t corresponds to the value of time series smoothed at time t by the simple exponential smoothing). Therefore, it holds according to (3.77)

$$S_t = \alpha y_t + (1-\alpha)S_{t-1} \tag{3.95}$$

(again we put $\alpha = 1 - \beta$). Similarly one introduces so-called *double smoothing statistics*

$$S_t^{[2]} = (1-\beta)\sum_{j=0}^{\infty}\beta^j S_{t-j} \tag{3.96}$$

(the relation (3.96) is analogous to (3.94), but the values y_t are replaced by S_t). Hence the analogy to the recursive relation (3.95) gives

$$S_t^{[2]} = \alpha S_t + (1-\alpha)S_{t-1}^{[2]}. \tag{3.97}$$

The introduced smoothing statistics enable to rewrite the system of normal equation (3.93) to the form

$$\beta_0 - \frac{\beta}{1-\beta}\cdot\beta_1 = S_t, \quad \beta\cdot\beta_0 - \frac{\beta(1+\beta)}{1-\beta}\cdot\beta_1 = S_t^{[2]} - (1-\beta)S_t. \tag{3.98}$$

Solving these equations, one gets the corresponding estimates as

$$b_0(t) = 2S_t - S_t^{[2]}, \qquad b_1(t) = \frac{1-\beta}{\beta}\left(S_t - S_t^{[2]}\right). \tag{3.99}$$

Then the prediction of value $y_{t+\tau}$ constructed at time t has the natural form

$$\widehat{y}_{t+\tau}(t) = b_0(t) + b_t(t)\cdot\tau = \left(2+\frac{\alpha\cdot\tau}{1-\alpha}\right)S_t - \left(1+\frac{\alpha\cdot\tau}{1-\alpha}\right)S_t^{[2]}. \tag{3.100}$$

The special case $\tau = 0$ delivers the smoothed value of time series, i.e.,

$$\widehat{y}_t = 2S_t - S_t^{[2]}. \tag{3.101}$$

The statistics S_t and $S_t^{[2]}$ are calculated recursively according to (3.95) and (3.97).

Again the practical realization of the recursive formulas of double exponential smoothing assumes the choice of initial values S_0 and $S_0^{[2]}$ and smoothing constant α :

1. The initial values S_0 and $S_0^{[2]}$ can be set up using the relations (3.99) for $t = 0$, where $b_0(0)$ and $b_1(0)$ are chosen as the classical regression estimates of parameters β_0 and β_1 fitting the line through a short initial segment of time series (e.g., y_1, ..., y_6). Then the values S_0 and $S_0^{[2]}$ can be expressed from (3.99) explicitly as

$$S_0 = b_0(0) - \frac{1-\alpha}{\alpha} b_1(0), \qquad S_0^{[2]} = b_0(0) - \frac{2(1-\alpha)}{\alpha} b_1(0). \tag{3.102}$$

2. For the choice of smoothing constant α in practice, one again recommends the interval $0 < \alpha \leq 0.3$, in which (similarly to the simple exponential smoothing) we can use:

   (a) The fixed choice $\alpha = 0.1$ or $\alpha = 0.2$.
   (b) The choice

$$\alpha = \sqrt{\frac{1}{m+1}}, \tag{3.103}$$

   where $2m + 1$ is the length of arithmetic moving averages which would be adequate in this case (it follows again by comparing the mean age of weights of arithmetic moving averages and the mean age of weights of double exponential smoothing)

   (c) The estimate of α : one admits the grid points 0.01; 0.02; ..., 0.30 as possible values of the smoothing constant and chooses such a value of α from this grid that gives the best predictions with minimum *SSE*.

Remark 3.8 Using the same denotation as in Remark 3.6, the $(1-p)\cdot 100\%$ prediction interval can be constructed in the form

$$\left(\widehat{y}_{n+\tau}(n) - u_{1-p/2}\cdot d_\tau \cdot MAE, \widehat{y}_{n+\tau}(n) + u_{1-p/2}\cdot d_\tau \cdot MAE\right), \tag{3.104}$$

where for any $\tau > 0$ one substitutes

$$d_\tau \approx 1.25\cdot \left(\frac{1+\frac{1-\beta}{(1+\beta)^3}\left((1+4\beta+5\beta^2)+2(1-\beta)(1+3\beta)\tau+2(1-\beta)^2\tau^2\right)}{1+\frac{1-\beta}{(1+\beta)^3}\left((1+4\beta+5\beta^2)+2(1-\beta)(1+3\beta)+2(1-\beta)^2\right)}\right)^{1/2}. \tag{3.105}$$

◇

Remark 3.9 A natural extension of simple and double exponential smoothing is the *triple exponential smoothing* (the local quadratic trend necessitates to introduce the triple smoothing statistics $S_t^{[3]}$ in addition to S_t and $S_t^{[2]}$). Even though the exponential smoothing of a general order r is possible, the order $r = 3$ is the highest, which is used in practice.

◇

3.3.3 Holt's Method

Holt's method generalizes the double exponential smoothing by introducing two smoothing constants: α (it smoothes the level L_t of time series) and γ (it smoothes the slope T_t of the same time series). Both constants lie between zero and one ($0 < \alpha$, $\gamma < 1$):

$$L_t = \alpha y_t + (1-\alpha)(L_{t-1} + T_{t-1}), \tag{3.106}$$

$$T_t = \gamma(L_t - L_{t-1}) + (1-\gamma)T_{t-1}, \tag{3.107}$$

$$\widehat{y}_t = L_t, \tag{3.108}$$

$$\widehat{y}_{t+\tau}(t) = L_t + T_t \cdot \tau \quad (\tau > 0). \tag{3.109}$$

The initial values are recommended as $L_0 = y_1$ and $T_0 = y_2 - y_1$. It is interesting that Holt's method has been suggested *ad hoc* by a logic consideration. For example, the elimination of level L_t of given time series according to (3.106) is constructed as a convex combination of the present value y_t at time t and the estimate of this value which was calculated in previous time $t - 1$ (simply as $L_{t-1} + T_{t-1} \times 1$). Later one has proved that the double exponential smoothing with smoothing constant α is a special case of Holt's method with smoothing constants α_{Holt} and γ_{Holt} of the form

$$\alpha_{Holt} = \alpha(2-\alpha), \qquad \gamma_{Holt} = \frac{\alpha}{2-\alpha}. \tag{3.110}$$

Example 3.9 Table 3.15 and Fig. 3.14 present double exponential smoothing (with the fixed choice of smoothing constant $\alpha = 0.15$) and Holt's method (with the fixed choice of smoothing constants $\alpha = 0.1$ and $\gamma = 0.2$) in the time series y_t of the US nominal short-term interest rates (in % p.a.) for particular years 1961–2015 ($t = 1, ..., 55$). The smoothed values and the one-step-ahead prediction for year 2016 have been obtained by EViews and can be compared with the corresponding results by moving averages in Example 3.7 for the same data (see Fig. 3.11).

◇

Table 3.15 Annual data 1961–2015 and smoothing and predicting by double exponential smoothing and Holt's method in Example 3.9 (*US nominal short-term interest rates in % p.a.*)

Year	t	y_t	Doub. exp. $\alpha = 0.15$	Holt $\alpha = 0.1, \gamma = 0.2$	Year	t	y_t	Doub. exp. $\alpha = 0.15$	Holt $\alpha = 0.1, \gamma = 0.2$
1961	1	2.37	2.37	2.26	1989	29	9.28	8.95	7.98
1962	2	2.77	2.77	2.88	1990	30	8.28	8.24	7.26
1963	3	3.17	3.17	3.30	1991	31	5.98	5.98	6.05
1964	4	3.57	3.53	3.62	1992	32	3.83	3.93	5.05
1965	5	3.97	4.18	3.91	1993	33	3.30	3.51	4.35
1966	6	4.86	4.43	4.46	1994	34	4.75	4.71	4.43
1967	7	4.30	4.67	5.21	1995	35	6.04	5.68	5.00
1968	8	5.35	5.43	5.63	1996	36	5.51	5.83	5.49
1969	9	6.74	6.52	5.43	1997	37	5.74	5.60	6.03
1970	10	6.28	6.03	5.34	1998	38	5.56	5.49	6.06
1971	11	4.32	4.49	5.79	1999	39	5.41	5.96	5.60
1972	12	4.18	4.76	5.97	2000	40	6.53	5.69	4.83
1973	13	7.19	6.77	5.95	2001	41	3.77	4.12	3.62
1974	14	7.89	7.49	5.94	2002	42	1.79	1.88	2.57
1975	15	5.77	6.15	5.99	2003	43	1.22	1.13	2.08
1976	16	5.00	4.93	6.07	2004	44	1.62	1.83	2.40
1977	17	5.33	5.47	6.19	2005	45	3.56	3.51	3.48
1978	18	7.37	7.45	7.80	2006	46	5.20	5.18	4.09
1979	19	10.11	9.75	9.90	2007	47	5.30	4.99	3.94
1980	20	11.56	12.33	11.14	2008	48	2.91	2.99	3.13
1981	21	13.97	12.77	11.70	2009	49	0.69	0.97	1.88
1982	22	10.60	11.10	11.39	2010	50	0.34	0.23	0.74
1983	23	8.67	9.20	10.24	2011	51	0.34	0.35	0.11
1984	24	9.54	8.99	8.87	2012	52	0.43	0.37	0.22
1985	25	8.38	8.32	7.66	2013	53	0.27	0.30	0.42
1986	26	6.83	7.15	7.71	2014	54	0.23	0.21	0.48
1987	27	7.19	7.06	8.03	2015	55	0.32	0.32	0.16
1988	28	7.98	8.23	8.09	2016	56	–	0.83	−0.76

Source: AMECO (European Commission Annual Macro-Economic Database) (https://ec.europa.eu/economy_finance/ameco/user/serie/SelectSerie.cfm)

3.4 Exercises

Exercise 3.1 Repeat the analysis from Example 3.1 (the linear trend in the Swiss gross national income) only for data since 1990 (*hint:* 343.348 1 + 12.566 40 t, $\widehat{y}_{27} =$ 682.6, (652.5; 712.8)).

Exercise 3.2 Repeat the analysis from Example 3.2 (the exponential trend in US gross national income) only for data since 1970 (*hint:* 2 466.88·1.048 05^t).

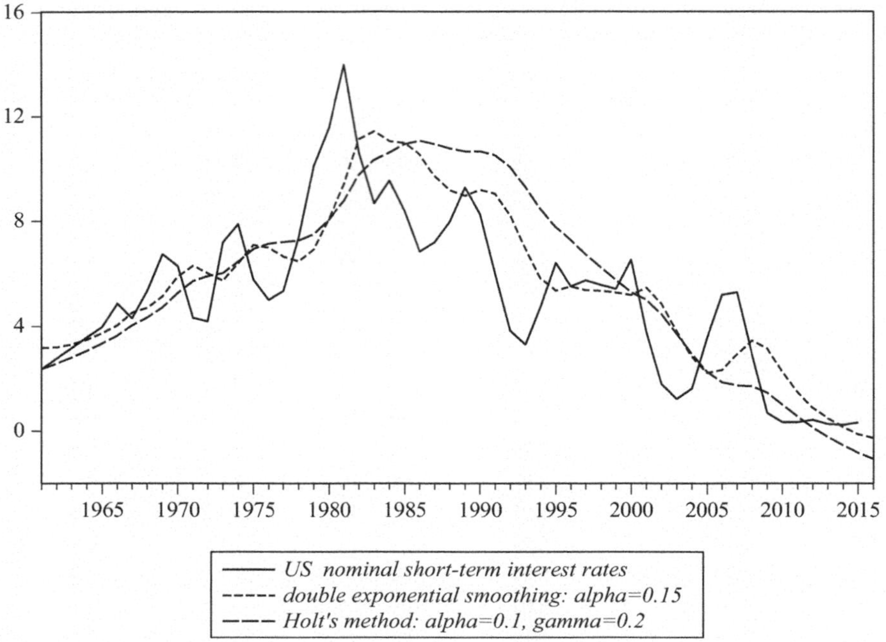

Fig. 3.14 Annual data 1961–2015 and smoothing and predicting by double exponential smoothing and Holt's method in Example 3.9 (*US nominal short-term interest rates in % p.a.*)

Exercise 3.3 Repeat the analysis from Example 3.3 (the modified exponential trend in Japan gross national income) only for data since 1970 (*hint:* 511 354-770 662· 0,862 404*t*, *saturation in* 511 360).

Exercise 3.4 Eliminate numerically the logistic trend in Example 3.4 (the Japan gross national income).

Exercise 3.5 Eliminate numerically Gompertz trend in Example 3.5 (the Japan gross national income).

Exercise 3.6 Repeat the analysis from Example 3.7 (the moving averages for the US nominal short-term interest rates) only for data since 1981.

Exercise 3.7 Derive the formulas (3.66) and (3.67) in the case arithmetic moving averages of length 5.

Exercise 3.8 Repeat the analysis from Example 3.8 (the simple exponential smoothing for the annual averages of exchange rates USD/EUR(ECU)) only for data since 1980 (these data are not presented numerically in the monograph; *hint:* $\alpha = 0.999$).

Exercise 3.9 Repeat the analysis from Example 3.9 (the double exponential smoothing and Holt's method for the US nominal short-term interest rates) only for data since 1981.

Chapter 4
Seasonality and Periodicity

4.1 Seasonality in Time Series

This deals with the elimination of *seasonal component* describing periodic changes in time series which pass off during one calendar year and repeat themselves each year. Even if the moving averages from Sect. 3.2 are capable of eliminating the seasonality significantly (e.g., the monthly centered moving averages (3.68) have such an effect in the case of monthly seasonal observations), an effective seasonal analysis should moreover deliver so-called *seasonal indices* $I_1, I_2, \ldots, I_s$ (s denotes the length of season, i.e., $s = 12$ in the case of monthly observations). These indices model the seasonality in particular seasons, and, moreover, they can be used not only to eliminate the seasonal phenomenon but also to construct predictions. However, they have sense only under the assumption that the seasonality is really regular so that its modeling by repeating seasonal indices is justified for the given time series.

The seasonal indices have the following properties:

- The units, in which the seasonal indices are measured, depend on the type of decomposition. When the decomposition is
  - *additive* (i.e., $y_t = Tr_t + Sz_t + E_t$): I_t is measured in the same units as the corresponding time series y_t (e.g., the December seasonal index of a retail sale amounting to EUR 45m means that the seasonality manifests itself by the December increase of time series by EUR 45m above the average trend behavior);
  - *multiplicative* (i.e., $y_t = Tr_t\, Sz_t\, E_t$): I_t is a relative variable (e.g., the December seasonal index of a retail sale amounting to 1.38 means that the seasonality manifests itself by the December increase of time series by 38% above the trend).
- It is typical for the multiplicative decomposition that the seasonal fluctuations increase (decrease) with increasing (decreasing) trend, respectively, even if the seasonal indices repeat themselves regularly in particular seasons (in the case of

T. Cipra, *Time Series in Economics and Finance*,
https://doi.org/10.1007/978-3-030-46347-2_4

Fig. 4.1 Characteristic shape of seasonal fluctuations for additive and multiplicative decomposition

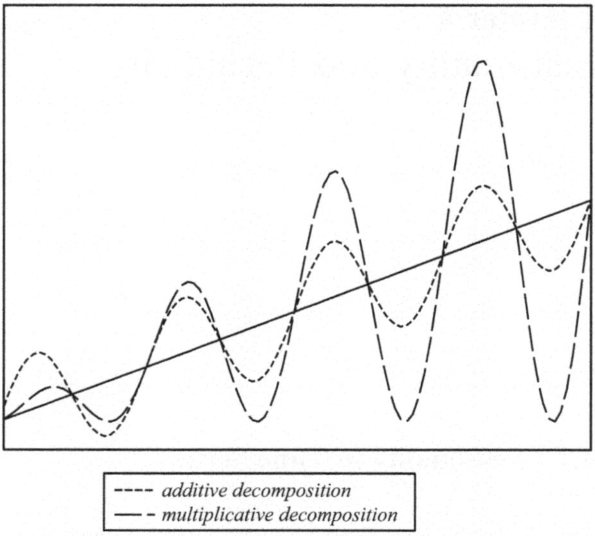

additive decomposition, the seasonal fluctuation does not depend on the trend monotonicity; see Fig. 4.1).

- The relation of trend and seasonal component is not determined unambiguously: one of them can be shifted upward in an arbitrary way, if it is offset by shifting the second one downward, and vice versa. This ambiguity is removed when the seasonal indices are normalized. Such a *normalization of seasonal indices* differs again according to the type of decomposition. When the decomposition is:
  - *additive*: then the usual normalization rule demands that the sum of seasonal indices over each season must be equal to zero; e.g., monthly observations must fulfill for each $i \geq 0$

$$I_{1+12i} + I_{2+12i} + \ldots + I_{12+12i} = 0; \tag{4.1}$$

  - *multiplicative*: then the usual normalization rule demands that the product of seasonal indices over each season must be equal to one; e.g., monthly observations must fulfill for each $i \geq 0$

$$I_{1+12i} \cdot I_{2+12i} : \ldots \cdot I_{12+12i} = 1 \tag{4.2}$$

(obviously after taking logarithm, this rule transfers to the form (4.1)), or occasionally

$$I_{1+12i} + I_{2+12i} + \ldots + I_{12+12i} = 12. \tag{4.3}$$

4.1.1 *Simple Approaches to Seasonality*

In practice (especially in various software systems) one prefers such approaches to seasonality that are as simple as possible from the calculation point of view. If the seasonal indices are (nearly) fixed for different seasons, then it can be recommended (e.g., in EViews) to apply the following algorithms (they will be described here only for the case of monthly observations, but they are quite analogous for quarterly observations):

4.1.1.1 Additive Decomposition

1. One constructs the centered moving averages $\overline{y}_t^{(12)}$ (in the case of quarterly observations, one should construct the centered moving averages $\overline{y}_t^{(4)}$). At the beginning and at the end of time series, one can repeat the first and the last calculable centered moving average, respectively (if it is necessary).
2. The centered moving averages can be looked upon as a raw estimate of trend component that enables to eliminate the trend from data

$$y_t^* = y_t - \overline{y}_t^{(12)}. \tag{4.4}$$

3. One constructs the (non-normalized) seasonal indices $I_1^*, I_2^*, \ldots, I_{12}^*$, where the seasonal index I_j^* for the jth month is estimated as the arithmetic average of all values y_t^*, which correspond to the jth month over all years included in time series ($j = 1, \ldots, 12$).
4. One normalizes the values $I_1^*, I_2^*, \ldots, I_{12}^*$ by subtracting their arithmetic mean

$$I_j = I_j^* - \overline{I}^* = I_j^* - \frac{I_1^* + \ldots + I_{12}^*}{12}, \qquad j = 1, \ldots, 12, \tag{4.5}$$

so that the normalization rule (4.1) is fulfilled.
5. One accomplishes the final seasonal elimination obtaining

$$\widehat{y}_t^{(12)} = y_t - I_j, \tag{4.6}$$

where the index t corresponds to the jth month of year.

4.1.1.2 Multiplicative Decomposition

1. One constructs the centered moving averages $\overline{y}_t^{(12)}$ similarly as in the case of additive decomposition (see above).
2. One eliminates the trend from data

$$y_t^* = \frac{y_t}{\overline{y}_t^{(12)}}. \tag{4.7}$$

3. One constructs the (non-normalized) seasonal indices $I_1^*, I_2^*, \ldots, I_{12}^*$, where the seasonal index I_j^* for the jth month is estimated as the arithmetic average of all values y_t^*, which correspond to the jth month over all years included in time series ($j = 1, \ldots, 12$).
4. One normalizes the values $I_1^*, I_2^*, \ldots, I_{12}^*$ by dividing by their geometric mean

$$I_j = \frac{I_j^*}{\widehat{I}^*} = \frac{I_j^*}{\sqrt[12]{I_1^* \cdot \ldots \cdot I_{12}^*}}, \qquad j = 1, \ \ldots, \ 12, \tag{4.8}$$

so that the normalization rule (4.2) is fulfilled.
5. One accomplishes the final seasonal elimination obtaining

$$\widehat{y}_t^{(12)} = \frac{y_t}{I_j}, \tag{4.9}$$

where the index t corresponds to the jth month of year.

Remark 4.1 A well-known software for seasonal time series denoted as X-ARIMA (e.g., X-12-ARIMA or X-13ARIMA-SEATS) has been developed by the U.S. Census Bureau (see, e.g., Dagum and Bianconcini (2016) or EViews). It consists in gradual application of several special moving averages in the form of compound filters. Moreover, it includes also procedures for calendar irregularities, seasonal Box–Jenkins methodology, and the like.

◇

Example 4.1 Table 4.1 and Fig. 4.2 present the additive elimination of seasonality in the time series y_t of the Czech construction production index for particular quarters 2009Q1-2016Q4 ($t = 1, \ldots, 32$). Table 4.1 shows also numerically the estimated seasonal indices $I_1, \ldots, I_4$ according to formulas (4.4)–(4.6).

◇

Example 4.2 Figure 4.3 presents the multiplicative elimination of seasonality in the time series y_t of the job applicants kept in the Czech labor office register for particular months 2005M1–2016M12 ($t = 1, \ldots, 144$; see also Table 4.4). The

Table 4.1 Quarterly data 2009Q1–2016Q4 and the simple approach to additive seasonal elimination in Example 4.1 (*Czech construction production index*)

Quarter	y_t	I_j	$\widehat{y}_t^{(4)}$	*Quarter*	y_t	I_j	$\widehat{y}_t^{(4)}$
2009Q1	72.06	−37.50	109.55	2013Q1	47.02	−37.50	84.52
2009Q2	110.55	−0.89	111.44	2013Q2	79.61	−0.89	80.50
2009Q3	124.02	15.95	108.07	2013Q3	98.65	15.95	82.70
2009Q4	125.56	22.44	103.13	2013Q4	107.21	22.44	84.77
2010Q1	55.70	−37.50	93.19	2014Q1	53.28	−37.50	90.77
2010Q2	101.24	−0.89	102.14	2014Q2	84.08	−0.89	84.97
2010Q3	120.37	15.95	104.41	2014Q3	101.47	15.95	85.51
2010Q4	122.69	22.44	100.25	2014Q4	107.91	22.44	85.48
2011Q1	58.85	−37.50	96.35	2015Q1	58.35	−37.50	95.85
2011Q2	95.91	−0.89	96.81	2015Q2	94.39	−0.89	95.29
2011Q3	109.37	15.95	93.42	2015Q3	108.92	15.95	92.97
2011Q4	121.59	22.44	99.15	2015Q4	109.53	22.44	87.10
2012Q1	52.94	−37.50	90.44	2016Q1	53.17	−37.50	90.67
2012Q2	90.20	−0.89	91.10	2016Q2	84.53	−0.89	85.43
2012Q3	102.61	15.95	86.65	2016Q3	99.26	15.95	83.31
2012Q4	110.69	22.44	88.25	2016Q4	105.93	22.44	83.50

Source: Czech Statistical Office (https://www.czso.cz/csu/czso/stavebnictvi-casove-rady-archiv-baze-2010)

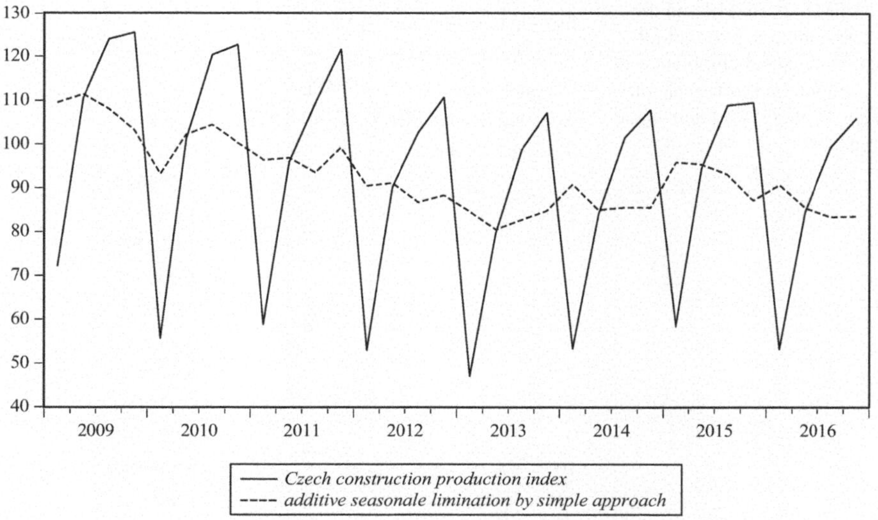

Fig. 4.2 Quarterly data 2009Q1–2016Q4 and the simple approach to the additive seasonal elimination in Example 4.1 (*Czech construction production index*)

output of EViews in Table 4.2 shows the estimated seasonal indices $I_1, \ldots, I_{12}$ according to formulas (4.7)–(4.9).

◇

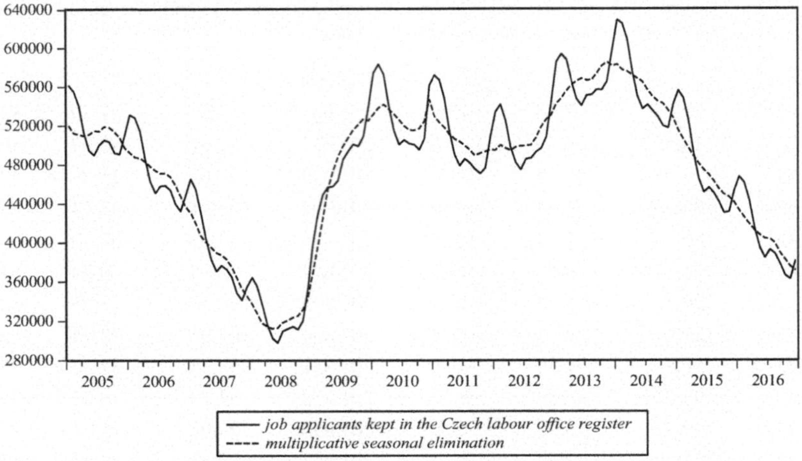

Fig. 4.3 Monthly data 2005M1–2016M12 and the simple approach to multiplicative seasonal elimination in Example 4.2 (*job applicants kept in the Czech labor office register*); see Table 4.4. Source: Czech Statistical Office

Table 4.2 Monthly data 2005M1–2016M12 and the simple approach to multiplicative seasonal elimination in Example 4.2 (*job applicants kept in the Czech labor office register*)

Sample: 2005M01–2016M12	
Included observations: 144	
Ratio to Moving Average	
Original Series: CZ_UNEMPL	
Adjusted Series: SEASON_MULTIPL	
Scaling Factors:	
1	1.079644
2	1.083429
3	1.057297
4	1.006022
5	0.968020
6	0.951058
7	0.972616
8	0.973159
9	0.971287
10	0.957589
11	0.965881
12	1.026710

Source: calculated by EViews

4.1.2 Regression Approaches to Seasonality

The regression approaches differ from the simple approaches of Sect. 4.1.1 only by estimating the seasonal indices using more sophisticated regression models.

4.1.2.1 Seasonality Modeled by Dummies

The additive seasonality can be conveniently modeled by means of regression with $s-1$ zero-one regressors denoted in econometrics usually as *dummies* (s is the length of season). As an example, let us consider the additive seasonal decomposition with linear trend for a quarterly time series y_t (i.e., $s = 4$). Then the corresponding regression model can be chosen as

$$y_t = \beta_0 + \beta_1 t + \alpha_2 x_{t2} + \alpha_3 x_{t3} + \alpha_4 x_{t4} + \varepsilon_t, \tag{4.10}$$

where the dummies x_2, x_3, x_4 are defined by the following table:

t	x_{t2}	x_{t3}	x_{t4}
1	0	0	0
2	1	0	0
3	0	1	0
4	0	0	1
5	0	0	0
6	1	0	0
7	0	1	0
8	0	0	1
$\vdots$	$\vdots$	$\vdots$	$\vdots$

The estimated model with OLS estimates b_0, b_1, a_2, a_3, a_4 can be used to construct the point and interval predictions, for which the future values of dummies are obtained by natural extension of the previous table till the corresponding prediction horizon. Moreover, if one needs the seasonal indices explicitly, then their normalization (4.3) is possible in the form

$$\widehat{Tr}_t = (b_0 + \overline{a}) + b_1 t,\ \widehat{I}_1 = -\overline{a},\ \widehat{I}_2 = a_2 - \overline{a},\ \widehat{I}_3 = a_3 - \overline{a},\ \widehat{I}_4 = a_4 - \overline{a}, \tag{4.11}$$

where

$$\overline{a} = \frac{a_2 + a_3 + a_4}{4}. \tag{4.12}$$

Table 4.3 Quarterly data 2009Q1–2016Q4 and the regression approach to additive seasonal elimination in Example 4.3 (*Czech construction production index*); see Table 4.1

Quarter	y_t	I_j	$\widehat{y}_t^{(4)}$	Quarter	y_t	I_j	$\widehat{y}_t^{(4)}$
2009Q1	72.06	−37.38	109.44	2013Q1	47.02	−37.38	84.40
2009Q2	110.55	−0.47	111.02	2013Q2	79.61	−0.47	80.08
2009Q3	124.02	15.73	108.29	2013Q3	98.65	15.73	82.92
2009Q4	125.56	22.12	103.44	2013Q4	107.21	22.12	85.09
2010Q1	55.70	−37.38	93.08	2014Q1	53.28	−37.38	90.66
2010Q2	101.24	−0.47	101.71	2014Q2	84.08	−0.47	84.55
2010Q3	120.37	15.73	104.64	2014Q3	101.47	15.73	85.74
2010Q4	122.69	22.12	100.57	2014Q4	107.91	22.12	85.79
2011Q1	58.85	−37.38	96.23	2015Q1	58.35	−37.38	95.73
2011Q2	95.91	−0.47	96.38	2015Q2	94.39	−0.47	94.86
2011Q3	109.37	15.73	93.64	2015Q3	108.92	15.73	93.19
2011Q4	121.59	22.12	99.47	2015Q4	109.53	22.12	87.41
2012Q1	52.94	−37.38	90.32	2016Q1	53.17	−37.38	90.55
2012Q2	90.20	−0.47	90.67	2016Q2	84.53	−0.47	85.00
2012Q3	102.61	15.73	86.88	2016Q3	99.26	15.73	83.53
2012Q4	110.69	22.12	88.57	2016Q4	105.93	22.12	83.81

Source: Czech Statistical Office

Example 4.3 Table 4.3 and Fig. 4.4 present the additive elimination of seasonality in the time series y_t of the Czech construction production index for particular quarters 2009Q1–016Q4 ($t = 1, \ldots, 32$) using the regression approach to seasonality (see also Example 4.1 for the same data but applying the simple approach to additive seasonality from Sect. 4.1.1). Again Table 4.3 shows numerically the seasonal indices $I_1, \ldots, I_4$ estimated according to formulas (4.10)–(4.12) with the only difference that the model using quadratic trend has been applied instead of linear trend used in (4.10), i.e.,

$$y_t = \beta_0 + \beta_1 t + \beta_2 t^2 + \alpha_2 x_{t2} + \alpha_3 x_{t3} + \alpha_4 x_{t4} + \varepsilon_t.$$

This model has been estimated as

$$\widehat{y}_t = 74.85 - 2.15 \cdot t + 0.04t^2 + 36.91x_{t2} + 53.11x_{t3} + 59.50x_{t4},$$

so that normalization (4.11) with $\overline{a} = 37.38$ gives

$$\widehat{Tr}_t = 112.23 - 2.15 \cdot t + 0.04t^2;\ \widehat{I}_1 = -37.38;\ \widehat{I}_2 = -0.47;\ \widehat{I}_3 = 15.73;\ \widehat{I}_4 = 22.12.$$

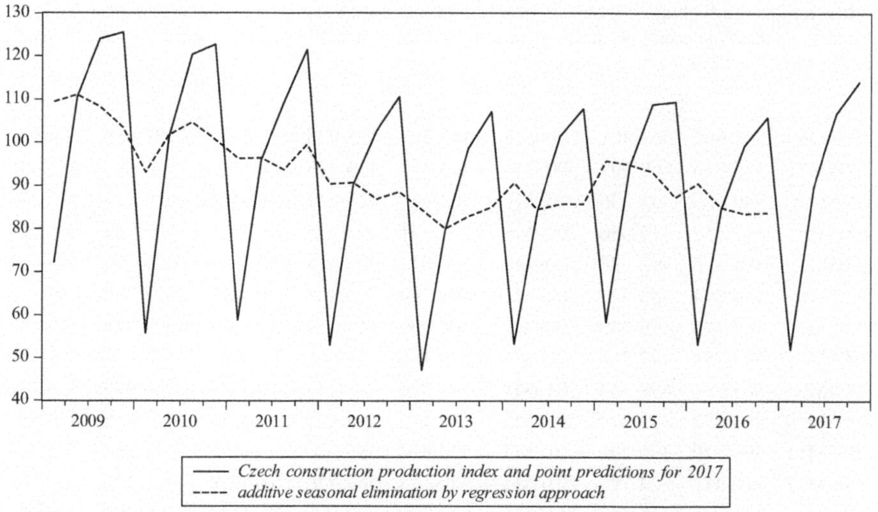

Fig. 4.4 Quarterly data 2009Q1–2016Q4 and the regression approach to the additive seasonal elimination including predictions for data 2017Q1–2017Q4 in Example 4.3 (*Czech construction production index*)

Table 4.3 and Fig. 4.4 again present the values $\widehat{y}_t^{(4)}$ after seasonal elimination (4.6). Finally, the predictions for year 2017 are shown in Fig. 4.4, e.g., for 2017Q1 calculated as

$$\begin{aligned}\widehat{y}_{33}(32) &= 74.85 - 2.15 \cdot 33 + 0.04 \cdot 33^2 + 36.91 \cdot 0 + 53.11 \cdot 0 + 59.50 \cdot 0 \\ &= 51.98.\end{aligned}$$

◇

4.1.2.2 Seasonality Modeled by Goniometric Functions

The goniometric functions enable us to model the seasonality with length of season s explicitly by means of models of the form

$$y_t = \beta_0 + \beta_1 \cdot t + \beta_2 \sin\left(\frac{2\pi t}{s}\right) + \beta_3 \cos\left(\frac{2\pi t}{s}\right) + \varepsilon_t \tag{4.13}$$

(*linear trend with additive seasonality*);

$$y_t = \beta_0 + \beta_1 \cdot t + \beta_2 \cdot t \cdot \sin\left(\frac{2\pi t}{s}\right) + \beta_3 \cdot t \cdot \cos\left(\frac{2\pi t}{s}\right) + \varepsilon_t \tag{4.14}$$

(*linear trend with multiplicative seasonality*) and the like.

Table 4.4 Monthly data 2005M1–2016M12 and the multiplicative seasonal elimination by Holt–Winters' method including predictions for data 2017M1–2017M12 in Example 4.4 (*job applicants kept in the Czech labor office register*)

obs	y_t	$\widehat{y}_t$	*obs*	y_t	$\widehat{y}_t$	*obs*	y_t	$\widehat{y}_t$
2005M1	561 662	561 401	2009M5	457 561	415 928	2013M9	557 058	560 632
2005M2	555 046	560 999	2009M6	463 555	436 915	2013M10	556 681	557 446
2005M3	540 456	543 236	2009M7	485 319	471 992	2013M11	565 313	566 554
2005M4	512 557	514 050	2009M8	493 751	490 025	2013M12	596 833	606 851
2005M5	494 576	492 728	2009M9	500 812	500 613	2014M1	629 274	636 620
2005M6	489 744	483 847	2009M10	498 760	500 469	2014M2	625 390	639 169
2005M7	500 325	496 289	2009M11	508 909	512 454	2014M3	608 315	619 771
2005M8	505 254	497 258	2009M12	539 136	554 066	2014M4	574 908	585 569
2005M9	503 396	498 453	2010M1	574 226	586 434	2014M5	549 973	560 130
2005M10	491 878	492 879	2010M2	583 135	594 621	2014M6	537 179	546 537
2005M11	490 779	496 296	2010M3	572 824	585 490	2014M7	541 364	554 316
2005M12	510 416	524 791	2010M4	540 128	562 720	2014M8	535 225	549 561
2006M1	531 235	549 598	2010M5	514 779	539 075	2014M9	529 098	545 301
2006M2	528 154	539 345	2010M6	500 500	526 087	2014M10	519 638	533 373
2006M3	514 759	519 664	2010M7	505 284	534 095	2014M11	517 508	531 787
2006M4	486 163	489 954	2010M8	501 494	526 799	2014M12	541 914	556 343
2006M5	463 042	468 218	2010M9	500 481	517 533	2015M1	556 191	579 989
2006M6	451 106	456 462	2010M10	495 161	502 021	2015M2	548 117	567 196
2006M7	458 270	461 290	2010M11	506 640	503 480	2015M3	525 315	540 863
2006M8	458 729	458 063	2010M12	561 551	536 489	2015M4	491 585	502 268
2006M9	454 182	453 279	2011M1	571 853	590 852	2015M5	465 689	473 425
2006M10	439 788	442 873	2011M2	566 896	589 323	2015M6	451 395	456 122
2006M11	432 573	441 332	2011M3	547 762	568 253	2015M7	456 341	456 930
2006M12	448 545	460 103	2011M4	513 842	533 396	2015M8	450 666	450 923
2007M1	465 458	479 713	2011M5	489 956	504 445	2015M9	441 892	446 274
2007M2	454 737	469 777	2011M6	478 775	488 996	2015M10	430 432	435 493
2007M3	430 474	448 495	2011M7	485 584	495 685	2015M11	431 364	431 793
2007M4	402 932	414 079	2011M8	481 535	491 332	2015M12	453 118	451 977
2007M5	382 599	388 422	2011M9	475 115	485 599	2016M1	467 403	469 766
2007M6	370 791	374 060	2011M10	470 618	471 305	2016M2	461 254	463 410
2007M7	376 608	374 935	2011M11	476 404	473 334	2016M3	443 109	444 681
2007M8	372 759	370 801	2011M12	508 451	505 004	2016M4	414 960	415 571
2007M9	364 978	363 714	2012M1	534 089	531 607	2016M5	394 789	393 370
2007M10	348 842	351 060	2012M2	541 685	532 930	2016M6	384 328	381 009
2007M11	341 438	345 060	2012M3	525 180	522 306	2016M7	392 667	384 149
2007M12	354 878	356 754	2012M4	497 322	496 235	2016M8	388 474	381 810
2008M1	364 544	370 824	2012M5	482 099	476 241	2016M9	378 258	379 133
2008M2	355 033	361 174	2012M6	474 586	469 617	2016M10	366 244	370 435
2008M3	336 297	342 850	2012M7	485 597	482 137	2016M11	362 755	367 779
2008M4	316 118	317 381	2012M8	486 693	483 795	2016M12	381 373	382 580
2008M5	302 507	298 817	2012M9	493 185	483 997	2017M1	–	393 632

(continued)

Table 4.4 (continued)

obs	y_t	$\hat{y}_t$	*obs*	y_t	$\hat{y}_t$	*obs*	y_t	$\hat{y}_t$
2008M6	297 880	289 208	2012M10	496 762	481 615	2017M2	–	388 287
2008M7	310 058	293 601	2012M11	508 498	493 515	2017M3	–	372 641
2008M8	312 333	295 350	2012M12	545 311	534 816	2017M4	–	348 250
2008M9	314 558	295 268	2013M1	585 809	564 588	2017M5	–	329 616
2008M10	311 705	291 838	2013M2	593 683	579 507	2017M6	–	318 508
2008M11	320 299	296 627	2013M3	587 768	572 841	2017M7	–	320 201
2008M12	352 250	321 184	2013M4	565 228	553 061	2017M8	–	313 932
2009M1	398 061	347 520	2013M5	547 463	541 313	2017M9	–	306 686
2009M2	428 848	368 210	2013M6	540 473	538 308	2017M10	–	297 872
2009M3	448 912	385 651	2013M7	551 096	555 683	2017M11	–	295 628
2009M4	456 726	398 596	2013M8	551 731	558 583	2017M12	–	308 814

Source: Czech Statistical Office (https://www.czso.cz/csu/czso/casove-rady-zakladnich-ukazatelu-statistiky-prace-leden-2020)

If the form of seasonality is more complex one can add components of the form

$$\beta_4 \sin\left(\frac{4\pi t}{s}\right) + \beta_5 \cos\left(\frac{4\pi t}{s}\right). \tag{4.15}$$

4.1.3 *Holt–Winters' Method*

This method extends Holt's method from Sect. 3.3.3 to include in the adaptive way not only the local linear trend but also the seasonality. Therefore, both versions of Holt–Winters' method (additive and multiplicative) exploit even three smoothing constants: α to smooth the level L_t, γ to smooth the slope T_t, and δ to smooth the seasonal index I_t of given time series with length of season s ($0 < \alpha, \gamma, \delta < 1$); see, e.g., Abraham and Ledolter (1983), Bowerman and O'Connell (1987), Montgomery and Johnson (1976), and others.

4.1.3.1 Additive Holt–Winters' Method

Recursive formulas of *additive Holt–Winters' method* have the form

$$L_t = \alpha(y_t - I_{t-s}) + (1 - \alpha)(L_{t-1} + T_{t-1}), \tag{4.16}$$

$$T_t = \gamma(L_t - L_{t-1}) + (1 - \gamma)T_{t-1}, \tag{4.17}$$

$$I_t = \delta(y_t - L_t) + (1 - \delta)I_{t-s}, \tag{4.18}$$

$$\widehat{y}_t = L_t + I_t, \tag{4.19}$$

$$\begin{aligned} \widehat{y}_{t+\tau}(t) &= L_t + T_t \cdot \tau + I_{t+\tau-s} \quad \text{for} \quad \tau = 1, \ \ldots, \ s, \\ &= L_t + T_t \cdot \tau + I_{t+\tau-2s} \quad \text{for} \quad \tau = s+1, \ \ldots, \ 2s, \\ &\vdots \end{aligned} \tag{4.20}$$

This method similarly as Holt's method in Sect. 3.3.3 has been initially suggested ad hoc using logical arguments only. For example, the eliminated seasonal index I_t of given time series in time t is constructed according to (4.18) as a convex combination of two items, namely (1) an estimation of this seasonal index constructed in time t by removing the trend component from the observed value y_t (i.e., $y_t - L_t$) and (2) an estimation of this seasonal index constructed in time $t - 1$ using the most actual estimated value I_{t-s} from the previous season. One proceeds in a similar way also in (4.16) to remove the seasonal component from the observed value y_t (i.e., $y_t - I_{t-s}$) using again the most actual estimated value I_{t-s} from the previous season, and in (4.20) to predict (in the case of predictions, one must distinguish in (4.20) particular future seasons respecting the fact that forecasting in prediction horizons, which are too remote in future, may be unreliable).

To start the recursive formulas of additive Holt–Winters' method. one must choose initial values L_0, T_0, I_{-s+1}, I_{-s+2}, …, I_0 and smoothing constants α, γ, δ:

1. Suitable initial values can be found simply if one models the seasonality by dummies as in Sect. 4.1.2 (the normalization is not here necessary)

$$y_t = \beta_0 + \beta_1 t + \alpha_2 x_{t2} + \ldots + \alpha_s x_{ts} + \varepsilon_t \tag{4.21}$$

   with dummy variables x_2, …, x_s, so that using OLS estimates b_0, b_1, a_2, …, a_s one can put

$$L_0 = b_0, \ T_0 = b_1, \ I_{-s+1} = 0, \ I_{-s+2} = a_2, \ \ldots \ I_0 = a_s. \tag{4.22}$$

2. Suitable smoothing constants α, γ, δ can be found using
   (a) Fixed choice: in routine situations one recommends $\alpha = \delta = 0.4$ and $\gamma = 0.1$.
   (b) Estimates α, γ, δ: one proceeds in a similar way as for the exponential smoothing by minimizing *SSE* (see, e.g., Sect. 3.3.1).

4.1.3.2 Multiplicative Holt–Winters' Method

Recursive formulas of *multiplicative Holt–Winters' method* have the form

$$L_t = \alpha(y_t/I_{t-s}) + (1-\alpha)(L_{t-1} + T_{t-1}), \tag{4.23}$$

$$T_t = \gamma(L_t - L_{t-1}) + (1-\gamma)T_{t-1}, \tag{4.24}$$

$$I_t = \delta(y_t/L_t) + (1-\delta)I_{t-s}, \tag{4.25}$$

$$\widehat{y}_t = L_t \cdot I_t, \tag{4.26}$$

$$\begin{aligned} \widehat{y}_{t+\tau}(t) &= (L_t + T_t \cdot \tau) \cdot I_{t+\tau-s} \quad \text{for} \quad \tau = 1, \ \ldots, \ s, \\ &= (L_t + T_t \cdot \tau) \cdot I_{t+\tau-2s} \quad \text{for} \quad \tau = s+1, \ \ldots, \ 2s, \quad \vdots \end{aligned} \tag{4.27}$$

In comparison with the previous additive Holt–Winters' method, one only replaces sums and differences within brackets in (4.16)–(4.20) by products and quotients, respectively.

To start the recursive formulas of multiplicative Holt–Winters' method, one must again choose initial values L_0, T_0, I_{-s+1}, I_{-s+2}, ..., I_0 and smoothing constants α, γ, δ:

1. Suitable initial values can be found by means of simple formulas

$$T_0 = \frac{\overline{y}_m - \overline{y}_1}{(m-1)s}, \quad L_0 = \overline{y}_1 - \frac{s+1}{2} T_0,$$

$$I_{j-s} = \frac{1}{m} \sum_{i=0}^{m-1} \frac{y_{j+s \cdot i}}{\overline{y}_{i+1} - \left(\frac{s+1}{2} - j\right) T_0}, \quad j = 1, \ \ldots, s, \tag{4.28}$$

   where $\overline{y}_i$ is the arithmetic average of observations over the ith season (of length s) and m is the total number of these seasons.

2. Suitable smoothing constants α, γ, δ can be found in the same way as in the case of additive Holt–Winters' method.

Remark 4.2 Using a similar denotation as in Remark 3.6, one recommends to construct the $(1-p)\cdot 100\%$ prediction interval in the form

$$\left(\widehat{y}_{n+\tau}(n) - u_{1-p/2} \cdot d_\tau \cdot MAE, \widehat{y}_{n+\tau}(n) + u_{1-p/2} \cdot d_\tau \cdot MAE\right), \tag{4.29}$$

where one substitutes for any $\tau > 0$

$$d_\tau \approx 1,25 \cdot \left(\frac{1 + \frac{\theta}{(1+\nu)^3}\left((1+4\nu+5\nu^2) + 2\theta(1+3\nu)\tau + 2\theta^2\tau^2\right)}{1 + \frac{\theta}{(1+\nu)^3}\left((1+4\nu+5\nu^2) + 2\theta(1+3\nu) + 2\theta^2\right)} \right)^{1/2} \tag{4.30}$$

(one denotes $\theta = \max\{\alpha, \gamma, \delta\}$ and $\nu = 1 - \theta$), and for additive Holt–Winters' method

$$MAE = \frac{1}{n-s} \sum_{t=s+1}^{n} \left| y_t - L_{t-1} - T_{t-1} - I_{t-s} \right|, \tag{4.31}$$

or for multiplicative Holt–Winters' method

$$MAE = \frac{1}{n-s} \sum_{t=s+1}^{n} \left| \frac{y_t}{I_{t-s}} - L_{t-1} - T_{t-1} \right|. \tag{4.32}$$

◊

Example 4.4 Table 4.4 and Fig. 4.5 present the multiplicative elimination of seasonality by Holt–Winters' method (with the fixed choice of smoothing constant $\alpha = \delta = 0.4$ and $\gamma = 0.1$) in the time series y_t of job applicants kept in the Czech labor office register for particular months 2005M1–2016M12 ($t = 1, \ldots, 144$) including predictions for data 2017M1–2017M12. The smoothed values and predictions have been obtained by EViews and can be compared with the corresponding results by the simple approach to multiplicative seasonal elimination in Example 4.2.

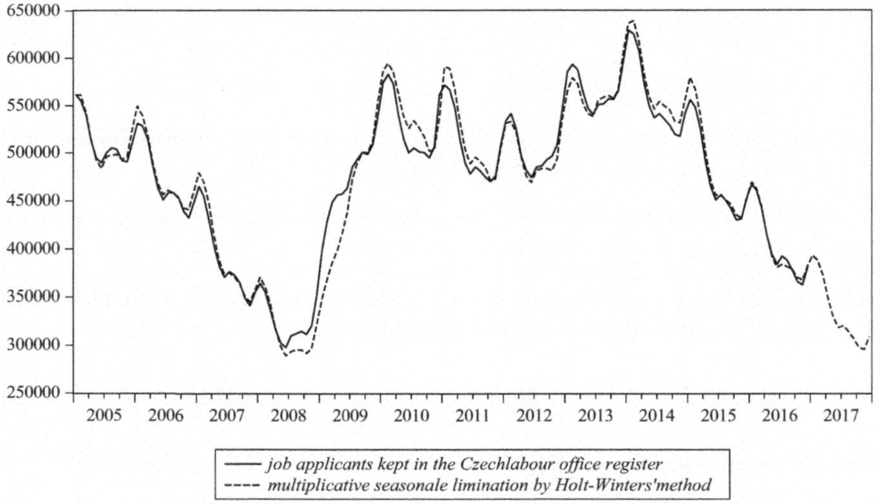

Fig. 4.5 Monthly data 2005M1–2016M12 and the multiplicative seasonal elimination by Holt–Winters' method in Example 4.4 (*job applicants kept in the Czech labor office register*), see Table 4.4. Source: Czech Statistical Office

4.1.4 Schlicht's Method

Schlicht's method approaches the problem of additive decomposition by optimizing specific criteria. If **y** is the vector of observed values of given time series, then these criteria characterize the vector of trend values **x**, the vector of seasonal values **z**, and the vector of residual values **u**

$$\mathbf{y} = \mathbf{x} + \mathbf{z} + \mathbf{u} \tag{4.33}$$

(all these vectors are column vector of type $T \times 1$). The criteria that characterize the particular decomposition components are recommended in the following form (see Schlicht (1982)):

1. Criterion minimizing the *measure of trend smoothness* f: $\mathrm{R}^T \to \mathrm{R}$ (over $\mathbf{x} \in \mathrm{R}^T$):

$$f(\mathbf{x}) = \alpha \cdot \sum_{t=3}^{T} (\Delta^2 x_t)^2 = \alpha \cdot \sum_{t=3}^{T} (x_t - 2x_{t-1} + x_{t-2})^2. \tag{4.34}$$

2. Criterion minimizing the *measure of seasonal stability* g: $\mathrm{R}^T \to \mathrm{R}$ with length of season s (over $\mathbf{z} \in \mathrm{R}^T$):

$$g(\mathbf{z}) = \beta \cdot \sum_{t=s+1}^{T} (z_t - z_{t-s})^2 + \gamma \cdot \sum_{t=s}^{T} \left(\sum_{\tau=0}^{s-1} z_{t-\tau} \right)^2. \tag{4.35}$$

3. Criterion minimizing the *measure SSE of residual component* h: $\mathrm{R}^T \to \mathrm{R}$ (over $\mathbf{u} \in \mathrm{R}^T$):

$$h(\mathbf{u}) = \mathbf{u}'\mathbf{u} = \sum_{t=1}^{T} u_t^2. \tag{4.36}$$

The positive coefficients α, β, γ control the relative intensity of particular components in the decomposition, e.g., the choice of higher α will result in a higher smoothness of trend. One must also respect the fact that $\mathbf{x} + \mathbf{z} + \mathbf{u} = \mathbf{y}$. Therefore, the decomposition of **y** is transferred to the optimization problem

$$\min_{\mathbf{x}, \mathbf{z} \in \mathrm{R}^T} \{ f(\mathbf{x}) + g(\mathbf{z}) + h(\mathbf{y} - \mathbf{x} - \mathbf{z}) \}. \tag{4.37}$$

It is equivalent to solving the following system of $2T$ linear equations with $2T$ unknowns **x** and **z**:

$$\mathbf{H}\begin{pmatrix} \mathbf{x} \\ \mathbf{z} \end{pmatrix} = \begin{pmatrix} \mathbf{y} \\ \mathbf{y} \end{pmatrix}, \tag{4.38}$$

where

$$\mathbf{H} = \begin{pmatrix} \alpha \cdot \mathbf{P}'\mathbf{P} + \mathbf{I} & \mathbf{I} \\ \mathbf{I} & \beta \cdot \mathbf{Q}'\mathbf{Q} + \gamma \cdot \mathbf{R}'\mathbf{R} + \mathbf{I} \end{pmatrix} \tag{4.39}$$

with matrices **P** of type $(T{-}2) \times T$, **Q** of type $(T{-}s) \times T$, and **R** of type $(T{-}s + 1) \times T$ defined as

$$\mathbf{P} = \begin{pmatrix} 1 & -2 & 1 & & \\ & 1 & -2 & 1 & \\ & & \ddots & \ddots & \ddots \\ & & 1 & -2 & 1 \end{pmatrix}, \mathbf{Q} = \begin{pmatrix} 1 & 0 & \cdots & 0 & -1 & & \\ & 1 & 0 & \cdots & 0 & -1 & \\ & & \ddots & \ddots & \ddots & \ddots & \\ & & 1 & 0 & \cdots & 0 & -1 \end{pmatrix}, \mathbf{R} = \begin{pmatrix} 1 & 1 & \cdots & 1 & & \\ & 1 & 1 & \cdots & 1 & \\ & & \ddots & \ddots & & \ddots \\ & & 1 & 1 & \cdots & 1 \end{pmatrix}$$

(these are band matrices that have zero elements with the exception of main diagonal and several upper diagonals that are formed by the same elements, e.g., by −2 and 1 in the case of **P**; here only nonzero bands are shown in **P**, **Q**, and **R**).

4.2 Tests of Periodicity

In the decomposition model (2.2.2), the cyclical component C_t can sometimes play an important role. It is the periodic component with periodicity longer than one year, e.g., the five-year business cycle from Sect. 2.2.2 (the annual periodicity is classified as seasonality). Sometimes there are even several such periodicities compounded in the given time series. Their elimination is complex since one must decide on the number and length of corresponding periodicities (e.g., quarterly one in combination with annual and five-year periodicities using monthly observations). As objective instruments in such situations one can apply various *tests of periodicity* which are usually based on a so-called *periodogram*.

The periodogram (as well as the spectral density) is an important instrument of *spectral analysis of time series* (see Sect. 2.2.2). Spectral analysis transfers the *time domain* (which looks upon the given time series as a sequence of observations in time) to the *spectral domain* (which looks upon the given time series as an (infinite) mixture of periodic components and calculates their intensities in this mixture).

More specifically, the *periodogram* $I(\omega)$ of time series $y_1, y_2, \ldots, y_n$ is a function of the frequency ω (such functions are typical for the spectral domain, while functions of time are used in the time domain). The *frequency* ω is usually measured by radians per time unit (this time unit corresponds to the time interval between neighboring observations, e.g., one year for an annual time series). Then $\omega/2\pi$ is the number of cycles per one time unit. For example, the five-year periodicity in an annual time series, where one-fifth of cycle occurs per one year, has the frequency $2\pi/5$. It is obvious that by observing a given time series one is capable of recognizing statistically only the frequencies ranging maximally to π radians per time unit, i.e., to one-half cycle per time unit (this upper limit is called *Nyquist frequency*); the "quicker" frequencies remain hidden from the point of view of the grid of observed values (e.g., the "quick" frequency 2π radians per time unit of the time series $y_t = \sin(2\pi t)$ observed at times $t = 1, \ldots, n$ with one cycle per each time unit obviously cannot be identified from the observed zero values $y_t = 0$ for $t = 1, \ldots, n$).

Numerically, the periodogram $I(\omega)$ is defined as

$$I(\omega) = \frac{1}{4\pi}\left(a^2(\omega) + b^2(\omega)\right), \qquad 0 \le \omega \le \pi, \tag{4.40}$$

where

$$a(\omega) = \sqrt{\frac{2}{n}}\sum_{t=1}^{n} y_t \cos(\omega \cdot t), \qquad b(\omega) = \sqrt{\frac{2}{n}}\sum_{t=1}^{n} y_t \sin(\omega \cdot t). \tag{4.41}$$

The key property of periodogram, which motivates the introduction of this instrument, is its behavior when the periodogram is applied for a periodic time series of the form

$$\begin{aligned} y_t &= \mu + \sum_{i=1}^{k} \delta_i \cos(\omega_i t + \phi_i) + \varepsilon_t \\ &= \mu + \sum_{i=1}^{k} (\alpha_i \cos(\omega_i t) + \beta_i \sin(\omega_i t)) + \varepsilon_t, \qquad t = 1, \quad \ldots, \quad n, \end{aligned} \tag{4.42}$$

where μ denotes the level of this time series, $\omega_1, \ldots, \omega_k$ are the mutually different (unknown in general) frequencies from the interval $(0, \pi)$ for k periodic components contained in (4.42), $\varphi_1, \ldots, \varphi_k$ are the corresponding phases (i.e., the shifts of cosinusoids from the origin for particular periodic components), and ε_t is the residual component in the form of white noise with variance σ^2. Then the periodogram of time series (4.42) fluctuates around the constant $\sigma^2/2\pi$ with the exception of frequencies $\omega_1, \ldots, \omega_k$, in which the periodogram rockets to local extremes comparable with the size of n (i.e., they are of order $O(n)$). Therefore, the periodogram can indicate by its "bursts" the position of frequencies $\omega_1, \ldots, \omega_k$.

In practice, a graphical search for the local extremes of periodogram may be subjective. Moreover, a practical realization of periodogram typically highly fluctuates because it is not consistent estimator of the spectrum (i.e., its variance may not decrease with the increasing length of time series). Therefore, in practice one should prefer suitable statistical tests. The best-known test of this type is *Fisher's test of periodicity*, which tests the null hypothesis

$$H_0 : y_t = \mu + \varepsilon_t, \qquad t = 1, \ldots, n \tag{4.43}$$

with the normally distributed white noise $\{\varepsilon_t\}$ against the alternative hypothesis (4.42) with a given significance level α. The test statistics is constructed using the periodogram values over the grid of frequencies

$$\omega_j^* = \frac{2\pi j}{n}, \qquad j = 1, \ldots, m, \tag{4.44}$$

where $m = \left\lfloor \frac{n-1}{2} \right\rfloor$ is the integer part of $\frac{n-1}{2}$, and has the form

$$W = \max_{j=1,\ldots,m} Y_j = Y_{j^*} = \max_{j=1,\ldots,m} \frac{I\left(\omega_j^*\right)}{I\left(\omega_1^*\right) + \ldots + I\left(\omega_m^*\right)} \tag{4.45}$$

(i.e., the test statistics equals to the maximum standardized value of periodogram over the grid (4.44) achieved for the index value denoted as j^*).

The critical region of Fisher's test of periodicity with significance level α is then

$$W \geq g_\alpha, \tag{4.46}$$

where g_α is the critical value of this test (see the tabulated values in Table 4.5). When the inequality (4.46) occurs, then we have found simultaneously the frequency of the periodic component that causes the rejection of null hypothesis (4.43): it is the grid point (4.44) for $j = j^*$. Repeating the test after removing this frequency (i.e., for $m - 1$), we can find further frequencies in the analyzed time series (see Example 4.5). The practical realization of Fisher's test of periodicity is described in Example 4.5.

Table 4.5 Critical values of Fisher's test

m	$g_{0.05}$	$g_{0.01}$
5	0.684	0.789
10	0.445	0.536
15	0.335	0.407
20	0.270	0.330
25	0.228	0.278
30	0.198	0.241
35	0.175	0.216
40	0.157	0.192
50	0.131	0.160

Remark 4.3 The distribution of test statistics under the null hypothesis in Fisher's test is complex. However, for $m \leq 50$ a simple approximation holds, namely

$$P(W > x) \approx m \cdot (1 - x)^{m-1}, \qquad 0 \leq x \leq 1. \tag{4.47}$$

There exist various modification of Fisher's test, which have been suggested to improve the power of this test, in particular when the periodicity is compounded $k > 1$ (e.g., Siegel's test (1980), Bølviken's test (1983), and others).

◇

If in time series $y_1, \ldots, y_n$ the frequencies $\widehat{\omega}_1, \ldots, \widehat{\omega}_k \in \{\omega_1^*, \ldots, \omega_m^*\}$ has been indicated by Fisher's test, then the OLS estimates in the resulting model (4.42) are very simple due to the orthogonality of its regressors, namely

$$\widehat{\mu} = \frac{1}{n} \sum_{t=1}^{n} y_t,$$

$$\widehat{\alpha}_j = \frac{2}{n} \sum_{t=1}^{n} y_t \cos\left(\widehat{\omega}_j t\right), \widehat{\beta}_j = \frac{2}{n} \sum_{t=1}^{n} y_t \sin\left(\widehat{\omega}_j t\right), \qquad j = 1, \ldots, k. \tag{4.48}$$

Example 4.5 Table 4.6 presents observed numbers of defective pieces (in thousands) in daily production of a production unit during 21 days (i.e., three weeks). One investigates whether this time series in Fig. 4.6 involves periodicities (e.g., the weekly periodicity due to the "bad" days in the beginning and end of particular weeks, or the three-day periodicity due to the rotation in shifts of workers, or others).

Since $n = 21$ and $m = 10$, one calculates (in Table 4.7) ten values of periodograms $I(\omega_j^*)$ in the grid points $\omega_j^* = 2\pi j/21$ ($j = 1, \ldots, 10$) according to (4.40) and (4.41). Then the value of the corresponding test statistics (4.45) is

Table 4.6 Data for Example 4.5 (numbers of defective pieces in daily production in thousands)

t	y_t	t	y_t	t	y_t
1	3.69	8	3.34	15	3.35
2	4.05	9	2.20	16	3.88
3	1.40	10	2.39	17	2.75
4	2.53	11	2.49	18	0.77
5	1.87	12	1.53	19	3.92
6	2.57	13	4.18	20	2.13
7	5.16	14	4.59	21	4.24

Table 4.7 Periodogram values in Example 4.5

j	$I(\omega_j^*)$
1	0.028 5
2	0.021 9
3	1.083 8
4	0.019 0
5	0.034 4
6	0.020 3
7	0.513 6
8	0.017 7
9	0.164 6
10	0.260 7
Σ	2.164 5

$$W = \max_{j=1,\ldots,10} Y_j = Y_3 = \frac{I(\omega_3^*)}{I(\omega_1^*) + \ldots + I(\omega_{10}^*)} = \frac{1.083\ 8}{2.164\ 5} = 0.500\ 7. \quad (4.49)$$

Using the tabulated critical values from Table 4.5, it holds for $m = 10$

$$W = 0.500\ 7 > g_{0.05} = 0.445, \quad (4.50)$$

so that Fisher's test of periodicity confirms with significance level of 5% the presence of the periodic component with frequency $\omega_3^* = 2\pi \cdot 3/21 = 2\pi/7$, which can be interpreted in the given time series of daily observations as a weekly periodicity.

Since the null hypothesis (4.43) on non-periodicity has not been rejected in the previous step, one should continue to look for further potential periodicities and repeat Fisher's test with periodogram values from Table 4.7 for $m = 9$ deleting the value $I(\omega_3^*)$, i.e.,

$$W = \frac{I(\omega_7^*)}{I(\omega_1^*) + I(\omega_2^*) + I(\omega_4^*) + I(\omega_5^*) + \ldots + I(\omega_{10}^*)} = \frac{0.513\ 6}{1.080\ 7}$$
$$= 0.475\ 2. \quad (4.51)$$

The approximation (4.47) can replace the tabulated critical value (since $m = 9 \leq 50$)

$$\mathrm{P}(W > 0.475\ 2) \approx 9 \cdot (1 - 0.475\ 2)^8 = 0.051\ 8, \quad (4.52)$$

so that the presence of further periodic component with frequency $\omega_7^* = 2\pi \cdot 7/21 = 2\pi/3$ cannot be confirmed with significance level of 5% (obviously, this component would model a 3-day periodicity).

Finally by (4.48), the model (4.42) can be estimated in the form (see also Fig. 4.6)

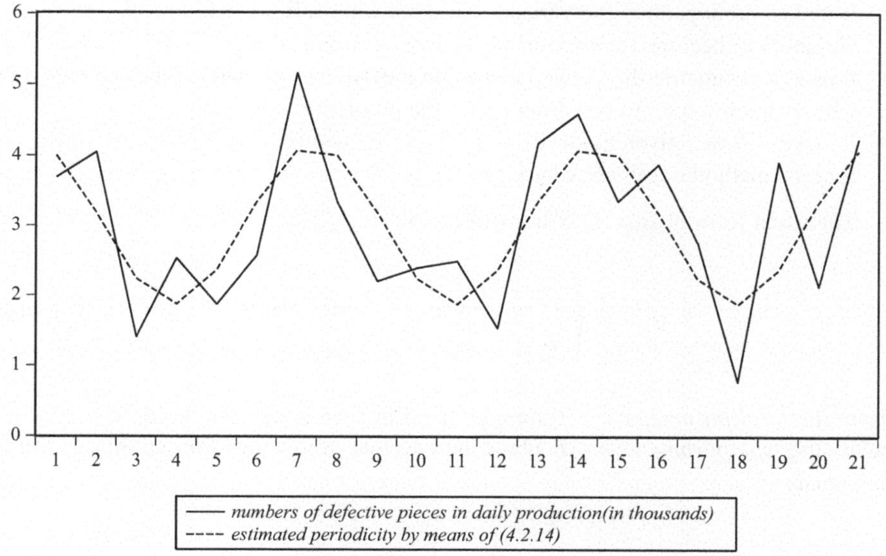

Fig. 4.6 Numbers of defective pieces in daily production (in thousands) and estimated periodicity by means of (4.53) in Example 4.5

$$\begin{aligned}\widehat{y}_t &= 3.01 + 1.06\cos\left(\frac{2\pi}{7}t\right) + 0.42\sin\left(\frac{2\pi}{7}t\right)\\ &= 3.01 + 1.14\cos\left(\frac{2\pi}{7}t - 0.377\ 2\right).\end{aligned} \tag{4.53}$$

◇

4.3 Transformations of Time Series

In practice, the analyzed time series are sometimes transformed in a suitable way to simplify the decomposition of the time series after transformation. Moreover, such transformations may be useful not only in the framework of decomposition. Two examples will be given here: Box–Cox transformation and the transformation based on differencing.

4.3.1 Box–Cox Transformation

Box–Cox transformation has some appealing properties:

- It makes homogenous the variance of given time series (including the seasonal variance) to become (approximately) constant in time.
- It makes symmetric the skewed distribution of given time series (or even normal, which enables, e.g., to construct easily the prediction intervals).
- It makes linear a given model of time series (frequently in the framework of Box–Jenkins methodology; see Chap. 6).

The usual form of Box–Cox transformation is

$$y_t^{(\lambda)} = \begin{cases} \dfrac{(y_t + c)^\lambda - 1}{\lambda} & \text{for } \lambda \neq 0, \\ \ln(y_t + c) & \text{for } \lambda = 0. \end{cases} \tag{4.54}$$

Here the *level parameter* $c > 0$ may be fixed in such a way that holds $y_t + c > 0$, while the *type parameter* $\lambda \in \mathrm{R}$ plays the key role in this transformation, e.g., it is obviously

$$\lim_{\lambda \to 0} \frac{(y_t + c)^\lambda - 1}{\lambda} = \ln(y_t + c) \tag{4.55}$$

(therefore the index λ participates in the symbol $y_t^{(\lambda)}$ denoting the transformed time series). Even though the parameter value λ that makes homogenous the variance of given time series can be estimated by the maximum likelihood method, practical applications frequently prefer more subjective approaches based on considerations of the following type. Since it holds

$$\operatorname{var}\left(y_t^{(\lambda)}\right) \approx \left(\frac{dy_t^{(\lambda)}}{dy_t}\right)^2 \Bigg|_{y_t = \bar{y}} \cdot \operatorname{var}(y_t) \tag{4.56}$$

and since we want to achieve the variance homogeneity

$$\operatorname{var}\left(y_t^{(\lambda)}\right) = k^2 = \text{const}, \tag{4.57}$$

one obtains from (4.56) and (4.57) the following important relation for the sample standard deviation s_y of time series y_t:

$$s_y \approx k \cdot \bar{y}^{\,1-\lambda}. \tag{4.58}$$

For example, the logarithmic transformation (4.55) with $\lambda = 0$ will make the considered time series homogenous if the relation $s_y \approx k \cdot \bar{y}$ holds approximately between the sample standard deviation s_y and the sample mean $\bar{y}$, and similarly for other values of λ.

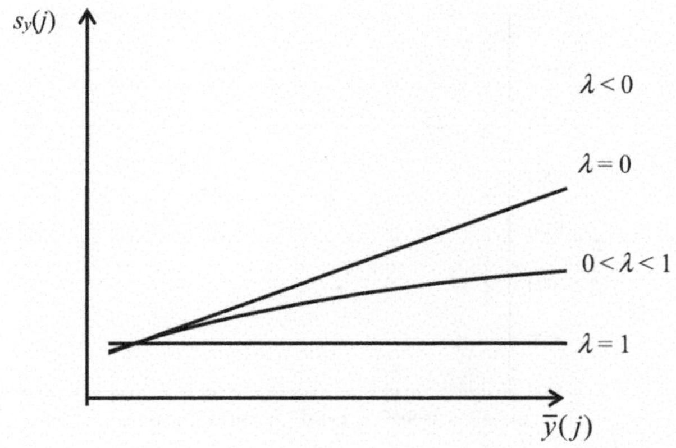

Fig. 4.7 Choice of type parameter λ in Box-Cox transformation

Table 4.8 Choice of type parameter λ in Box–Cox transformation

Shape of curve	Values of λ	Simplified choice of λ	Type of transformation
Constant	$\lambda = 1$	$\lambda = 1$	$y_t^{(\lambda)} = y_t$
Concave	$0 < \lambda < 1$	$\lambda = 1/2$	$y_t^{(\lambda)} = \sqrt{y_t}$
Line	$\lambda = 0$	$\lambda = 0$	$y_t^{(\lambda)} = \ln y_t$
Convex	$\lambda < 0$	$\lambda = -1/2$	$y_t^{(\lambda)} = 1/\sqrt{y_t}$

Therefore, in practice one recommends to divide the given time series into short segments of the same length (logically the length of segments may be 4 for quarterly time series and 12 for monthly time series). In each segment, the sample mean $\bar{y}(j)$ and the sample standard deviation $s_y(j)$ are calculated (j denotes the jth segment), and one plots a point with these coordinates in the plane (i.e., one has a system of points corresponding to particular segments). Finally, a smooth curve is fitted subjectively to this system of points in the plane (see Fig. 4.7). According to the

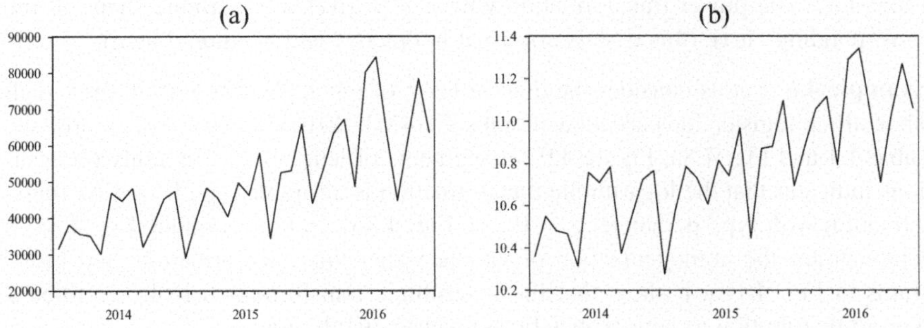

Fig. 4.8 Monthly data 2014M1–2016M12 in Example 4.6 (*job applicants kept in the Czech labor office register*): (**a**) before transformation; (**b**) after logarithmic transformation

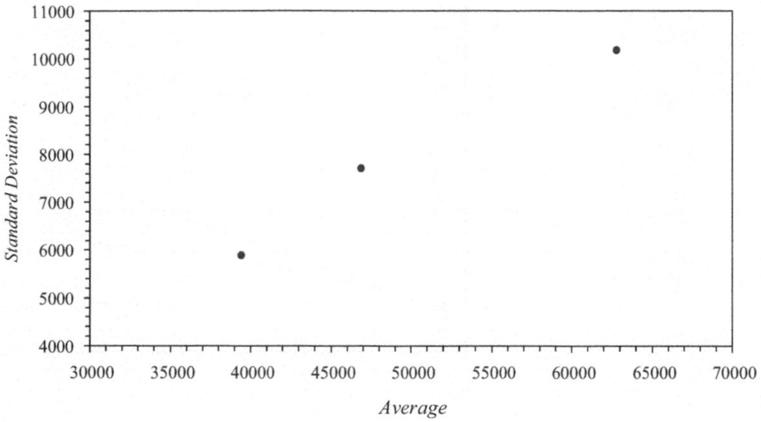

Fig. 4.9 Choice of type parameter λ in Box–Cox transformation in Example 4.6

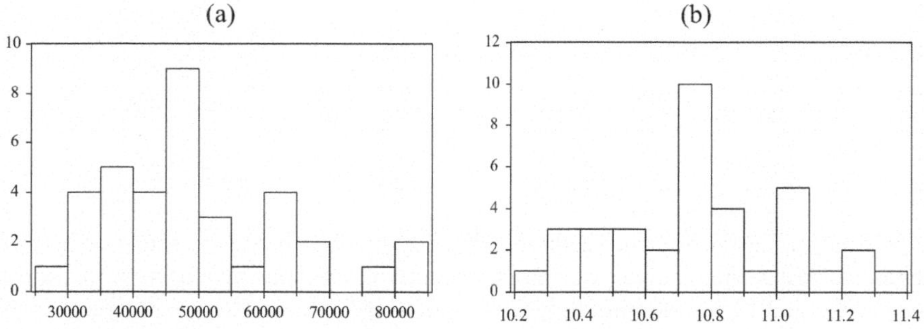

Fig. 4.10 Histogram of job applicants kept in the Czech labor office register in Example 4.6: (**a**) before transformation; (**b**) after logarithmic transformation

shape of such a curve, one selects a value for the parameter λ and decides in this way on the corresponding form of Box–Cox transformation for given time series (see Table 4.8). The power transformation with $\lambda > 1$ gives a hyperbolic shape of the corresponding curve (this case is not usual in practice and is ignored here).

Example 4.6 Let us consider the time series y_t of job applicants kept in the Czech labor office register for particular months 2014M1–2016M12 ($t = 1, \ldots, 36$); see Table 4.4 and Fig. 4.8a. Figure 4.9 for segments of length of 12 monthly observations indicates that the logarithmic transformation is desirable (i.e., Box–Cox transformation with type parameter $\lambda = 0$; see Fig. 4.8b). In this example, Fig. 4.8a, b demonstrates the homogenization of variance after this transformation, and histograms in Fig. 4.10a, b show that the logarithmic transformation really rectified a skewed distribution to approximately symmetric distribution.

◇

4.3.2 *Transformation Based on Differencing*

Other transformations frequently used for time series consist in a suitable differencing (see Remark 3.4) that simplifies decomposition components of the original time series (such transformations can be looked upon as special cases of moving averages from Sect. 3.2). Usually, a constant trend remains in the transformed time series only, as it is the case in the following examples with various decomposition structure:

(a) *Linear trend* $y_t = \beta_0 + \beta_1 \cdot t + \varepsilon_t$:

$$(1 - B)y_t = \Delta y_t \approx \beta_1. \tag{4.59}$$

(b) *Polynomial trend* $y_t = \beta_0 + \beta_1 \cdot t + \ldots + \beta_k \cdot t^k + \varepsilon_t$:

$$(1 - B)^k y_t = \Delta^k y_t \approx \beta_k. \tag{4.60}$$

(c) *Additive seasonality* $y_t = Sz_t + \varepsilon_t$ (s is the length of season):

$$(1 - B^s)y_t = \Delta_s y_t \approx 0. \tag{4.61}$$

(d) *Polynomial trend and additive seasonality* $y_t = \beta_0 + \beta_1 \cdot t + \ldots + \beta_k \cdot t^k + Sz_t + \varepsilon_t$ (s is the length of season):

$$(1 - B)^{k-1}(1 - B^s)y_t = \Delta^{k-1}\Delta_s y_t \approx konst. \tag{4.62}$$

(e) *Polynomial trend and multiplicative seasonality* $y_t = (\beta_0 + \beta_1 \cdot t + \ldots + \beta_k \cdot t^k) \cdot Sz_t + \varepsilon_t$ (s is the length of season):

$$(1 - B^s)^{k+1} y_t = (\Delta_s)^{k+1} y_t \approx 0. \tag{4.63}$$

4.4 Exercises

Exercise 4.1 Repeat the analysis from Example 4.1 (the simple approach to additive seasonal elimination for the Czech construction production index) only for data since 2013 (*hint:* $I_1 = -33.76$, $I_2 = -1.00$, $I_3 = 15.01$, $I_4 = 19.75$).

Exercise 4.2 Repeat the analysis from Example 4.2 (the multiplicative elimination of seasonality for the job applicants kept in the Czech labor office register) only for data since 2013 (*hint:* $I_1 = 1.084$, $I_2 = 1.082$, $I_3 = 1.053$, $I_4 = 1.000$, $I_5 = 0.962$, $I_6 = 0.948$, $I_7 = 0.970$, $I_8 = 0.969$, $I_9 = 0.970$, $I_{10} = 0.964$, $I_{11} = 0.976$, $I_{12} = 1.035$).

Exercise 4.3 In Example 4.3 (the additive elimination of seasonality for the Czech construction production index using the regression approach with dummies) construct the prediction intervals for the year 2017.

Exercise 4.4 Repeat the analysis from Example 4.4 (the multiplicative Holt–Winters' method for the job applicants kept in the Czech labor office register) only for data since 2010 (*hint: predictions for 2017*: 395068; 390412; 375287; 350837; 332122; 320818; 322058; 315104; 307204; 298323; 296419; 310548).

Exercise 4.5 Apply Fisher's test of periodicity for the time series in Table 5.1 (*hint: no significant periodicities with significance level of 5%,* $\widehat{\mu} = 0.167$).

Chapter 5
Residual Component

5.1 Tests of Randomness

Sometimes it seems from the visual point of view that the analyzed time series does not indicate the presence of any systematic component, so that it is white noise only (even if this white noise can be shifted to a nonzero level). For example, the graphical record of monthly time series in Table 5.1 plotted for 2015–2017 ($t = 1, \ldots, 36$) in Fig. 5.1 seems to be white noise. Moreover, sometimes one must assess whether the elimination of systematic components from a decomposed time series has been perfect, i.e., whether some reminders of systematic behavior do not persist in the estimated residuals (e.g., patterns of trend, seasonality, and the like).

However, a visual decision can be subjective so that objective statistical tests with fixed significance levels are desirable to test the null hypothesis

$$H_0 : y_t \sim iid. \tag{5.1}$$

This hypothesis is stronger than a test of white noise since it requires the independence and the identical distribution (*iid*). On the other hand, (5.1) does not require the level in zero (as is the case of white noise).

In general, the tests of this type are denoted as *tests of randomness* and they are mostly nonparametric. We will describe some of them briefly. For all of them one recommends before initiating the test procedure to arrange the given time series in such a way that in each group of equal neighboring observations (equal approximately in the sense of applied rounding) one keeps only one observation (the other equal observations in the group are deleted). In each of following tests, let $y_1, \ldots, y_n$ denote the tested time series after this adjustment (i.e., $y_t \neq y_{t+1}$ for all $t = 1, \ldots, n - 1$). To be on the safe side, we remind that we deal exclusively with time series with continuous states.

T. Cipra, *Time Series in Economics and Finance*,
https://doi.org/10.1007/978-3-030-46347-2_5

Table 5.1 Time series to be analyzed by tests of randomness (see also Fig. 5.1)

2015		2016		2017	
t	y_t	t	y_t	t	y_t
1	4	13	5	25	−2
2	0	14	3	26	8
3	5	15	4	27	−5
4	−13	16	−1	28	5
5	5	17	−6	29	−21
6	−4	18	−4	30	−3
7	7	19	−14	31	4
8	6	20	8	32	−4
9	3	21	0	33	11
10	2	22	4	34	2
11	−5	23	12	35	−5
12	−9	24	−4	36	8

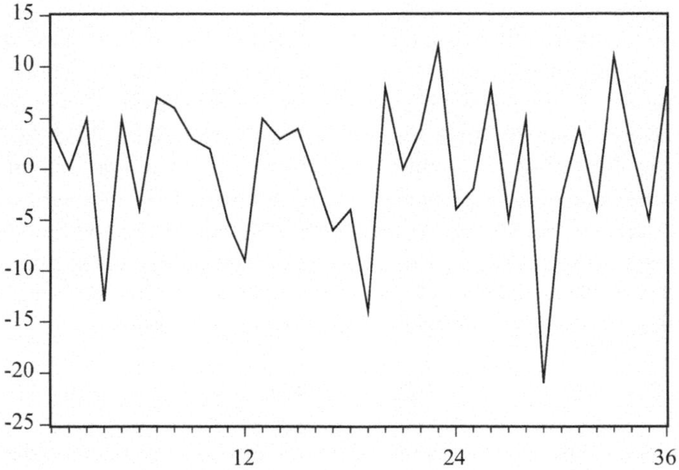

Fig. 5.1 Time series to be analyzed by tests of randomness (see also Table 5.1)

5.1.1 Test Based on Signs of Differences

This test is based on the number of positive first differences of given time series, i.e., on the number of points in which this time series grows (so-called *points of growth*; see also the growth function in Sect. 3.1.2.4).

Let V_t be random variables defined as

$$V_t = \begin{cases} 1 & \text{for } y_t < y_{t+1}, \\ 0 & \text{for } y_t > y_{t+1} \end{cases} \tag{5.2}$$

(the case $y_t = y_{t+1}$ is excluded due to the preliminary adjustment). The mean value of the number of positive first differences k (or equivalently the number of points of growth) is then obviously under the null hypothesis (5.1) equal to

$$\mathrm{E}(k) = \mathrm{E}\left(\sum_{t=1}^{n-1} V_t\right) = \sum_{t=1}^{n-1}\left(\frac{1}{2}\cdot 1 + \frac{1}{2}\cdot 0\right) = \frac{n-1}{2}, \tag{5.3}$$

since the relations between values of two neighboring values y_t and y_{t+1} have under the null hypothesis the same probabilities 1/2. One can derive analogously that the variance of k under the null hypothesis (5.1) fulfills

$$\mathrm{var}(k) = \frac{n+1}{12}. \tag{5.4}$$

Even if under (5.1) one can tabulate the (non-asymptotic) distribution of random variable k, in practice one prefers the asymptotic version of the test which is acceptable for higher n. Its critical region is

$$\frac{|\ k - (n-1)/2\ |}{\sqrt{(n+1)/12}} \geq u_{1-\alpha/2}, \tag{5.5}$$

where $u_{1-\alpha/2}$ is the $(1-\alpha/2)$-quantile of standard normal distribution $N(0, 1)$.

5.1.2 *Test Based on Turning Points*

Let r denote the total number of upper and lower turning points in the tested time series (see Sect. 3.1.1). Analogously as in the previous test one can derive that under the null hypothesis (5.1) it holds

$$\mathrm{E}(r) = \frac{2(n-2)}{3}, \quad \mathrm{var}(r) = \frac{16n-29}{90}. \tag{5.6}$$

In practice, one again applies the asymptotic version of the corresponding test with the critical region

$$\frac{|\ r - 2(n-2)/3\ |}{\sqrt{(16n-29)/90}} \geq u_{1-\alpha/2}. \tag{5.7}$$

5.1.3 Test Based on Kendall Rank Correlation Coefficient τ

This test makes use of *Kendall rank correlation coefficient* τ (or briefly *Kendall's tau*), which was originally suggested as a measure of ordinal association between two observed quantities. In our context, it has the form

$$\tau = \frac{4v}{n(n-1)} - 1, \tag{5.8}$$

where v denotes such a number of pairs y_s and y_t in the given time series $y_1, \ldots, y_n$ fulfilling $y_s < y_t$ for $s < t$. The formula (5.8) for τ standardizes v in such a way that $-1 \leq \tau \leq 1$ and under the null hypothesis (5.1) it holds

$$\mathrm{E}(\tau) = 0, \qquad \mathrm{var}(\tau) = \frac{2(2n+5)}{9n(n-1)}. \tag{5.9}$$

In practice, one applies mainly the asymptotic version of the corresponding test with critical region

$$\frac{|\,\tau\,|}{\sqrt{\frac{2(2n+5)}{9n(n-1)}}} \quad \geq \quad u_{1-\ \alpha/2}. \tag{5.10}$$

5.1.4 Test Based on Spearman Rank Correlation Coefficient ρ

Let $q_1, \ldots, q_n$ denote the *ranks* of values of given time series. For example, if it is $y_1 = 10$, $y_2 = -6$, $y_3 = 2$, $y_4 = -6$, then $q_1 = 4$, $q_2 = 1$, $q_3 = 3$, $q_4 = 2$ (sometimes one uses *fractional ranks* with rank averages for equal values, i.e., $q_1 = 4$, $q_2 = 1.5$, $q_3 = 3$, $q_4 = 1.5$). Then the *Spearman rank correlation coefficient* ρ (or briefly *Spearman's rho*), suggested similarly as τ to measure statistical dependence between the ranking of two observed variables, can be calculated as

$$\rho = 1 - \frac{6}{n(n^2-1)} \sum_{i=1}^{n} (i - q_i)^2 \tag{5.11}$$

(in our context, the one of rankings is obviously the natural one $1, 2, \ldots, n$). Even if the tabulated critical values $r_{1-\alpha/2}$ fulfilling $\mathrm{P}(|\rho| \geq r_{1-\alpha/2}) \leq \alpha$ under the null hypothesis (5.1) are easily available nowadays, in practice again the asymptotic version of the corresponding test is preferred with critical region

$$\sqrt{n-1}\ |\rho| \quad \geq \quad u_{1-\ \alpha/2}. \tag{5.12}$$

5.1.5 Test Based on Numbers of Runs Above and Below Median

In this test, one must construct the sample median M of observations in given time series (therefore one calls it sometimes *median test*). Graphically it means that we look for such a line parallel with the time axis that the numbers of observations above and below it are the same (see Fig. 5.2).

Sometimes several observations of the given time series must lie on this line. In other situations (see, e.g., Fig. 5.2), such a line cannot be even constructed: then one recommends to shift arbitrary observations from the line to the region (above or below) with smaller number of observations to make of the line the correct median (in Fig. 5.2, we have to shift one observation downward). Now we ignore all observations on the line and pool the others into groups called *runs* in such a way that all neighboring observations lying above or below the line create one particular run (see Fig. 5.2). Let us denote the number of runs by u and the number of observations above (or equivalently below) the median line by m.

Even if the critical values of this test are tabulated, in practice one usually makes use of the asymptotic version of this test with the following critical region:

$$\frac{|\ u - (m+1)\ |}{\sqrt{\frac{m(m-1)}{2m-1}}} \geq u_{1-\alpha/2}. \tag{5.13}$$

Example 5.1 The time series from Table 5.1 and Fig. 5.1 has the length $n = 36$ (obviously the preliminary adjustment recommended in the previous text is not

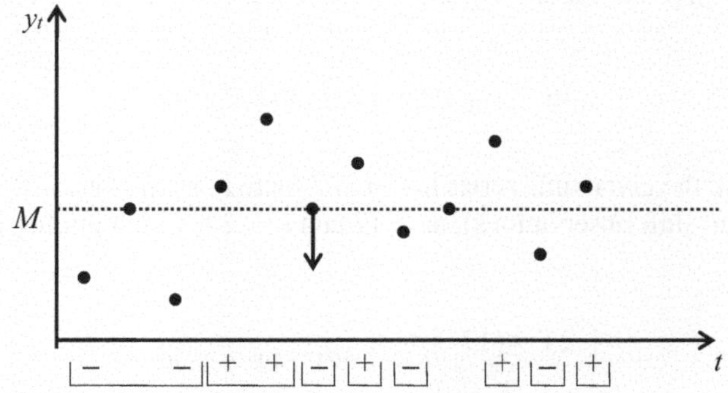

Fig. 5.2 Median test

necessary). Particular tests of randomness (each of them with significance level of 5% and corresponding critical value $u_{0.975} = 1.96$) give the following results:

- Test based on signs of differences: one obtains according to (5.5) with $k = 16$

$$\frac{|\ 16 - (36 - 1)/2\ |}{\sqrt{(36 + 1)/12}} = 0.854 < 1.96;$$

- Test based on turning points: one obtains according to (5.7) with $r = 25$

$$\frac{|\ 25 - 2(36 - 2)/3\ |}{\sqrt{(16 \cdot 36 - 29)/90}} = 0.946 < 1.96;$$

- Test based on τ: here $v = 297$ gives

$$\tau = \frac{4 \cdot 297}{36 \cdot 35} - 1 = -0.057,$$

so that according to (5.10) one obtains

$$\frac{|\ -0.057\ |}{\sqrt{\frac{2(2 \cdot 36+5)}{9 \cdot 36(36-1)}}} = 0.489 < 1.96;$$

- Test based on ρ: here it is $\rho = 0.051$, so that according to (5.12) one obtains

$$\sqrt{36 - 1}\ |\ 0.051| = 0.302 < 1.96;$$

- Median test: the given time series has $M = 2$ (here it is $y_{10} = y_{34} = 2$, so that there is no need to shift observations), $m = 17$ and $u = 23$, which implies according to (5.13)

$$\frac{|\ 23 - (17 + 1)\ |}{\sqrt{\frac{17(17-1)}{2 \cdot 17-1}}} = 1.742 < 1.96.$$

Obviously the null hypothesis that the observations in Table 5.1 are *iid* could not be rejected with significance level of 5% by any of the applied tests of randomness. ◇

Remark 5.1 The recommended choice of a test may be subjective depending on our suspicion that a systematic behavior survives in residuals, e.g.:

- If we suspect that a linear trend remains in residuals, then one recommends the test based on signs of differences, test based on τ ,and test based on ρ.
- If we suspect that a periodicity remains in residuals, then one recommends the test based on turning points and median test (e.g., the test based on signs of differences is not suitable for residuals with remaining periodicity since in such a case obviously $k \sim n/2$ so that the test statistics (5.5) lies close to zero and the test has a low power).

◇

5.2 Exercises

Exercise 5.1 Derive the formula (5.4) for the variance of test statistics k in the test of randomness based on signs of differences. *Hint: under* H_0 *it holds* $\mathrm{var}(V_t) = 1/4$ *and* $\mathrm{cov}(V_t, V_{t+1}) = -1/12$ *in the variance of k:*

$$\mathrm{var}(k) = \mathrm{var}\left(\sum_{t=1}^{n-1} V_t\right) = \sum_{t=1}^{n-1} \mathrm{var}(V_t) + 2\sum_{t=1}^{n-2} \mathrm{cov}(V_t, V_{t+1}).$$

Exercise 5.2 Simulate white noise $N(0, 1)$ with length of 100 and apply five tests of randomness from Sect. 5.1 to it.

Part III
Autocorrelation Methods for Univariate Time Series

Chapter 6
Box–Jenkins Methodology

This chapter is devoted to so-called *Box–Jenkins methodology* applying special stochastic models (ARMA, ARIMA, SARIMA, and others) to time series analysis (e.g., to time series predictions). It enables us to model satisfactorily time series with general courses that cannot be handled by the classical decomposition approach (see also Sect. 2.2.2). The methodology is entitled according to the well-known monograph by Box and Jenkins (1970). The authors summarized the temporary knowledge on this issue and transferred theoretical results to algorithmic form. The methodology has some typical features: in particular, it prefers the (auto)correlation analysis as the main instrument of time series analysis, it models the trend and seasonality in a stochastic way, and other particularities can be stressed. It implies that time series with strongly (auto)correlated observations can be studied using this approach. Indeed, the linear models such as ARMA offer the most popular approach to the routine correlatedness among observations in time (however, the financial time series require specific nonlinear modifications of linear models; see, e.g., models GARCH in Sect. 8.3, even if the basic principles are the same). In this chapter, we describe the given issue in a systematic way. We start introducing some pros and cons of Box–Jenkins methodology:

(+) The stochastic models of the type ARMA are flexible enough to model time series with general courses.
(+) One can document plenty of successful applications of this approach.
(+) The software based on Box–Jenkins methodology is easily available in most econometric and statistical packets nowadays.
(+) Meanwhile there do not exist better routine instruments for analysis of time-dependent observations.
(–) Box–Jenkins methodology requires longer time series (minimal length of fifty observations is recommended, which is not usually a problem in the case of financial time series).
(–) Box–Jenkins methodology cannot be used for real data without disposing of relevant software and instructions.

T. Cipra, *Time Series in Economics and Finance*,
https://doi.org/10.1007/978-3-030-46347-2_6

(–) The interpretation of constructed models is not mostly easy; typically, laymen ask how it is possible that their data are modeled combining random shocks; numerical outputs (e.g., predictions) may serve as acceptable arguments in such cases.

References for more comprehensive study are, e.g., Brockwell and Davis (1993, 1996), Hamilton (1994), and others.

6.1 Autocorrelation Properties of Time Series

6.1.1 Stationarity

Generally speaking, the stationarity of a time series $\{y_t\}$ means that the behavior of this series is stable in a specific way. One usually distinguishes two cases:

- *Strict stationarity* means that the probability behavior of corresponding stochastic process is invariant to shifts in time, i.e., the probability distribution of random vector $\left(y_{t_1}, \ldots, y_{t_k}\right)$ is the same as the distribution of vector $\left(y_{t_1+h}, \ldots, y_{t_k+h}\right)$ for arbitrary h.
- *(Weak) Stationarity* is not so restrictive as the strict stationarity since the invariance to time shifts suffices only for the first and second moments, i.e., it must hold for each s and t

$$\mathrm{E}(y_t) = \mu = const; \tag{6.1}$$

$$\mathrm{cov}(y_s, y_t) = \mathrm{E}(y_s - \mu)(y_t - \mu) = \mathrm{cov}\left(y_{s+h}, y_{t+h}\right) \text{ for arbitrary } h, \tag{6.2}$$

i.e., particularly

$$\mathrm{var}(y_t) = \sigma_y^2 = const. \tag{6.3}$$

In other words, the level and variance of stationary time series are constant in time. A trend, seasonality or non-constant variance (volatility) is incompatible with stationarity and should be removed from time series to make it stationary. Also the covariance structure of stationary time series must be invariable in time (e.g., the character of dependence between the first and second quarter of stationary quarterly series must be the same in all years).

Remark 6.1 If finite second moments of a given process exist, then obviously the strict stationarity implies the weak one. Moreover, if such a process is *normal* (i.e., each finite sample from this process has joint normal distribution), then the both types of stationarity are equivalent.

◇

This text deals only with the weak stationarity that will be addressed simply as *stationarity*. If introducing Box–Jenkins methodology it is suitable to start just with models of stationary time series. The text respects this methodological recommendation which will be valid until being canceled. The concept of autocovariance and autocorrelation functions will be introduced only for stationary time series as well.

6.1.2 Autocovariance and Autocorrelation Function

The typical feature of time series is frequently a strong correlatedness among observations in time. For instance, if the value of 3-month LIBOR for a given day was 1.45 % p.a., then it will range probably between 1.40 and 1.50 % in next days (and not somewhere around 3 %). The common instruments to describe quantitatively this phenomenon are autocovariance and autocorrelation functions:

Autocovariance function for lag k (simply *autocovariance* for lag k) is defined as

$$\gamma_k = \text{cov}(y_t, y_{t-k}) = \text{E}(y_t - \mu)(y_{t-k} - \mu) \ , \qquad k = \ldots, -1, 0, 1, \ldots . \tag{6.4}$$

Analogously *autocorrelation function* for lag k (simply *autocorrelation* for lag k abbreviated in software systems usually as ACF) is defined as

$$\rho_k = \frac{\gamma_k}{\gamma_0} = \frac{\gamma_k}{\sigma_y^2} \ , \qquad k = \ldots, -1, 0, 1, \ldots . \tag{6.5}$$

Remark 6.2 The term "autocorrelation" for ρ_k in (6.5) is correct, since one can write due to stationarity

$$\rho_k = \frac{\gamma_k}{\sigma_y^2} = \frac{\text{cov}(y_t, y_{t-k})}{\sqrt{\text{var}(y_t)}\sqrt{\text{var}(y_{t-k})}} = corr(y_t, y_{t-k}). \tag{6.6}$$

Further the autocovariance and autocorrelation functions are obviously even functions (i.e., $\gamma_k = \gamma_{-k}$ and $\rho_k = \rho_{-k}$), so that their domain of definition can be curtailed to $k \geq 0$). It is always $\rho_0 = 1$ and $|\rho_k| \leq 1$. The graphical plot of ρ_k for particular k is called *correlogram* (see, e.g., Fig. 6.1). In fact, the correlogram describes by means of several values (e.g., ρ_1, ..., ρ_{10} from Fig. 6.1) the short-term dynamics of given time series (on the contrary, the long-term dynamics may be reflected by trend). There are just several last values y_{t-k} ($k \geq 1$), whose correlations ρ_k with y_t explain this value of time series in the stationary model. It stands in contrast to the classical model of linear regression, where the regressor y is explained by means of other (exogenous) variables x; the classical OLS estimate (see, e.g., Sect. 3.1.2) includes the correlation between x and y.

◇

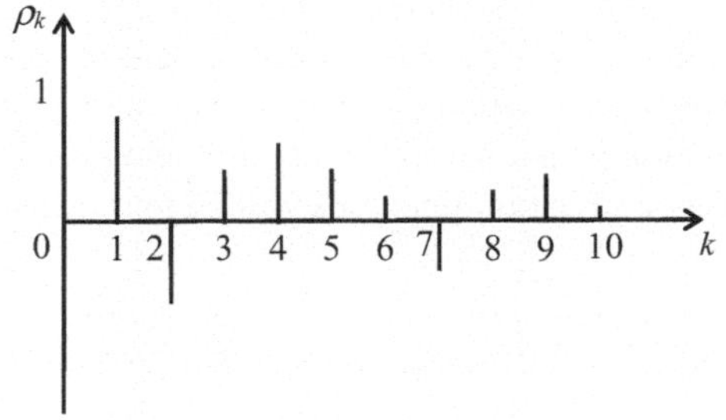

Fig. 6.1 Example of correlogram

6.1.3 Estimated Autocovariance and Autocorrelation Function

For a given stationary time series, one usually constructs the *estimated mean value*

$$\bar{y} = \frac{1}{n}\sum_{t=1}^{n} y_t, \tag{6.7}$$

the *estimated autocovariance function*

$$c_k = \frac{1}{n}\sum_{t=k+1}^{n} (y_t - \bar{y})(y_{t-k} - \bar{y}), \qquad k = 0, 1, \ldots, n-1 \tag{6.8}$$

and the *estimated autocorrelation function*

$$r_k = \frac{c_k}{c_0}, \qquad k = 0, 1, \ldots, n-1. \tag{6.9}$$

Remark 6.3 The estimates (6.7)–(6.9) are applicable if empirical recommendations are fulfilled, namely $n > 50$ and $k < n/4$ (however, these requirements are not usually respected in practice). Sometimes, the divisor n in formula (6.8) is replaced by $n - k$: in this way, one achieves a lower bias of estimate c_k (i.e., $E(c_k)$ will be closer to the theoretical value γ_k), but on the other hand, the mean squared error of c_k defined as $E(c_k - \gamma_k)^2$ will increase. In any case for n going to infinity, both $E(c_k)$ and $E(r_k)$ approach to γ_k and ρ_k, respectively, so that these estimates are unbiased asymptotically.

◇

In the framework of Box–Jenkins methodology, the behavior of autocorrelation function is an important instrument that helps to indicate which type of model is suitable for the given time series; one says that this behavior identifies the corresponding model. In addition, it is also important to find a value k_0 such that all ρ_k are zero for $k > k_0$ (k_0 is then called *truncation point*), or to conclude that such a point k_0 does not exist at all. For example in a model of the form

$$y_t = \varepsilon_t + \theta_1 \varepsilon_{t-1} \tag{6.10}$$

(ε_t is the white noise (see (2.1)) and θ_1 is a parameter (see Sect. 6.2)) it holds

$$\rho_1 = \frac{\theta_1}{1+\theta_1^2},\ \rho_k = 0 \ \text{for}\ k > 1, \tag{6.11}$$

so that $k_0 = 1$ (in Box–Jenkins methodology this truncation point identifies the model (6.10) denoted as MA(1)). However, the theoretical autocorrelations ρ_k are unknown for observed time series. Therefore, in practice one must replace ρ_k by r_k estimated easily according to (6.8) and (6.9). In particular, it is important to answer the question which distance of small r_k from zero justifies the null hypothesis $\rho_k = 0$ (with a given significance level). For such a test, so-called *Bartlett's approximation* is recommended: if $\rho_k = 0$ for $k > k_0$, then under specific assumptions it holds (asymptotically with growing length of time series n; see, e.g., Brockwell and Davis (1993), Theorem 7.2.2)

$$r_k \sim N\left(0\ ,\frac{1}{n}\left(1+2\sum\nolimits_{j=1}^{k_0} r_j^2\right)\right) \qquad \text{for}\ k > k_0. \tag{6.12}$$

6.1.4 Partial Autocorrelation Function and Its Estimate

In addition to the autocorrelation function ρ_k, Box–Jenkins methodology also makes use of the *partial autocorrelation function* denoted as ρ_{kk} (and abbreviated as PACF in software). The value ρ_{kk} is defined as the partial correlation coefficient between y_t and y_{t-k} under fixed values $y_{t-k+1}, \ldots, y_{t-1}$ (e.g., the symbol ρ_{22} could be replaced by more correct though tedious denotation $\rho_{13.2}$). Obviously, it holds $\rho_{00} = 1$ and $\rho_{11} = \rho_1$.

Due to the definition of partial autocorrelation ρ_{kk}, its logic estimate r_{kk} is the estimated parameter $\widehat{\varphi}_{kk}$ in the model

$$y_t = \delta + \varphi_{k1} y_{t-1} + \varphi_{k2} y_{t-2} + \ldots + \varphi_{kk} y_{t-k} + \varepsilon_t. \tag{6.13}$$

However, in practice one usually applies the following (Durbin-Levinson) recursive algorithm:

$$r_{11} = r_1, \qquad r_{kk} = \frac{r_k - \sum_{j=1}^{k-1} r_{k-1,j} \cdot r_{k-j}}{1 - \sum_{j=1}^{k-1} r_{k-1,j} \cdot r_j} \quad \text{for } k > 1, \tag{6.14}$$

where

$$r_{kj} = r_{k-1,j} - r_{kk} \cdot r_{k-1,k-j} \quad \text{for } j = 1, \ldots, k-1. \tag{6.15}$$

Similarly to the autocorrelation function, there can exist truncation points for the partial autocorrelation function as well (e.g., for autoregressive processes), so that ρ_{kk} is an important identifying instrument again. In this case, one can apply so-called *Quenouille's approximation*: if $\rho_{kk} = 0$ for $k > k_0$, then under specific assumptions again it holds (asymptotically with growing length n)

$$r_{kk} \sim N\left(0, \frac{1}{n}\right) \quad \text{for } k > k_0. \tag{6.16}$$

6.2 Basic Processes of Box–Jenkins Methodology

6.2.1 *Linear Process*

The theoretical ground of Box–Jenkins methodology (though not important for mastering it practically so that practitioners can skip this theoretical concept) is so-called *linear process* defined as

$$y_t = \varepsilon_t + \psi_1 \varepsilon_{t-1} + \psi_2 \varepsilon_{t-2} + \ldots = \left(1 + \psi_1 B + \psi_2 B^2 + \ldots\right)\varepsilon_t = \psi(B)\varepsilon_t, \tag{6.17}$$

where $\{\varepsilon_t\}$ is white noise (i.e., a sequence $\{\varepsilon_t\}$ of uncorrelated random variables with zero mean values and constant (finite) variances $\sigma^2 > 0$; see (2.1)) and B is lag operator (see Remark 3.4: the transcription of the models of Box–Jenkins methodology by means of the operators B and Δ is popular due to its simplicity, e.g., in (6.17) one constructs a power series $\psi(B)$ applying formally the operator B as if it is the variable z in the classical power series $\psi(z)$). Moreover, one assumes that

$$\psi(z) \text{ converges for } |\, z \,| \leq 1 \text{ (i.e., inside the unit circle in complex plane)} \tag{6.18}$$

(see Brockwell and Davis (1996)). One can show under this assumption that the infinite series of random variables (6.17) for particular times t converge in the sense of convergence in mean square and the limits form a stationary process with zero mean value ($E(y_t) = 0$). Some authors (see, e.g., Davidson (2000)) assume more strongly that $\varepsilon_t \sim iid\ (0, \sigma^2)$ in the linear process.

Another expression of the linear process (6.17), which can be useful especially when constructing predictions, is possible for so-called *invertible* process. In this case, one can rewrite (6.17) in the form

$$y_t = \pi_1 y_{t-1} + \pi_2 y_{t-2} + \ldots + \varepsilon_t, \text{i.e., } \varepsilon_t = y_t - \pi_1 y_{t-1} - \pi_2 y_{t-2} - \ldots = \pi(B) y_t. \quad (6.19)$$

The sufficient condition of invertibility is analogous to the assumption (6.18), namely the power series

$$\pi(z) \text{ converges for } |z| \leq 1 \text{ (i.e., inside the unit circle in complex plane).} \quad (6.20)$$

Remark 6.4 There is a lot of reasons why the models based on the principle of linear process are suitable to model reality. For instance, let us consider a stationary process $\{y_t\}$ with zero mean value and let us predict the value y_t on the basis of last values $Y_{t-1} = \{y_{t-1}, y_{t-2}, \ldots\}$. Then the optimal prediction (in the sense of minimal mean squared error *MSE* in (2.11)) is $E(y_t \mid Y_{t-1})$. The error of this prediction

$$e_t = y_t - E(y_t | \ Y_{t-1}) \quad (6.21)$$

has properties of white noise. One calls it *innovation* (this name is logic since the innovation process $\{e_t\}$ corresponds to unpredictable movements in values $\{y_t\}$). Moreover, if the process $\{y_t\}$ is normal, then the conditional mean value $E(y_t \mid Y_{t-1})$ has the form of linear combination of values $y_{t-1}, y_{t-2}, \ldots$, and (6.21) can be rewritten as

$$e_t = y_t - \pi_1 y_{t-1} - \pi_2 y_{t-2} - \ldots . \quad (6.22)$$

Obviously, it is just the inverted form (6.19) of linear process.

◊

Remark 6.5 Since $\varepsilon_t = \pi(B) y_t = \pi(B) \psi(B) \varepsilon_t$, it must hold

$$\psi(B) \cdot \pi(B) = 1, \quad (6.23)$$

i.e., $\psi_1 - \pi_1 = 0$, $\psi_2 - \psi_1 \pi_1 - \pi_2 = 0$, etc. These relations transform the parameters $\{\psi_j\}$ to $\{\pi_j\}$, and vice versa. Formally, one can also write $\pi(B) = \psi(B)^{-1}$.

◊

6.2.2 Moving Average Process MA

Here one must start warning that MA models have nothing to do with the method of moving averages for trend elimination (see Sect. 3.2). *Moving average process of order q* denoted as MA(q) has the form

$$y_t = \varepsilon_t + \theta_1 \varepsilon_{t-1} + \ldots + \theta_q \varepsilon_{t-q} = \theta(B)\varepsilon_t, \tag{6.24}$$

where $\theta_1, \ldots, \theta_q$ are parameters and $\theta(B) = 1 + \theta_1 B + \ldots + \theta_q B^q$ is *moving average operator* (obviously, MA(q) originates by truncating the linear process (6.17) behind the lag q).

The process MA(q) is always stationary with zero mean value and variance

$$\sigma_y^2 = \left(1 + \theta_1^2 + \ldots + \theta_q^2\right)\sigma^2 \tag{6.25}$$

and autocorrelation function

$$\rho_k = \begin{cases} \dfrac{\theta_k + \theta_1 \theta_{k+1} + \ldots + \theta_{q-k}\theta_q}{1 + \theta_1^2 + \ldots + \theta_q^2} & \text{for } k = 1, \ldots, q \\ 0 & \text{for } k > q \end{cases} \tag{6.26}$$

(apparently the autocorrelation function has the truncation point k_0 equal to the model order q). The partial autocorrelation function ρ_{kk} of the process MA(q) has no truncation point, but it is bounded by a linear combination of geometrically decreasing sequences and sinusoids with geometrically decreasing amplitudes.

The process MA(q) is invertible if all roots $z_1, \ldots, z_q$ of polynomial $\theta(z)$ lie outside the unit circle in complex plane (i.e., $|z_1|, \ldots, |z_q| > 1$, since then the assumption (6.20) is fulfilled).

Remark 6.6 The process MA(1) (see (6.10)) has the autocorrelation function (6.11) with truncation point $k_0 = 1$. Its partial autocorrelation function is (without truncation point)

$$\rho_{kk} = \frac{(-1)^{k-1}\theta_1^k (1 - \theta_1^2)}{1 - \theta_1^{2(k+1)}} \quad \text{for } k = 1, 2, \ldots, \tag{6.27}$$

so that in the case of process invertibility, ρ_{kk} is really bounded by a geometrically decreasing sequence ($|\rho_{kk}| < |\theta_1|^k$). Indeed, the invertibility condition (6.20) has a simple form $|\theta_1| < 1$ here. Since $\rho_1 = \theta_1/(1 + \theta_1^2)$, it must be $|\rho_1| < 1/2$ for invertible MA(1) process (this inequality holds even for all $|\theta_1| \neq 1$).

◇

Remark 6.7 The process MA(2)

$$y_t = \varepsilon_t + \theta_1 \varepsilon_{t-1} + \theta_2 \varepsilon_{t-2} \tag{6.28}$$

has the autocorrelation function

$$\rho_k = \begin{cases} \dfrac{\theta_1(1+\theta_2)}{1+\theta_1^2+\theta_2^2} & \text{for } k = 1 \\ \dfrac{\theta_2}{1+\theta_1^2+\theta_2^2} & \text{for } k = 2 \\ 0 & \text{for } k > 2 \end{cases} \tag{6.29}$$

with truncation point $k_0 = 2$. The invertibility condition (6.20) for MA(2) process has the form

$$\theta_1 + \theta_2 > -1, \qquad \theta_2 - \theta_1 > -1, \qquad -1 < \theta_2 < 1, \tag{6.30}$$

so that the invertibility region of MA(2) (in the plane with horizontal axis for values θ_1 and vertical axis for values θ_2) is the interior of triangle with vertices (–2, 1), (0, –1), and (2, 1).

◇

6.2.3 Autoregressive Process AR

Autoregressive process of order p denoted as AR(p) has the form

$$y_t = \varphi_1 y_{t-1} + \ldots + \varphi_p y_{t-p} + \varepsilon_t, \text{ i.e. } y_t - \varphi_1 y_{t-1} - \ldots - \varphi_p y_{t-p} = \varphi(B) y_t = \varepsilon_t, \tag{6.31}$$

where φ_1, ..., φ_p are parameters and $\varphi(B) = 1 - \varphi_1 B - \ldots - \varphi_p B^p$ is *autoregressive operator* (it originates by truncating the inverted linear process (6.19) behind the lag p).

The process AR(p) is stationary, if all roots z_1, ..., z_p of polynomial $\varphi(z)$ lie outside the unit circle in complex plane (i.e., $|z_1|$, ..., $|z_p| > 1$, since then the assumption (6.18) is fulfilled). In such a case the process has zero mean value and variance

$$\sigma_y^2 = \frac{\sigma^2}{1 - \varphi_1 \rho_1 - \ldots - \varphi_p \rho_p} \tag{6.32}$$

and its autocorrelation function fulfills the following difference equation:

$$\rho_k = \varphi_1 \rho_{k-1} + \varphi_2 \rho_{k-2} + \ldots + \varphi_p \rho_{k-p} \quad \text{for } k > 0 \tag{6.33}$$

(to derive (6.33) it suffices to multiply all terms in the equality (6.31) by the value y_{t-k}/σ_y^2 and to calculate mean values at the both sides; moreover, it is $\mathrm{E}(y_{t-k} \cdot \varepsilon_t) = 0$ for $k > 0$ since the stationary process AR(p) can be expressed as the linear process). Due to the theory of difference equations (see Brockwell and Davis (1993), Section 3.6) the solution of (6.33) can be expressed in the form

$$\rho_k = \alpha_1 z_1^{-k} + \alpha_2 z_2^{-k} + \ldots + \alpha_p z_p^{-k} \quad \text{for } k \geq 0, \tag{6.34}$$

where $z_1, \ldots, z_p$ are mutually distinct roots of the polynomial $\varphi(z)$ ($|z_1|, \ldots, |z_p| > 1$; see above) and $\alpha_1, \ldots, \alpha_p$ are fixed coefficients (if the roots z_i and z_j are complex conjugate, then they can be replaced by a single term of the type $\alpha \cdot d^{\,k} \cdot \sin(\lambda k + \varphi)$ with $0 < d < 1$; similarly if the roots $z_1, \ldots, z_p$ are not mutually distinct, then all terms with a multiple root z_i of multiplicity r must be replaced in (6.34) by a more complex term $(\beta_0 + \beta_1 k + \ldots + \beta_{r-1} k^{r-1}) \cdot z_i^{-k}$, which is always significantly overlapped by the behavior of the term z_i^{-k} for higher k). In any case, the autocorrelation function of process AR(p) can be approximated by a linear combination of geometrically decreasing sequences and sinusoids with geometrically decreasing amplitudes (see, e.g., Fig. 6.4).

Remark 6.8 If we write (6.33) only for $k = 1, \ldots, p$, then we obtain so-called system of *Yule-Walker equations* for unknown parameters $\varphi_1, \ldots, \varphi_p$ by means of autocorrelations $\rho_1, \ldots, \rho_p$ (or vice versa)

$$\begin{array}{lclclclcl}
\rho_1 & = & \varphi_1 & + & \varphi_2 \rho_1 & + & \ldots & + & \varphi_p \rho_{p-1}, \\
\rho_2 & = & \varphi_1 \rho_1 & + & \varphi_2 & + & \ldots & + & \varphi_p \rho_{p-2}, \\
\vdots & & \vdots & & \vdots & & & & \vdots \\
\rho_p & = & \varphi_1 \rho_{p-1} & + & \varphi_2 \rho_{p-2} & + & \ldots & + & \varphi_p.
\end{array} \tag{6.35}$$

◇

The partial autocorrelation function ρ_{kk} of the process AR(p) has the truncation point k_0 equal to the model order p (it follows directly from the definition of partial autocorrelation function of an autoregressive process of order p fulfilling $\rho_{kk} = 0$ for all $k > p$; see (6.13)). This property makes of the partial autocorrelation function an important instrument for the identification of autoregressive processes.

The process AR(p) is always invertible since (6.31) is directly the invertible form of this model.

Remark 6.9 The process AR(1)

$$y_t = \varphi_1 y_{t-1} + \varepsilon_t \tag{6.36}$$

is stationary for $|\varphi_1| < 1$. In such a case, it has zero mean value and variance

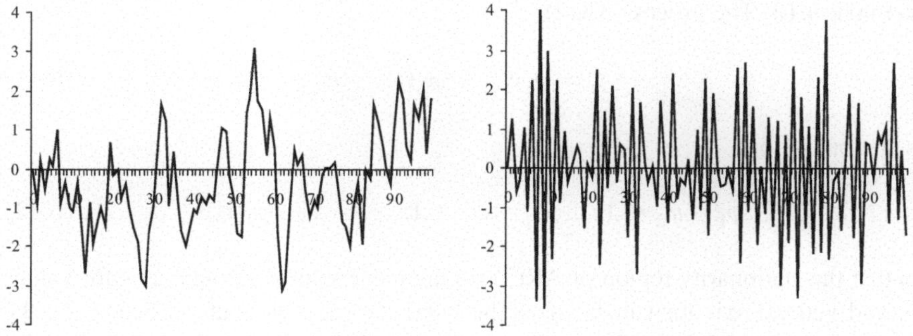

Fig. 6.2 (**a**) Positive correlatedness for $y_t = 0.8y_{t-1} + \varepsilon_t$ ($\rho > 0$) and (**b**) negative correlatedness for $y_t = -0.8y_{t-1} + \varepsilon_t$ ($\rho < 0$)

$$\sigma_y^2 = \frac{\sigma^2}{1 - \varphi_1^2} \tag{6.37}$$

and autocorrelation function

$$\rho_k = \varphi_1^k \quad \text{for } k \geq 0 \tag{6.38}$$

in the form of geometrically decreasing sequence (oscillating for negative φ_1) without any truncation point. In particular, it holds for $k = 1$

$$\rho_1 = \varphi_1, \tag{6.39}$$

i.e., the first autocorrelation of the process AR(1) equals its autoregressive parameter. Hence the sign of parameter φ_2 plays an important role here: the positive $\varphi_1 > 0$ (so-called *positive correlatedness*) induces the inertia for the signs of neighboring values of the corresponding time series (see Fig. 6.2(a) with a relatively rare crossing of time axis), while on the contrary the negative $\varphi_1 < 0$ (so-called *negative correlatedness*) induces frequent changes of the signs of neighboring values of the corresponding time series (see Fig. 6.2(b) with a relatively dense crossing of time axis).

The partial autocorrelation function of the process AR(1) has the form

$$\rho_{11} = \varphi_1, \qquad \rho_{kk} = 0 \quad \text{for } k > 1 \tag{6.40}$$

with truncation point $k_0 = 1$.

◇

Remark 6.10 The process AR(2)

$$y_t = \varphi_1 y_{t-1} + \varphi_2 y_{t-2} + \varepsilon_t \tag{6.41}$$

is stationary for

$$\varphi_1 + \varphi_2 < 1, \qquad \varphi_2 - \varphi_1 < 1, \qquad -1 < \varphi_2 < 1, \tag{6.42}$$

so that the stationarity region of AR(2) (in the plane with horizontal axis for values φ_1 and vertical axis for values φ_2) is the interior of triangle with vertices (–2, –1), (0, 1), and (2, –1). In such a case, the process AR(2) has zero mean value and variance

$$\sigma_y^2 = \frac{\sigma^2}{1 - \varphi_1 \rho_1 - \varphi_2 \rho_2} \tag{6.43}$$

and autocorrelation function

$$\rho_k = \frac{z_1^{-1}\left(1 - z_2^{-2}\right) z_1^{-k} - z_2^{-1}\left(1 - z_1^{-2}\right) z_2^{-k}}{\left(z_1^{-1} - z_2^{-1}\right)\left(1 + z_1^{-1} z_2^{-1}\right)} \quad \text{for } k \geq 0, \tag{6.44}$$

where z_1 and z_2 are mutually distinct roots of the polynomial $\varphi(z)$ ($|z_1|, |z_2| > 1$, in the case of double root the form of autocorrelation function is analogous); ρ_k is without any truncation point and has the form of a linear combination of two geometrically decreasing sequences or the form of a sinusoid with geometrically decreasing amplitude.

The partial autocorrelation function of the process AR(2) has the truncation point $k_0 = 2$.

◇

6.2.4 *Mixed Process ARMA*

Mixed process of order p and q denoted as ARMA(p, q) has the form

$$y_t = \varphi_1 y_{t-1} + \ldots + \varphi_p y_{t-p} + \varepsilon_t + \theta_1 \varepsilon_{t-1} + \ldots + \theta_q \varepsilon_{t-q}, \text{ i.e. } \varphi(B) y_t = \theta(B) \varepsilon_t, \tag{6.45}$$

where the operators $\varphi(B)$ and $\theta(B)$ have been defined in the context of processes AR(p) and MA(q), respectively. The condition of stationarity and the condition of invertibility of the process ARMA(p, q) correspond with the condition of stationarity of AR(p) and the condition of invertibility of MA(q), respectively.

The stationary process ARMA(p, q) has zero mean value, and its autocorrelation function fulfills the following difference equation:

$$\rho_k = \varphi_1 \rho_{k-1} + \varphi_2 \rho_{k-2} + \ldots + \varphi_p \rho_{k-p} \quad \text{for } k > q \tag{6.46}$$

with solution of the form

$$\rho_k = \alpha_1 z_1^{-k} + \alpha_2 z_2^{-k} + \ldots + \alpha_p z_p^{-k} \quad \text{for } k \geq \max(0, q - p + 1), \tag{6.47}$$

where z_1, ..., z_p are mutually distinct roots of the polynomial $\varphi(z)$ ($|z_1|$, ..., $|z_p| > 1$). Hence the autocorrelation function of process ARMA(p, q) is without any truncation point and can be approximated by a linear combination of geometrically decreasing sequences and sinusoids of various frequencies with geometrically decreasing amplitudes excepting the initial values ρ_0, ρ_1, ..., ρ_{q-p} (this exception is non-empty only in the case of $q \geq p$).

The partial autocorrelation function of the process ARMA(p, q) has no truncation point as well and it is bounded by a linear combination of geometrically decreasing sequences and sinusoids of various frequencies with geometrically decreasing amplitudes excepting the initial values ρ_{00}, ..., $\rho_{p-q,\ p-q}$ (this exception is non-empty only in the case of $p \geq q$).

Remark 6.11 The process ARMA(1, 1)

$$y_t = \varphi_1 y_{t-1} + \varepsilon_t + \theta_1 \varepsilon_{t-1} \tag{6.48}$$

is stationary for $|\varphi_1| < 1$. In such a case, it has zero mean value and variance

$$\sigma_y^2 = \frac{1 + \theta_1^2 + 2\varphi_1\theta_1}{1 - \varphi_1^2} \sigma^2 \tag{6.49}$$

and autocorrelation function

$$\rho_1 = \frac{(1 + \varphi_1\theta_1)(\varphi_1 + \theta_1)}{1 + \theta_1^2 + 2\varphi_1\theta_1}, \qquad \rho_k = \varphi_1 \rho_{k-1} \quad \text{for } k > 1 \tag{6.50}$$

in the form of geometrically decreasing sequence excepting ρ_0 without any truncation point.

The condition of invertibility of the process ARMA(1, 1) is $|\theta_1| < 1$.

The partial autocorrelation function of ARMA(1, 1) is bounded (in absolute values) by a geometrically decreasing sequence starting from ρ_{11}.

◇

Remark 6.12 The stationary processes introduced in this section have the zero mean value. It is natural to generalize them to the case of *nonzero* value (constant in time). For example, the process MA(q) with the mean value μ has the form

$$y_t = \mu + \varepsilon_t + \theta_1 \varepsilon_{t-1} + \ldots + \theta_q \varepsilon_{t-q}. \tag{6.51}$$

More generally, the process ARMA(p, q) with the mean value μ has the form

$$y_t - \mu = \varphi_1 (y_{t-1} - \mu) + \ldots + \varphi_p (y_{t-p} - \mu) + \varepsilon_t + \theta_1 \varepsilon_{t-1} + \ldots + \theta_q \varepsilon_{t-q}, \tag{6.52}$$

or equivalently

$$y_t = \alpha + \varphi_1 y_{t-1} + \ldots + \varphi_p y_{t-p} + \varepsilon_t + \theta_1 \varepsilon_{t-1} + \ldots + \theta_q \varepsilon_{t-q}, \quad \text{where}$$
$$\alpha = (1 - \varphi_1 - \ldots - \varphi_p) \cdot \mu. \tag{6.53}$$

In special cases, nonlinear modifications of ARMA models with *thresholds* can be also useful; see, e.g., the so-called SETAR (or TAR) model (9.18).

◇

6.3 Construction of Models by Box–Jenkins Methodology

The construction of models by Box–Jenkins methodology is recommended in three steps:

1. *Identification of model*: e.g., for time series y_1, ..., y_n one identifies the model AR(1).
2. *Estimation of model*: e.g., the estimated model is $y_t = 0.68\ y_{t-1} + \varepsilon_t$ with $\widehat{\sigma} = 11.24$.
3. *Verification of model*: e.g., the model from the step (2) is verified with significance level of 5 % or one assesses its prediction competences.

If the diagnostic results from the step (3) are not satisfying, one should repeat all three steps for another competitive model (in such a case, one frequently corrects or modifies the rejected model, and the previous identification offers instructions how to do it).

6.3.1 *Identification of Model*

6.3.1.1 Identification Based on Autocorrelation and Partial Autocorrelation Function

General findings on the form of autocorrelation and partial autocorrelation function of stationary and invertible processes AR(p), MA(q), and ARMA(p, q) described in Sect. 6.2 are summarized in Table 6.1.

Then the corresponding identifying method consists in examining the graphical plot of the estimated correlogram and partial correlogram of modeled time series striving to find the most relevant model for this time series just according to characteristics from Table 6.1. If there are any doubts, we can test the potential truncation point k_0 by means of Bartlett's approximation (6.12) with the approximate (asymptotic) critical region (applying the significance level of 5 %)

$$|r_k| \geq 2\sqrt{\frac{1}{n}\left(1 + 2\sum\nolimits_{j=1}^{k_0} r_j^2\right)} \quad \text{for some } k > k_0 \tag{6.54}$$

or by means of Quenouille's approximation (6.16) with the critical region (applying again the significance level of 5 %)

$$|r_{kk}| \geq 2\sqrt{\frac{1}{n}} \quad \text{for some } k > k_0. \tag{6.55}$$

Example 6.1 Table 6.2 presents values y_t of 3-month interbank interest rate (in % p. a.) in Germany (*Dreimonatsgeld*; see Deutsche Bundesbank) for particular years 1960–1999 ($t = 1, \ldots, 40$). Since the corresponding graph in Fig. 6.3 can be regarded as stationary in this time period (see also Example 6.4 in Sect. 6.3.3), one has estimated the corresponding correlogram and partial correlogram (see Table 6.3 and Fig. 6.4).

Applying the characteristics from Table 6.1, the most suitable model for this time series seems to be the process AR(4): the correlogram r_k corresponds to a sinusoid

Table 6.1 Form of autocorrelation and partial autocorrelation function of stationary and invertible processes AR(p), MA(q), and ARMA(p, q) (U denotes the curve in the form of linear combination of geometrically decreasing sequences and sinusoids with geometrically decreasing amplitudes)

	AR(p)	MA(q)	ARMA(p, q)
ρ_k	Non-existent k_0 ; ρ_k in form of curve U	$k_0 = q$	Non-existent k_0 ; ρ_k in form of curve U excepting values $\rho_0, \rho_1, \ldots, \rho_{q-p}$
ρ_{kk}	$k_0 = p$	Non-existent k_0 ; ρ_{kk} bounded by curve U	Non-existent k_0 ; ρ_{kk} bounded by curve U excepting values $\rho_{00}, \ldots, \rho_{p-q,\, p-q}$

Table 6.2 Annual data 1960–1999 in Example 6.1 (*3-month interbank interest rate in Germany in % p.a.—Dreimonatsgeld*)

t	Year	y_t	t	Year	y_t	t	Year	y_t	t	Year	y_t
1	1960	5.10	11	1970	9.41	21	1980	9.54	31	1990	8.43
2	1961	3.59	12	1971	7.15	22	1981	12.11	32	1991	9.18
3	1962	3.42	13	1972	5.61	23	1982	8.88	33	1992	9.46
4	1963	3.98	14	1973	12.14	24	1983	5.78	34	1993	7.24
5	1964	4.09	15	1974	9.90	25	1984	5.99	35	1994	5.31
6	1965	5.14	16	1975	4.96	26	1985	5.44	36	1995	4.48
7	1966	6.63	17	1976	4.25	27	1986	4.60	37	1996	3.27
8	1967	4.27	18	1977	4.37	28	1987	3.99	38	1997	3.30
9	1968	3.81	19	1978	3.70	29	1988	4.28	39	1998	3.52
10	1969	5.79	20	1979	6.69	30	1989	7.07	40	1999	2.94

Source: OECD (https://data.oecd.org/interest/short-term-interest-rates.htm#indicator-chart)

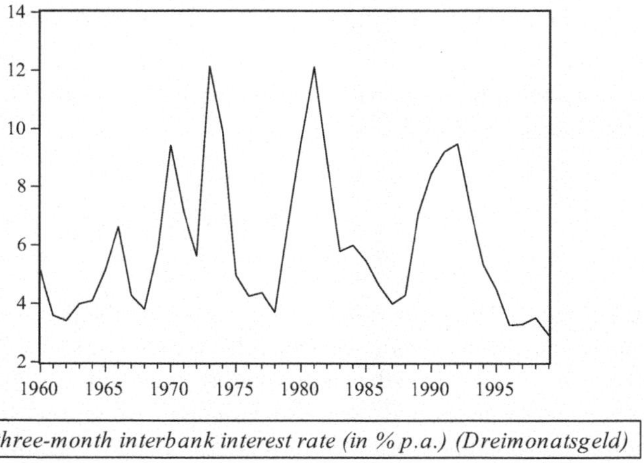

Fig. 6.3 Annual data 1960–1999 in Example 6.1 (*3-month interbank interest rate in Germany in % p.a.—Dreimonatsgeld*)

with geometrically decreasing amplitude and the partial correlogram r_{kk} has evidently the truncation point $k_0 = 4$; the statistical test (6.55) gives really

$$|r_{kk}| \quad < 2\sqrt{\frac{1}{n}} \;=\; 2\sqrt{\frac{1}{40}} = 0.316 \qquad \text{for } k > 4$$

(and it is not true for $k = 4$). One could try as an alternative the process MA(1) with the truncation point $k_0 = 1$ in the correlogram r_k, since it holds according to (6.54)

Table 6.3 Correlogram and partial correlogram in Example 6.1 (*Dreimonatsgeld*) estimated by means of EViews

	AC	PAC		AC	PAC
1	0.612	0.612	11	0.256	−0.083
2	0.150	−0.360	12	0.004	−0.128
3	−0.028	0.124	13	−0.195	−0.004
4	−0.228	−0.392	14	−0.241	−0.119
5	−0.400	−0.082	15	−0.193	0.154
6	−0.318	0.023	16	−0.139	−0.102
7	−0.037	0.208	17	−0.137	−0.206
8	0.147	−0.015	18	−0.062	−0.053
9	0.162	−0.048	19	0.039	0.018
10	0.249	0.189	20	0.048	0.037

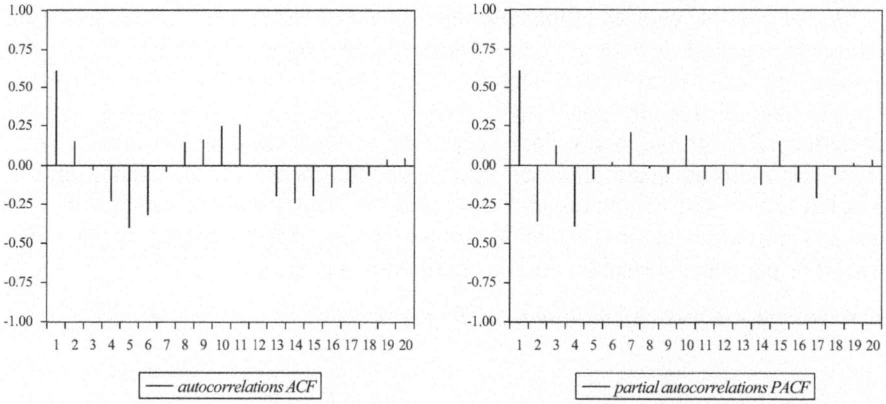

Fig. 6.4 Correlogram and partial correlogram in Example 6.1 (*Dreimonatsgeld*) estimated by means of EViews

$$|r_k| \quad < 2\sqrt{\frac{1}{n}\ \left(1+2r_1^2\right)} = 2\sqrt{\frac{1}{40}\ \left(1+2\cdot 0.612^2\right)} = 0.418 \qquad \text{for } k > 1,$$

but the value $|r_5| = 0.400$ is relatively close to this border (moreover, it would be evidently difficult to look for a curve U bounding the partial correlogram r_{kk}).

◇

To check the correctness of identified model, one sometimes makes use of the inequalities for estimated autocorrelations r_k that should hold theoretically under the assumption of stationarity and invertibility of process (see Table 6.5: e.g., according to Remark 6.6 it holds $|\rho_1| < 1/2$ in the process MA(1)).

6.3.1.2 Identification Based on Information Criteria

This advanced approach to the model identification enables (at least theoretically) a fully automatic identification excluding any subjective interference of analysts. The problem of identification of process ARMA(p, q) for a given time series is addressed here as the problem of estimation of unknown parameters p and q by means of optimization

$$(\widehat{p}, \widehat{q}) = \arg\min_{(k,\,l)} A(k, l), \tag{6.56}$$

where $A(k,\ l)$ is a suitable criterion constructed by estimating the process ARMA $(k,\ l)$ for the given time series. The minimization is performed over a grid of values $k = 0, 1, ..., K$ and $l = 0, 1, ..., L$ chosen a priori.

One of possible choices of the criterion A could be to set $A(k,\ l) = \widehat{\sigma}_{k,l}$ using the estimated standard deviation of white noise of the process ARMA($k,\ l$), i.e., for a given time series one prefers the model with the lowest residual error in such a case (thus in Example 6.1 one would prefer the process AR(4) with the residual standard deviation 1.756 to the process MA(1) with the residual standard deviation 1.782). However, more adequate approach is to apply the theory of information and to penalize unnecessarily high orders k and l (simultaneously with the minimization of residual standard deviation) achieving the consistency of the estimates (6.56) in this way. The popular information criteria based on this idea are:

1. *Criterion AIC* (*Akaike information criterion*):

$$AIC(k, l) = \ln \widehat{\sigma}^2_{k,l} + \frac{2(k+l+1)}{n}. \tag{6.57}$$

2. *Criterion BIC* (*Bayes information criterion* or *Schwarz information criterion*):

$$BIC(k, l) = \ln \widehat{\sigma}^2_{k,l} + \frac{(k+l+1)\ln n}{n}. \tag{6.58}$$

The value $\widehat{\sigma}^2_{k,l}$ in (6.57) and (6.58) denotes the estimated variance of white noise in the process ARMA(k, l) (more correctly one should use the minimal value of logarithmic likelihood multiplied by coefficient ($-2/n$) instead of the first term in (6.57) and (6.58); see, e.g., EViews), the numerator of the second term contains obviously the number of estimated parameters (including the level parameter μ to penalize unnecessarily high orders k and l), and n is the length of given time series. The criterion AIC produces the strongly consistent estimator of the model order (i.e., this estimator converges to the true order with probability one for increasing n), but it may have a high variance (i.e., it lacks of efficiency). The properties of the criterion BIC are just opposite: the corresponding estimator of model order is not consistent, but it is efficient.

Table 6.4 Values of information criteria *AIC* and *BIC* from Example 6.2 (*Dreimonatsgeld*) calculated by means of EViews (the minimal values are in bold figures)

	AIC	BIC
White noise	4.678	4.721
AR(1)	4.257	4.342
AR(2)	4.170	**4.299**
AR(3)	4.225	4.399
AR(4)	**4.093**	4.313
AR(5)	4.159	4.426
AR(6)	4.251	4.565

Example 6.2 The values of information criteria AIC and BIC calculated by means of EViews for the time series from Example 6.1 (*Dreimonatsgeld*) are shown in Table 6.4, where one examines autoregressions up to the order six. The identified process is AR(4), since the process AR(2) according to BIC is nested into the process AR(4) according to AIC.

◇

Remark 6.13 If two models are acceptable in the identification step and one of them is nested into the second one (e.g., AR(2) is nested into AR(4); see above), then one can decide on proper identification by means of statistical tests (*F*-test or Lagrange Multiplier (LM) test), which test whether the parameters distinguishing the both models are zero (see, e.g., EViews).

◇

6.3.2 *Estimation of Model*

Simple models of Box–Jenkins methodology (up to the orders two) can be estimated by means of the moment estimates making use of relations among the parameters of the identified model and its autocorrelations (see Table 6.5: e.g., according to (6.39) it holds $\varphi_1 = \rho_1$ in the process AR(1)). However, such estimates are usually perceived as preliminary ones and are used in practice as initial values for more complex (iterative) procedure.

The estimation procedures for construction of the final estimates (not only the initial ones) in particular models are software matters definitely. For the process AR(p)

$$y_t = \alpha + \varphi_1 y_{t-1} + \ldots + \varphi_p y_{t-p} + \varepsilon_t \tag{6.59}$$

one can use the classical OLS estimation (including the classical estimation of its variance matrix). Under the stationarity assumption, this estimate is consistent due to the orthogonality of regressors to residuals in (6.59) (the orthogonality $\text{cov}(y_{t-1}, \varepsilon_t) = \ldots = \text{cov}(y_{t-p}, \varepsilon_t) = 0$ can be shown by expressing the process AR(p) in the form of linear process (6.17)).

Table 6.5 Moment estimates of simple stationary and invertible models of Box–Jenkins methodology and check inequalities for estimated autocorrelations

Model	Moment estimates	Check inequalities for r_k
AR(1)	$\widehat{\varphi}_1 = r_1, \quad \widehat{\sigma}^2 = \widehat{\sigma}_y^2(1 - \widehat{\varphi}_1 r_1)$	$\|r_1\| < 1$
AR(2)	$\widehat{\varphi}_1 = \frac{r_1(1-r_2)}{1-r_1^2}, \quad \widehat{\varphi}_2 = \frac{r_2 - r_1^2}{1-r_1^2},$ $\widehat{\sigma}^2 = \widehat{\sigma}_y^2(1 - \widehat{\varphi}_1 r_1 - \widehat{\varphi}_2 r_2)$	$\|r_2\| < 1, \quad r_1^2 < \frac{1+r_2}{2}$
MA(1)	$\widehat{\theta}_1 = \frac{1-\sqrt{1-4r_1^2}}{2r_1}, \quad \widehat{\sigma}^2 = \frac{\widehat{\sigma}_y^2}{1+\widehat{\theta}_1^2}$	$\|r_1\| < 1/2$
MA(2)	$\widehat{\theta}_1 = \widehat{\theta}_2 \approx 0.1\ , \quad \widehat{\sigma}^2 = \frac{\widehat{\sigma}_y^2}{1+\widehat{\theta}_1^2+\widehat{\theta}_2^2}$	$r_1 + r_2 > -1/2, \quad r_2 - r_1 > -1/2$ $r_1^2 < 4r_2(1-2r_2)$
ARMA (1, 1)	$\widehat{\varphi}_1 = \frac{r_2}{r_1}, \quad \widehat{\theta}_1 = \frac{\widehat{b} \pm \sqrt{\widehat{b}^2 - 4}}{2}, \quad \left\|\widehat{\theta}_1\right\| < 1,$ $\widehat{b} = \frac{1 - 2r_2 + \widehat{\varphi}_1^2}{r_1 - \widehat{\varphi}_1}, \quad \widehat{\sigma}^2 = \frac{\widehat{\sigma}_y^{2*}}{1+\widehat{\theta}_1^2},$ $\widehat{\sigma}_y^{2*} = \sum_{t=1}^{n} \frac{(y_t^* - \bar{y}^*)^2}{n}, \quad y_t^* = y_t - \widehat{\varphi}_1 y_{t-1}$	$2r_1^2 - \|r_1\| < r_2 < \|r_1\|$

In the case of the stationary and invertible process ARMA(p, q) (with zero mean value for simplicity)

$$y_t = \varphi_1 y_{t-1} + \ldots + \varphi_p y_{t-p} + \varepsilon_t + \theta_1 \varepsilon_{t-1} + \ldots + \theta_q \varepsilon_{t-q} \tag{6.60}$$

one usually uses the NLS estimates (nonlinear least squares) which are realized by means of iterative algorithms (Gauss–Newton and others; see, e.g., EViews). The corresponding NLS procedures consist mostly in the minimization of sum of squares (which are nonlinear in parameters $\varphi_1, \ldots, \theta_q$)

$$\min_{\varphi_1, \ldots, \theta_q} \sum_{t=p+1}^{n} e_t(\varphi_1, \ \ldots, \theta_q)^2, \tag{6.61}$$

where residuals $e_t(\varphi_1, \ldots, \theta_q)$ are constructed recursively by means of the relation

$$e_t(\varphi_1, \ \ldots, \theta_q) = e_t = y_t - \varphi_1 y_{t-1} - \ldots - \varphi_p y_{t-p} - \theta_1 e_{t-1} - \ldots - \theta_q e_{t-q}$$
$$\text{for } t = p+1, \ \ldots, \ n. \tag{6.62}$$

with suitable initial values $e_{p-q+1}, \ldots, e_p$. Finally, the estimate of the variance of white noise σ^2 is obtained if dividing the minimal value of (6.61) by the length of time series n. Under the normality assumption and for higher n, these estimates are very

Table 6.6 Approximate standard deviations of estimated parameters of simple stationary and invertible models of Box–Jenkins methodology

Model	Standard deviations of estimated parameters
AR(1)	$\sigma(\widehat{\varphi}_1) \approx \left(\frac{1-\widehat{\varphi}_1^2}{n}\right)^{1/2}$
AR(2)	$\sigma(\widehat{\varphi}_1) \approx \sigma(\widehat{\varphi}_2) \approx \left(\frac{1-\widehat{\varphi}_2^2}{n}\right)^{1/2}$
MA(1)	$\sigma\left(\widehat{\theta}_1\right) \approx \left(\frac{1-\widehat{\theta}_1^2}{n}\right)^{1/2}$
MA(2)	$\sigma\left(\widehat{\theta}_1\right) \approx \sigma\left(\widehat{\theta}_2\right) \approx \left(\frac{1-\widehat{\theta}_2^2}{n}\right)^{1/2}$
ARMA(1, 1)	$\sigma(\widehat{\varphi}_1) \approx \left(\frac{\left(1-\widehat{\varphi}_1^2\right)\left(1+\widehat{\varphi}_1\widehat{\theta}_1\right)^2}{n\left(\widehat{\varphi}_1+\widehat{\theta}_1\right)^2}\right)^{1/2}$, $\sigma\left(\widehat{\theta}_1\right) \approx \left(\frac{\left(1-\widehat{\theta}_1^2\right)\left(1+\widehat{\varphi}_1\widehat{\theta}_1\right)^2}{n\left(\widehat{\varphi}_1+\widehat{\theta}_1\right)^2}\right)^{1/2}$

close to the ML estimates constructed by maximizing the logarithmic likelihood function

$$L(\varphi_1, \ldots, \theta_q, \sigma^2) = -\frac{n-p}{2}\ln(2\pi) - \frac{n-p}{2}\ln(\sigma^2) - \frac{1}{2\sigma^2}\sum_{t=p+1}^{n} e_t(\varphi_1, \ldots, \theta_q)^2. \quad (6.63)$$

Remark 6.14 Table 6.6 enables us to evaluate the errors of estimated parameters by means of their approximate standard deviations (the applicability and derivations can be found in Box and Jenkins (1970, Chapter 7)). For instance in the process AR(1), one can evaluate the corresponding error as

$$\widehat{\sigma}(\widehat{\varphi}_1) = \left(\frac{1-\widehat{\varphi}_1^2}{n}\right)^{1/2}. \quad (6.64)$$

◇

Example 6.3 In Table 6.7, the time series from Example 6.1 (*Dreimonatsgeld*) identified as the process AR(4) (see also Example 6.2) is estimated by means of EViews as

$$y_t = 6.203 + 0.950y_{t-1} - 0.726y_{t-2} + 0.542y_{t-3} - 0.452y_{t-4} + \varepsilon_t, \qquad \widehat{\sigma} = 1.756.$$

◇

Table 6.7 Estimation of the process AR(4) from Example 6.3 (*Dreimonatsgeld*) calculated by means of EViews

Variable	Coefficient	Std. Error	*t*-Statistic	Prob.
C	6.203253	0.427508	14.51027	0.0000
AR(1)	0.950188	0.163212	5.821793	0.0000
AR(2)	−0.725744	0.216765	−3.348060	0.0021
AR(3)	0.541991	0.215576	2.514154	0.0173
AR(4)	−0.451544	0.165381	−2.730320	0.0103
R-squared	0.567557	Mean dependent var		6.186667
Adjusted *R*-squared	0.511758	S.D. dependent var		2.513598
S.E. of regression	1.756359	Akaike info criterion		4.092609
Sum squared resid	95.62875	Schwarz criterion		4.312543
Log likelihood	−68.66697	*F*-statistic		10.17145
Durbin–Watson stat	2.057756	Prob(F-statistic)		0.000022
Inverted AR Roots	0.70−0.49i	0.70+0.49i	−0.22+0.76i	−0.22−0.76i

Source: Calculated by EViews

6.3.3 Verification of Model

The verification step of Box–Jenkins methodology is relatively elaborate. Applying various diagnostic instruments, the compatibility of the estimated model with the analyzed data should be verified. It is usually done by checking various properties of the constructed model:

6.3.3.1 Check of Stationarity

In this case, one checks whether the estimated model fulfills the condition of stationarity, i.e., whether the roots of estimated autoregressive polynomial lie outside the unit circle in complex plane (or equivalently, whether their inverted values, which are the roots of autoregressive polynomial written with the opposite order of powers $z^p - \varphi_1 z^{p-1} - \ldots - \varphi_p$, lie inside this circle; see Example 6.4). In particular, this check of stationarity is important in the cases in which the estimation method is strongly based on the stationarity assumption (e.g., for the estimates based on the Yule–Walker equations; see Remark 6.8). It is also possible to separate several segments in the given time series and to test the coincidence of estimated levels, variances, and autocorrelations (or higher moments such as skewness and others) among particular segments.

Another approach (so-called *impulse response*) consists in analyzing the response of an impulse that occurred in the estimated model either in a single time moment or repeatedly since this moment and that influences the consecutive values of the process (such an impulse is mostly standardized to the size of standard deviation of the corresponding white noise or to a multiple of this standard deviation). For example, the estimated ARMA structure is transferred to the form of linear process

(6.17), and hereinto one substitutes (since a given time moment) an "artificial" innovation process $\{\varepsilon_t\}$ either with the only nonzero value in this time or with fixed nonzero values since this moment. If the given time series is stationary, then by increasing the time distance from the initial moment of impulse (1) the response to a single impulse should fade away gradually to the zero level and (2) the response to repeated impulses should stabilize itself to an appropriate (non-zero) level (see Example 6.4).

6.3.3.2 Check of ARMA Structure

This check means first of all the coincidence of the correlation structure estimated from the data (i.e., the functions r_k and r_{kk}) with the correlation structure derived from the estimated model that is to be verified (see Example 6.4). Another check of structure of model is based on testing the uncorrelatedness (e.g., by means of Q-tests; see below) in the white noise that has been estimated using the tested model.

6.3.3.3 Graphical Examination of Estimated White Noise

An important diagnostic instrument is the white noise $\{\widehat{\varepsilon}_t\}$ constructed from the estimated model of given time series (similarly as the residuals calculated using an estimated regression model). The graphical record of this *estimated white noise* (and its estimated correlogram, histogram, etc.) can indicate eventual flaws of the model (in standard situations the estimated white noise is usually expected to show zero mean value, constant variance, uncorrelatedness and normality; see Example 6.4).

6.3.3.4 Tests of Uncorrelatedness for Estimated White Noise

The uncorrelatedness of the estimated white noise (see above) can be tested under the normality assumption directly by means of the test based on Bartlett's approximation (6.54), where we use the estimated autocorrelations of the estimated white noise $r_k(\widehat{\varepsilon}_t)$. Obviously, the null hypothesis has the critical region (applying the significance level of 5 %)

$$|r_k(\widehat{\varepsilon}_t)| \quad \geq 2\sqrt{\frac{1}{n}} \qquad \text{for } k = 1, 2, \ldots . \tag{6.65}$$

However, so-called *Q-tests* (or equivalently *portmanteau tests*) are also frequently used that test cumulatively the significance of the K initial autocorrelations of estimated white noise (the integer K must be chosen in advance with recommended size $K \approx \sqrt{n}$, where n is the length of given time series). In this way one verifies simultaneously the used structure ARMA(p, q) since the corresponding

Q-statistics of this test has the asymptotic distribution $\chi^2(K-p-q)$ (under the null hypothesis that the original time series admits to be modeled by ARMA(p, q)). As the Q-statistics are concerned, in practice one uses mainly *Box–Pierce statistics* with the critical region (applying the significance level α) of the form

$$Q = n\sum_{k=1}^{K}(r_k(\widehat{\varepsilon}_t))^2 \geq \chi^2_{1-\alpha}(K-p-q) \tag{6.66}$$

or statistically more powerful *Ljung–Box statistics* with the critical region (applying again the significance level α) of the form

$$Q^* = n(n+2)\sum_{k=1}^{K}\frac{1}{n-k}(r_k(\widehat{\varepsilon}_t))^2 \geq \chi^2_{1-\alpha}(K-p-q). \tag{6.67}$$

Example 6.4 The stationarity of model AR(4) estimated in Example 6.3 (*Dreimonatsgeld*) was checked at first in the framework of verification: in Table 6.7 and Fig. 6.5, one can see the inverted roots of estimated autoregressive polynomial which lie distinctly inside the unit circle in complex plane. Further in Fig. 6.6 one shows the response corresponding to impulse (standardized to the double size of estimated standard deviation of white noise) and to repeated impulses. In the first case, the response fades away gradually reaching the zero level finally and in the second case it stabilizes to the level of 2.56. Obviously, none of performed checks reject stationarity.

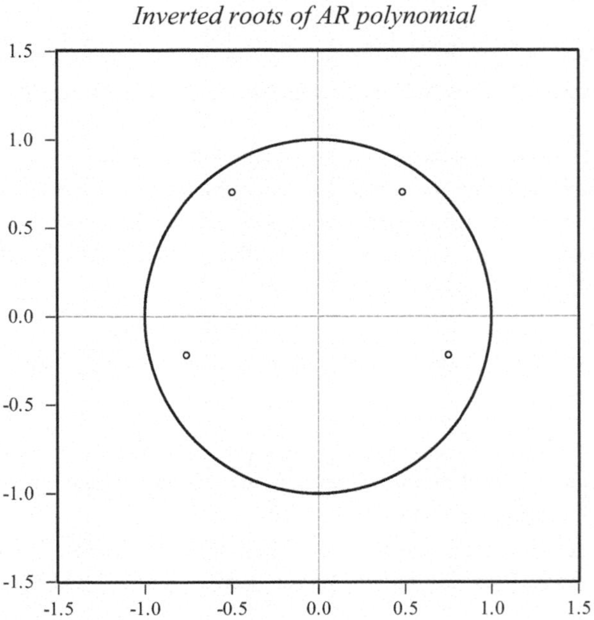

Fig. 6.5 Inverted roots of estimated autoregressive polynomial from Example 6.4 (*Dreimonatsgeld*) calculated by means of EViews

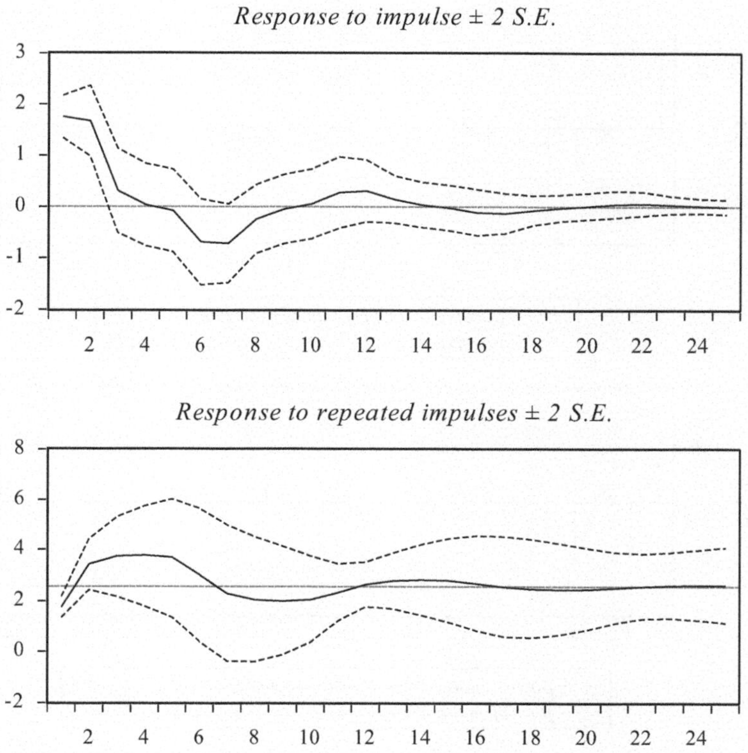

Fig. 6.6 Response to impulse standardized to the double size of (estimated) standard deviation of white noise in the case of (**a**) single impulse (see upper graph) and (**b**) repeated impulses (see lower graph) from Example 6.4 (*Dreimonatsgeld*) calculated by means of EViews

In Fig. 6.7, one compares the correlation structure estimated from data (i.e., the functions r_k and r_{kk}) with the correlation structure derived from estimated model (i.e., the functions ρ_k and ρ_{kk} corresponding to the estimated model). The achieved coincidence testifies to the adequacy of constructed model AR(4).

Finally, Table 6.8 shows the estimated autocorrelations of estimated white noise and the results of a Q-test. According to (6.65) one gets

$$|r_k(\widehat{\varepsilon}_t)| \quad \geq 2\sqrt{\frac{1}{40}} = 0,316 \qquad \text{for } k = 1, 2, \ldots . \tag{6.68}$$

In addition using the Q-test based on Ljung–Box statistics (6.67) for various $K = 5$, 6, ... , one cannot reject (applying the significance level 5 %) the null hypothesis on uncorrelatedness of white noise (i.e., null hypothesis on adequacy of the constructed model AR(4)).

◇

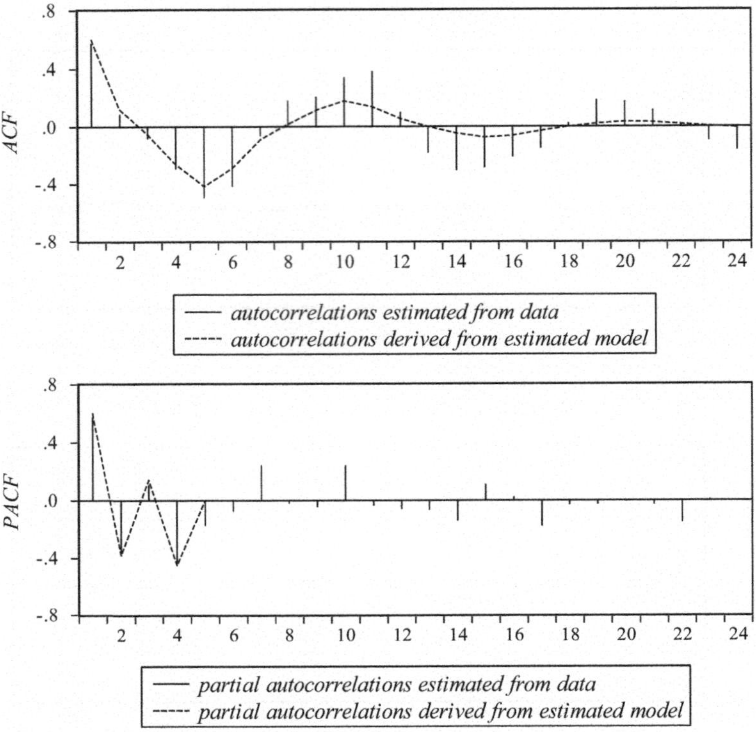

Fig. 6.7 Coincidence of the correlation structure estimated from data (i.e., the functions r_k and r_{kk}) with the correlation structure derived from estimated model (i.e., the functions ρ_k and ρ_{kk}) in Example 6.4 (*Dreimonatsgeld*) calculated by means of EViews

Table 6.8 Estimated autocorrelations of estimated white noise and results of Q-test based on Ljung–Box statistics in Example 6.4 (*Dreimonatsgeld*) calculated by means of EViews

	AC	Q-Stat	Prob
1	−0.056	0.1240	
2	−0.016	0.1344	
3	0.028	0.1663	
4	−0.064	0.3439	
5	−0.009	0.3475	0.556
6	−0.117	0.9681	0.616
7	−0.054	1.1053	0.776
8	0.147	2.1594	0.706
9	−0.107	2.7413	0.740
10	0.018	2.7579	0.839
11	0.281	7.0829	0.420
12	−0.068	7.3437	0.500
13	−0.060	7.5545	0.580
14	−0.200	10.050	0.436
15	−0.103	10.736	0.466
16	0.094	11.338	0.500

Source: Calculated by EViews

6.4 Stochastic Modeling of Trend

Majority of time series in practice (in particular, in economy and finance) is nevertheless nonstationary, i.e., such time series do not fulfill minimally the time invariance assumption for mean value, variance, and correlation structure. There are frequently significant differences in behavior of stationary and nonstationary time series with important impacts for their analysis and model instruments, e.g.,

- In Sect. 6.3.3.1, we have discussed the *response to impulse* analysis: while in the stationary environment the impulse influence expires gradually (i.e., the impulse realized in time t has mostly smaller "power" in time $t + \tau_2$ than in time $t + \tau_1$ for $0 < \tau_1 < \tau_2$), it does not hold in nonstationary environment where the persistence of the impulse can be unlimited with nondecreasing power of response.
- Presence of nonstationary data can lead to so-called *spurious regression*: if two stationary time series are mutually independent, then the regression of the first one on the second one (or on the contrary) does not usually exhibit attributes of a significant regression relation (there are usually nonsignificant t-ratios for the corresponding regression parameter and small coefficients of determination R^2). In the nonstationary situation, the regression may give highly significant results (particularly when the series contain trend components), but such regression relations may be only spurious without any rational reason.
- Moreover, in the previous point under nonstationarity one cannot rely on some routine properties according to which t-ratio has (asymptotically) t-distributions or F-statistics has (asymptotically) F-distribution.

In addition, one must distinguish two types of nonstationarity:

1. The first type of nonstationarity can be demonstrated simply by means of the following linear trend model:

$$y_t = \alpha + \beta \cdot t + \varepsilon_t, \tag{6.69}$$

where ε_t is a white noise (see, e.g., Fig. 3.2). It is the example of so-called *deterministic nonstationarity* caused for instance by *deterministic* trend (in our case by a linear line), and when it is eliminated the time series becomes stationary (e.g., white noise in our case).

2. The second type of nonstationarity is represented, e.g., by the model

$$y_t = \alpha + y_{t-1} + \varepsilon_t, \tag{6.70}$$

where ε_t is again the white noise with variance σ^2 (see, e.g., Fig. 6.8(a)), though here one usually assumes that in addition $\varepsilon_t \sim iid$. It is the example of the *stochastic nonstationarity*, which can be modeled in some specific situations by

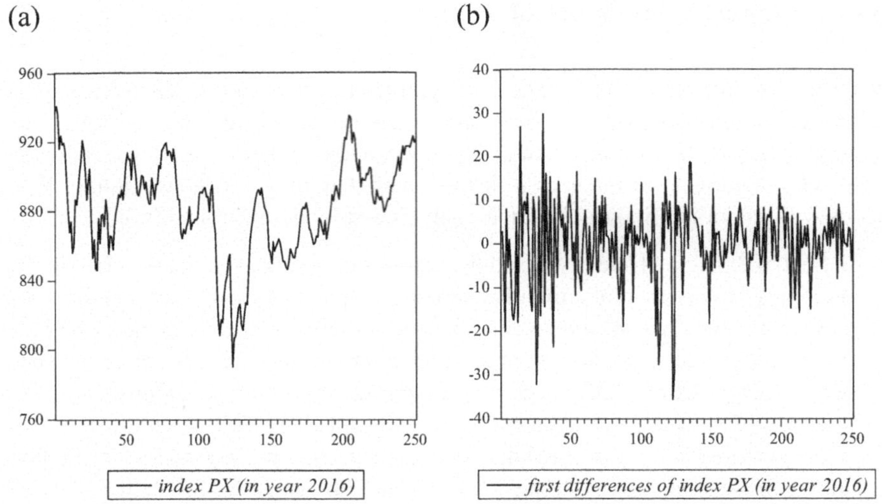

Fig. 6.8 (**a**) Index PX in the year 2016 (values for 251 trading days) from Example 6.5. (**b**) First differences of time series from (a)

special (stochastic) models and then also made stationary by exploiting these models in a suitable way. More specifically, the model (6.70) is so-called *random walk with drift*, and in this case, the corresponding time series can be "stationarized" simply by transferring it to the time series of first differences Δy_t, since from the model (6.70) one easily obtains

$$\Delta y_t = \alpha + \varepsilon_t. \tag{6.71}$$

The right-hand side of (6.71), i.e., the white noise shifted to level α, is trivially a stationary time series. The principle of stochastic nonstationarity in the case of model (6.70) can be presented better if one rewrite it in the form

$$y_t = y_1 + \alpha \cdot (t-1) + \sum_{\tau=2}^{t} \varepsilon_\tau. \tag{6.72}$$

The time series has obviously not only the deterministic trend (namely, the linear trend with the slope α), but also a stochastic trend consisting in progressive cumulation of white noise. From the interpretation point of view, the conditional values are also interesting (under the assumption of mutual independence of ε_t)

$$\mathrm{E}(\Delta y_t | y_{t-1}, \ldots, y_1) = \alpha, \tag{6.73}$$

$$\mathrm{E}(y_t|y_1) = y_1 + \alpha \cdot (t-1),$$

$$\mathrm{var}(y_t|y_1) = \sigma^2 \cdot (t-1), \quad \mathrm{corr}(y_{t-k}, y_t|y_1) = \sqrt{\frac{t-k-1}{t-1}}. \tag{6.74}$$

According to (6.73), this time series tends not to revert back to the original level, but on the contrary, it tends to higher (or lower) values for $\alpha > 0$ (or $\alpha < 0$), respectively, since the development rate of the mean value is $O(t)$, while for the standard deviation it is only $O(\sqrt{t})$. Even in the case of zero slope ($\alpha = 0$), the *random walk without drift* (in contrast to the white noise) intersects the horizontal axis (i.e., the zero level) only rarely. Moreover, the relations (6.74) imply that the mean level and variance (volatility) are unlimited, while the autocorrelation function has values near to one and decreases to zero in a slower rate than linearly.

Remark 6.15 Let us rewrite the relation (6.70) to a more general form

$$y_t = \alpha + \varphi_1 y_{t-1} + \varepsilon_t \tag{6.75}$$

(the relation (6.70) is a special case for $\varphi_1 = 1$). If it holds $|\varphi_1| < 1$, then (6.75) is obviously the stationary process AR(1) with nonzero mean value $\mu = \alpha/(1 - \varphi_1)$

$$y_t - \mu = \varphi_1(y_{t-1} - \mu) + \varepsilon_t \tag{6.76}$$

(see Remark 6.12), which can be rewritten as $\Delta y_t = (\varphi_1 - 1)(y_{t-1} - \mu) + \varepsilon_t$. Then the conditional mean value (6.73) of such a stationary process AR(1) obviously fulfills

$$\begin{aligned} \mathrm{E}(\Delta y_t|y_{t-1}, \ldots, y_1) < 0 \quad \text{for } y_{t-1} > \mu, \\ \mathrm{E}(\Delta y_t|y_{t-1}, \ldots, y_1) > 0 \quad \text{for } y_{t-1} < \mu, \end{aligned} \tag{6.77}$$

i.e., now in contrast to the random walk with drift, the process $\{y_t\}$ does not drift, but it reverts to the previous level (so-called *mean reverting*). Finally, the remaining case of $|\varphi_1| > 1$ is a very special one since then $\{y_t\}$ is the *explosive process* comparable with the powers $\varphi_1{}^k$ (e.g., the process $y_t = 2y_{t-1} + \varepsilon_t$ behaves since later times t as the deterministic sequence 2^t regardless of the size of white noise ε_t).

◊

Remark 6.16 Once more let us stress the distinction in stationarization for the model described above:

- The stationarity of the model (6.69) with deterministic trend can be achieved simply by regression methods eliminating trend. The stationarization based on differences should not be used since it may lead to models with residuals in the form of a (noninvertible) MA process

$$\Delta y_t = \beta + \varepsilon_t - \varepsilon_{t-1}. \tag{6.78}$$

- To achieve stationarity in the model (6.70) of random walk with drift it is sufficient to construct the (first) differences. As the possibility of a regression elimination of stochastic trend is concerned, it is not clear what we should eliminate. If we have extended the model (6.75) to the form

$$\Delta y_t = \alpha + \beta \cdot t + (\varphi_1 - 1)y_{t-1} + \varepsilon_t \quad \text{or} \quad y_t = \alpha + \beta \cdot t + \varphi_1 y_{t-1} + \varepsilon_t \tag{6.79}$$

with both deterministic and stochastic trend, then the effort to eliminate the trend by means of regression methods would face the problem that, e.g., t-ratio may not have (not even asymptotically) t-distribution. The model (6.79) can be also rewritten as

$$y_t - \beta_0 - \beta_1 t = \varphi_1 (y_{t-1} - \beta_0 - \beta_1 (t-1)) + \varepsilon_t, \quad \text{where } \beta_1 = \frac{\beta}{1-\varphi_1}, \quad \beta_0 = \frac{\alpha - \varphi_1 \beta_1}{1 - \varphi_1}, \tag{6.80}$$

i.e., if $|\varphi_1| < 1$, then (6.80) is in fact the stationary process AR(1) with linear trend.

◇

6.4.1 *Tests of Unit Root*

The possibility to stationarize the analyzed time series by means of differencing can be considered as the evidence of existence of (nearly) unit roots of the autoregressive operator for given model (e.g., the autoregressive operator of the model (6.70) has obviously the single root equal to one). In accord with the previous discussion, the decision on existence of such a unit root (or multiple unit roots) is often the key point of the corresponding analysis, even if the form of estimated correlogram can indicate the presence of such a root (namely, a very slow decline starting from unit to zero, since the particular estimated autocorrelations converge to 1 with the increasing length of nonstationary time series). However, a subjective checking of estimated correlograms cannot usually distinguish nonstationary models of the type $y_t = y_{t-1} + \varepsilon_t$ from stationary ones with nearly unit root of the type $y_t = 0.95 \cdot y_{t-1} + \varepsilon_t$ so that an application of a statistical test with a prescribed significance level is recommended here.

6.4.1.1 Dickey–Fuller Test

DF test (see Dickey and Fuller (1979, 1981)) was the pioneering one among the tests of the unit root. In particular, Dickey and Fuller suggested three variants all denoted as *τ-tests*:

(1) τ-test: H_0: $y_t = y_{t-1} + \varepsilon_t$ versus H_1: $y_t = \varphi_1 y_{t-1} + \varepsilon_t$ for $\varphi_1 < 1$, i.e., the one-tailed test of random walk versus stationary AR(1) process (the possible nonstationarity caused by $\varphi_1 \leq -1$ is not important in practice);
(2) τ_μ-test: H_0: $y_t = y_{t-1} + \varepsilon_t$ versus H_1: $y_t = \alpha + \varphi_1 y_{t-1} + \varepsilon_t$ pro $\varphi_1 < 1$, i.e., the one-tailed test of random walk versus stationary AR(1) process with (nonzero) level;
(3) τ_τ-test: H_0: $y_t = y_{t-1} + \varepsilon_t$ versus H_1: $y_t = \alpha + \beta \cdot t + \varphi_1 y_{t-1} + \varepsilon_t$ pro $\varphi_1 < 1$, i.e., the one-tailed test of random walk versus stationary AR(1) process with linear trend.

The null hypothesis in each of all three cases can be written simply as

$$H_0: \ \Delta y_t = \psi y_{t-1} + \varepsilon_t \quad \text{for } \psi = 0, \tag{6.81}$$

while the alternative generally as

$$H_1: \ \Delta y_t = \alpha + \beta \cdot t + \psi y_{t-1} + \varepsilon_t \quad \text{for } \psi < 0, \tag{6.82}$$

where $\psi = \varphi_1 - 1$ and (1) $\alpha = \beta = 0$ and (2) $\beta = 0$. One should stress that all alternatives consist in the inequality $\psi < 0$ only, and the equalities $\alpha = \beta = 0$ in the alternative (1) or $\beta = 0$ in the alternative (2) are not investigated at all (including the numerical values of the intercept α or the slope β whose correctness is not guaranteed under nonstationarity caused by $\psi = 0$ anyhow).

The test statistics in each of three variants of DF test is the classical t-ratio (we test simply the significance of regression parameter ψ in the model (6.81))

$$DF = \frac{\widehat{\psi}}{\widehat{\sigma}(\widehat{\psi})} \tag{6.83}$$

using the estimates constructed by means of the methodology from Sect. 6.3.2 and with the critical region

$$DF \leq t_\alpha^*(n). \tag{6.84}$$

However under the null hypothesis $\psi = 0$, the test statistics DF does not have t-distribution (not even asymptotically or under the assumption of $\varepsilon_t \sim iid$) as is the case of the classical t-ratio, but a nonstandard distribution, for which one must calculate the critical value in (6.84) by means of simulations separately for particular tests (1), (2), and (3) and for particular lengths n of time series (see selected critical

Table 6.9 Selected critical values of DF tests (for the asymptotic case of $n \to \infty$)

Significance level	10 %	5 %	1 %
Critical values for τ-test ($n \to \infty$)	–1.62	–1.95	–2.58
Critical values for τ_μ-test ($n \to \infty$)	–2.57	–2.86	–3.43
Critical values for τ_τ-test ($n \to \infty$)	–3.12	–3.41	–3.96

values for the asymptotic case $n \to \infty$ in Table 6.9). It holds generally that this distribution has heavier tails than the corresponding t-distribution so that its critical values are much more higher in the absolute value than for t-distribution (e.g., the critical value –3.41 for 5 % and $n \to \infty$ is in the absolute value twice as much as the critical value –1.645 for the classical t-test, i.e., one needs a more significant value of t-ratio to reject the null hypothesis of $\psi = 0$). The reason consists in the fact that one applies the nonstationary regressor (see also the introduction of Sect. 6.4). Even if the critical values were calculated by Dickey and Fuller, nowadays the software systems use more sophisticated algorithms delivering directly the corresponding p-values; see, e.g., MacKinnon (1996).

6.4.1.2 Augmented Dickey-Fuller Test

The previous DF test is applicable only in the case that the residual component ε_t has the form of independent white noise. However, if the model (6.81) explaining the dependent variable Δy_t includes autocorrelatedness, which is not reflected correctly, then the type-one error of DF test (i.e., the probability of rejection of valid H_0) is higher than the declared α. Mainly for this case, so-called *augmented DF test* (*ADF-test*) has been suggested which has the null hypothesis of the form

$$H_0 : \Delta y_t = \psi y_{t-1} + \sum_{i=1}^{p} \gamma_i \Delta y_{t-i} + \varepsilon_t \quad \text{for } \psi = 0 \tag{6.85}$$

instead of (6.81). The test statistics and the critical values for particular variants (1), (2), and (3) (i.e., for τ-test, τ_μ-test, and τ_τ-test) are the same as before the augmentation (the test concerns again the parameter ψ only). The added autoregressive terms in (6.85) absorb the dynamic structure explaining the dependent variable. For the identification of order p of added autoregressive terms, one recommends to apply the information criteria (see Sect. 6.3.1.2).

6.4.1.3 Phillips–Perron Test

PP test (see Phillips and Perron (1988)) is similar to ADF test, except that it models the possible autocorrelatedness of residuals not by adding autoregressive terms as in (6.85), but directly by correcting the estimated standard deviation in the denominator of the original DF statistics (6.83). Essentially, it is HAC approach (*heteroscedasticity and autoregression consistent covariances*) to the linear regression model with autocorrelated residuals; see, e.g., Newey–West (1987).

6.4.1.4 KPSS Test

KPSS test (see Kwaitkovski et al. (1992)) improves the resolving power of DF test which can be sometimes weaker. For instance, one should reject the null hypothesis of unit root for the theoretical model $y_t = 0.95 \cdot y_{t-1} + \varepsilon_t$. If it is not the case, then it means that either the model is really nonstationary or we do not have sufficient information to reject it (e.g., a short segment of time series $y_t = 0.95 \cdot y_{t-1} + \varepsilon_t$ is observed only). Therefore, KPSS test was suggested in such a way that the hypotheses H_0 and H_1 are just opposite than for ADF test (i.e., the null hypothesis H_0 represents stationarity versus nonstationarity in the alternative H_1). Moreover, one recommends to carry out ADF test and KPSS test always simultaneously with the following conclusions: (a) if H_0^{ADF} is rejected and simultaneously H_0^{KPSS} cannot be rejected, then the stationarity is confirmed; (b) if H_0^{ADF} cannot be rejected and simultaneously H_0^{KPSS} is rejected, then the nonstationarity is confirmed; and (c) both remaining combinations are regarded as inconclusive. To summarize the topic, the previous tests (and others) can be found in modern software systems recommended for time series analysis.

Nowadays the topic of the unit roots testing is very complex so that other references should be also addressed for a deeper understanding (see, e.g., Brockwell and Davis 1996, Section 6.3 or Heij et al. 2004, Section 7.3.3).

Example 6.5 Figure 6.8(a) and Table 6.10 show the values of index PX of Prague Exchange (i.e., the time series $\{PX_t\}$) for 251 trading days of the year 2016. This time series seems to be the random walk $PX_t = PX_{t-1} + \varepsilon_t$, (see (6.70) for $\alpha = 0$).

Table 6.11 shows the results of DF test of type τ (i.e., with H_1: $y_t = \varphi_1 y_{t-1} + \varepsilon_t$ for $\varphi_1 < 1$): the null hypothesis of nonstationarity with one unit root is not rejected even when applying the significance level of 10 %. This test is performed simultaneously with ADF test since according to Table 6.11 the system EViews performs the automatic choice of the order p of autoregressive terms in (6.85) by means of the information criterion SIC (so-called Schwarz information criterion also denoted sometimes as BIC; see Sect. 6.3.1.2).

After transferring to the first differences ΔPX_t, the previous DF test of type τ rejects the null hypothesis of nonstationarity (i.e., the existence of second unit root in the original time series PX_t) applying the significance level of 1 % (see Table 6.12), so that the construction of first differences is sufficient in order to make the series PX_t stationary.

Finally, Table 6.13 presents the results of KPSS test by means of EViews that rejects significantly the null hypothesis of stationarity of the (nondifferenced) time series PX_t even when applying the significance level of 1 %. This result together with the previous ADF test confirms unambiguously the nonstationarity of PX_t.

◇

Table 6.10 Index PX in year 2016 (values for 251 trading days written in columns) from Example 6.5 (see also Fig. 6.8(a))

	1	2	3	4	5	6	7	8	9	10
1	938.23	872.53	913.94	916.94	890.28	808.21	849.79	879.83	915.33	886.31
2	941.07	852.97	909.99	918.60	891.30	816.91	862.37	870.08	922.50	888.14
3	936.17	863.74	910.20	919.60	890.47	824.43	856.79	868.29	921.35	885.13
4	916.09	847.23	898.32	912.35	892.68	826.29	859.18	864.48	928.28	879.33
5	924.04	845.92	914.85	914.89	893.76	814.58	861.20	860.81	935.36	881.22
6	918.64	875.81	908.44	913.71	888.21	811.26	863.87	861.83	934.05	885.05
7	919.62	861.37	901.14	909.01	888.41	820.31	861.30	865.53	919.18	887.20
8	914.73	877.55	890.47	916.04	879.51	827.31	856.30	864.77	925.46	886.66
9	898.62	878.51	888.93	909.34	892.27	826.17	850.80	875.87	921.78	894.86
10	881.12	871.22	893.17	896.67	895.20	844.90	850.98	874.05	908.80	894.24
11	867.85	886.64	900.82	886.94	874.05	863.55	847.60	869.10	902.89	899.57
12	873.98	879.15	899.91	886.83	867.79	870.26	846.29	866.34	909.66	900.71
13	855.92	855.43	892.92	867.79	840.05	876.21	850.72	874.57	893.82	905.43
14	859.87	865.67	896.85	864.20	818.38	882.07	858.06	863.58	899.00	911.09
15	886.77	865.37	889.53	869.46	808.20	887.50	855.79	868.59	897.95	903.02
16	883.85	857.61	884.39	866.93	817.58	891.37	853.19	875.13	897.76	911.98
17	892.81	871.89	884.30	871.43	815.86	892.44	852.94	881.09	901.69	917.59
18	902.56	879.32	899.33	873.87	831.21	889.42	858.83	889.38	900.99	912.46
19	909.43	883.06	893.56	882.66	838.94	893.42	859.14	885.72	905.11	917.48
20	921.07	889.69	887.01	869.31	842.31	887.26	866.37	891.00	904.68	917.55
21	914.71	888.01	895.62	873.38	852.05	881.74	875.71	894.57	889.62	917.53
22	902.20	892.56	895.62	873.17	855.26	880.08	880.84	890.74	884.41	916.75
23	886.72	886.71	906.63	875.19	819.58	876.28	882.42	886.64	884.00	920.35
24	897.48	896.15	904.65	874.83	790.09	857.86	881.41	899.16	892.29	923.54
25	904.79	907.56	915.23	876.22	806.43	856.10	884.80	906.97	888.72	919.58
26										921.61

Source: Prague Stock Exchange (https://www.pse.cz/en/indices/index-values/detail/XC0009698371?tab=detail-history)

Table 6.11 DF test of time series of index PX from Example 6.5 by means of EViews

Null Hypothesis: PX2016 has a unit root			
Exogenous: None			
Lag Length: 0 (Automatic—based on SIC, maxlag=15)			
		t-Statistic	Prob.[a]
Augmented Dickey–Fuller test statistic		–0.202174	0.6126
Test critical values:	1% level	–2.574245	
	5% level	–1.942099	
	10% level	–1.615852	

Source: calculated by EViews
[a]MacKinnon (1996) one-sided *p*-values

Table 6.12 DF test of first differences of index PX from Example 6.5 by means of EViews

Null Hypothesis: D(PX2016) has a unit root			
Exogenous: None			
Lag Length: 0 (Automatic—based on SIC, maxlag=15)			
		t-Statistic	Prob.[a]
Augmented Dickey—Fuller test statistic		–14.49788	0.0000
Test critical values:	1% level	–2.574282	
	5% level	–1.942104	
	10% level	–1.615849	

Source: calculated by EViews
[a]MacKinnon (1996) one-sided *p*-values

Table 6.13 KPSS test of time series of index PX from Example 6.5 by means of EViews

Null Hypothesis: PX2016 is stationary		
Exogenous: Constant		
Bandwidth: 11 (Newey–West automatic) using Bartlett kernel		
Kwiatkowski–Phillips–Schmidt–Shin test statistic		0.284476
Asymptotic critical values:	1% level	0.739000
	5% level	0.463000
	10% level	0.347000

Source: calculated by EViews

6.4.2 Process ARIMA

The time series with stochastic trend of type (6.70), which can be stationarized by means of differencing, are modeled as processes ARIMA in Box–Jenkins methodology. *Integrated mixed process of order* p, d, q denoted as ARIMA(p, d, q) has the form

$$\varphi(B)w_t = \alpha + \theta(B)\varepsilon_t, \tag{6.86}$$

where

$$w_t = \Delta^d y_t \tag{6.87}$$

is the dth difference of the original time series y_t (see also (3.62)), which is modeled as a stationary (and invertible) process ARMA(p, q) in (6.86). In other words, the principle of ARIMA processes is as follows: at first (1) the modeled time series is stationarized by differencing it suitably and then (2) the corresponding stationary time series (denoted as w_t in (6.86) and (6.87)) is modeled by means of mixed process ARMA. Usually one writes it summarily as

$$\varphi(B)\Delta^d y_t = \alpha + \theta(B)\varepsilon_t. \tag{6.88}$$

An important special case is the *integrated process* I(d) presented mostly in a simple form

$$\Delta^d y_t = \varepsilon_t, \tag{6.89}$$

which can be constructed by "integrating" the white noise, e.g., for $d = 1$ it holds

$$y_t = y_1 + \sum_{\tau=2}^{t} \varepsilon_\tau. \tag{6.90}$$

Remark 6.17 The drift parameter α serves to model a possible nonzero level of the process w_t, i.e., a deterministic trend in the form of polynomial of the dth order for original time series $\{y_t\}$. If $d > 0$, then the model ARIMA of time series y_t is invariant when shifting the time series by an arbitrary constant. Therefore, obviously it makes no sense to center such series by subtracting the sample means before their analysis.

◊

Remark 6.18 The operator $\varphi(B)\Delta^d$ on the left-hand side of the model (6.88) is called *generalized autoregressive operator*. This operator is characterized by the property that the corresponding polynomial $\varphi(z)\Delta^d$ has p roots lying outside the unit circle in complex plane and, in addition, a unit root of multiplicity d. More general types are the processes ARUMA that have at least one root different from unit, but lying on the unit circle, and the explosive processes (see also Remark 6.15), which have at least one root inside the unit circle.

◊

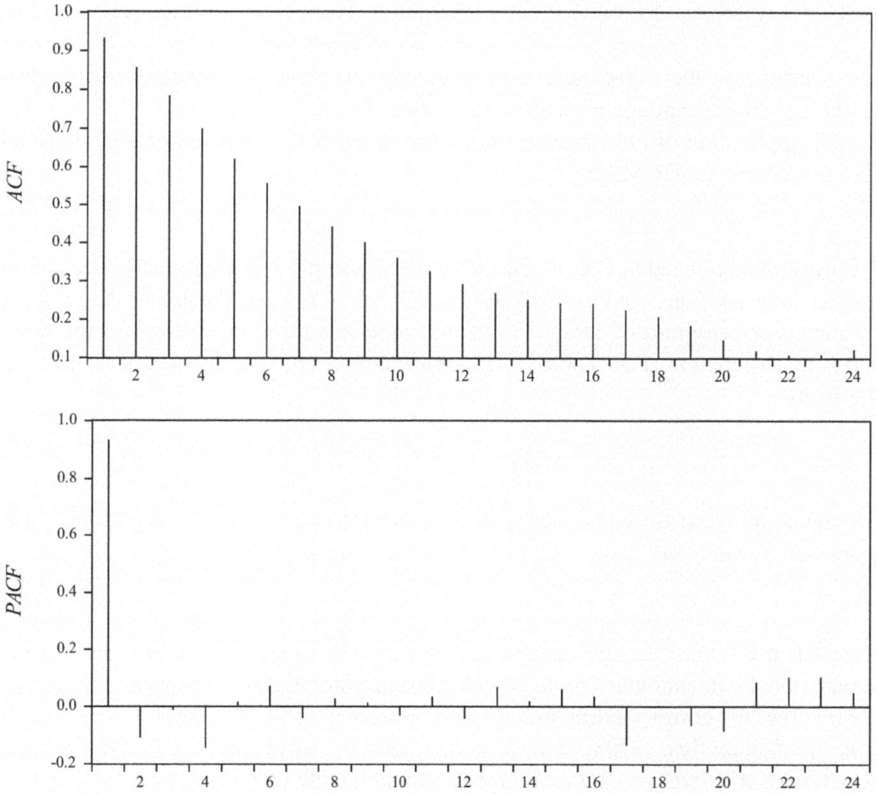

Fig. 6.9 Estimated correlogram and partial correlogram from Example 6.6 (*index PX in the year 2016*)

The construction of model ARIMA is based on the construction of stationary model ARMA for appropriate differences of original time series (possibly after an initial transformation which linearizes the given time series before differencing it; see Sect. 4.3). In practice, the order of differencing d does not exceed the order two (the routine time series of economic or financial character have usually $d = 1$, and particularly, the time series of consumer price indices or nominal salaries can demand $d = 2$). There are various possibilities how to find the order of differencing d for modeled time series, e.g.:

- The tests of unit root (see Sect. 6.4.1).
- A subjective examination of plots of time series y_t, Δy_t, $\Delta^2 y_t$, ... including their estimated correlograms and partial correlograms (in particular, a slow (linear) decrease of estimated autocorrelations indicates that a higher differencing is necessary; see Fig. 6.9).

- The comparison of sample standard deviations (volatilities) of time series y_t, Δy_t, $\Delta^2 y_t$, ... (one chooses such an order of differencing that shows the lowest volatility; on the other hand, one must pay attention to *overdifferencing* since the volatilities can again increase for higher d).
- The application of information criteria (see Sect. 6.3.1.2) which can be modified for the models ARIMA.

Example 6.6 For index PX in year 2016, the Example 6.5 identified by means of tests of unit root the random walk $PX_t = PX_{t-1} + \varepsilon_t$ (also the linear decrease of estimated correlogram of $\{PX_t\}$ in Fig. 6.9 indicates the need to transfer this time series by differencing). In Table 6.14, one estimated $\{PX_t\}$ by means of EViews in the form

$$\Delta PX_t = \varepsilon_t, \qquad \widehat{\sigma} = 9.26$$

(the intercept (or drift parameter) α in the model $\Delta PX_t = \alpha + \varepsilon_t$ is highly insignificant; see Table 6.14).

◊

Remark 6.19 One should understand correctly the meaning of constants in time series models. In stationary models, this constant interrelates to the mean value (i.e., the level) of the corresponding process. For example, for MA(1) process $y_t = \alpha + \varepsilon_t + \theta_1\varepsilon_{t-1}$ it is directly $\mu = \mathrm{E}(y_t) = \alpha$. For the stationary AR(1) process $y_t = \alpha + \varphi_1 y_{t-1} + \varepsilon_t$ it holds $\mu = \alpha/(1-\varphi_1)$. Finally for the random walk with drift $y_t = \alpha + y_{t-1} + \varepsilon_t$, the constant α presents the slope of process (even if this trend is loaded significantly by the integrated random walk).

◊

Table 6.14 Estimation of random walk from Example 6.6 (*index PX in the year 2016*) calculated by means of EViews

Dependent Variable: DPX2016				
Method: Least Squares				
DPX2016=C(1)				
	Coefficient	Std. Error	*t*-Statistic	Prob.
C(1)	–0.066480	0.585607	–0.113523	0.9097
R-squared	0.000000	Mean dependent var		–0.066480
Adjusted *R*-squared	0.000000	S.D. dependent var		9.259263
S.E. of regression	9.259263	Akaike info criterion		7.293118
Sum squared resid	21,347.75	Schwarz criterion		7.307204
Log likelihood	–910.6397	Hannan–Quinn criter.		7.298787
Durbin–Watson stat	1.834323			

Source: Calculated by EViews

Remark 6.20 In financial practice, we model frequently the time series of logarithmic rates of returns r_t (so-called *log returns*)

$$r_t = \ln\left(\frac{P_t}{P_{t-1}}\right) = \ln P_t - \ln P_{t-1} = p_t - p_{t-1} \tag{6.91}$$

for various financial assets (e.g., stocks or commodities) or price indices. These time series have usually a constant mean value of small positive size with added white noise

$$r_t = \mu + \varepsilon_t. \tag{6.92}$$

Then the corresponding time series of logarithmic prices $p_t = \ln P_t$ fulfill

$$p_t = \mu + p_{t-1} + \varepsilon_t, \tag{6.93}$$

which can be looked upon as a random walk with drift increasing approximately as $\mu \cdot t$.

◇

6.5 Stochastic Modeling of Seasonality

In addition to the stochastic modeling of trend, the methodology described in this chapter enables us to model also the seasonality in a stochastic way (by means of so-called *seasonal models of Box–Jenkins methodology*).

Let us consider, e.g., the seasonality in monthly observations y_t (i.e., $s = 12$). In such a case, one constructs at first the following model for the time series of January observations only:

$$\Phi(B^{12})\Delta_{12}^{D} y_t = \Theta(B^{12})\eta_t, \tag{6.94}$$

where the time index skips across the January periods. The symbols used in the formula (6.94) are following:

$$\Phi(B^{12}) = 1 - \Phi_1 B^{12} - \Phi_2 B^{24} - \ldots - \Phi_P B^{12P} \tag{6.95}$$

is the *seasonal autoregressive operator of order P*, and

$$\Theta(B^{12}) = 1 + \Theta_1 B^{12} + \Theta_2 B^{24} + \ldots + \Theta_Q B^{12Q} \tag{6.96}$$

is the *seasonal moving average operator of order Q*, and

$$\Delta_{12} = 1 - B^{12} \tag{6.97}$$

is the *seasonal difference operator*, e.g., it holds

$$\Delta_{12}^2 = \left(1 - B^{12}\right)^2 y_t = \left(1 - 2B^{12} + B^{24}\right) y_t = y_t - 2y_{t-12} + y_{t-24} \tag{6.98}$$

(see also (3.63)). The model (6.94) can be looked upon as the process ARIMA describing development of January observations. Similar models are constructed for the time series that skips across the February observations only, and so on. Let us suppose now that the models for particular months are approximately the same. However, the random components η_t in these models should be correlated mutually in time since there can exist, e.g., a relation between January and February values. Therefore, let us assume that also the time series $\{\eta_t\}$ is described by a model ARIMA of the form

$$\varphi(B)\Delta^d \eta_t = \theta(B)\varepsilon_t, \tag{6.99}$$

where ε_t is finally the white noise and the time index runs in the classical way. Obviously, the models (6.94) and (6.99) can be linked together to a single model of the form

$$\varphi(B)\Phi\left(B^{12}\right)\Delta^d \Delta_{12}^D y_t = \theta(B)\Theta\left(B^{12}\right)\varepsilon_t. \tag{6.100}$$

The model (6.100) is called the *multiplicative seasonal process of order* $(p, d, q) \times (P, D, Q)_{12}$ and is denoted usually by the acronym SARIMA (the adjective "multiplicative" expresses the fact that the operators of models (6.94) and (6.99) are multiplied mutually). For example, the process SARIMA $(0, 1, 1) \times (0, 1, 1)_{12}$ has the form

$$(1 - B)\left(1 - B^{12}\right) y_t = (1 + \theta_1 B)\left(1 + \Theta_1 B^{12}\right)\varepsilon_t \tag{6.101}$$

or equivalently

$$y_t - y_{t-1} - y_{t-12} + y_{t-13} = \varepsilon_t + \theta_1\varepsilon_{t-1} + \Theta_1\varepsilon_{t-12} + \theta_1\Theta_1\varepsilon_{t-13}. \tag{6.102}$$

Clearly, the number twelve is replaced by four in the case of quarterly seasonality. There are also the *additive seasonal processes* but they are applied in practice only rarely: if comparing with (6.104), a simple example of an additive seasonal processes is

$$y_t = \varepsilon_t + \theta_1\varepsilon_{t-1} + \theta_{12}\varepsilon_{t-12} + \theta_{13}\varepsilon_{t-13}. \tag{6.103}$$

The previous SARIMA models can be completed by a deterministic trend (see Example 6.7).

The construction of seasonal models is realized in three steps similarly as for ARIMA models (see Sect. 6.3). Here the first identification step is more difficult since the shapes of correlograms and partial correlograms may be more complex due to seasonality. For example the process SARIMA (0, 0, 1) × (0, 0, 1)$_{12}$, i.e.,

$$y_t = \varepsilon_t + \theta_1\varepsilon_{t-1} + \Theta_1\varepsilon_{t-12} + \theta_1\Theta_1\varepsilon_{t-13}, \tag{6.104}$$

has the following variance:

$$\sigma_y^2 = (1 + \theta_1^2 + \Theta_1^2 + \theta_1^2\Theta_1^2)\sigma^2 = (1 + \theta_1^2)(1 + \Theta_1^2)\sigma^2 \tag{6.105}$$

and the following autocorrelation function:

$$\rho_1 = \frac{\theta_1}{1+\theta_1^2}, \qquad \rho_{12} = \frac{\Theta_1}{1+\Theta_1^2}, \; \rho_{11} = \rho_{13} = \frac{\theta_1\Theta_1}{(1+\theta_1^2)(1+\Theta_1^2)},$$
$$\rho_k = 0 \quad \text{for } k \neq 1, 11, 12, 13. \tag{6.106}$$

In general, the process SARIMA (0, 0, q) × (0, 0, Q)$_{12}$ has nonzero autocorrelations only for arguments 1, ..., q, 12 – q, ..., 12 + q, 24 – q, ..., 24 + q, ..., 12Q – q, 12Q + q (and analogously for the partial autocorrelations of the process SARIMA (p, 0, 0) × (P, 0, 0)$_{12}$). In practice, one usually chooses among several alternatives offered by a suitable software (see, e.g., EViews).

Example 6.7 Figure 6.10 and Table 4.4 show the time series y_t (t = 1, ..., 96) of the job applicants kept in the Czech labor office register for particular months 2009M1-2016M12 (the same data with different time range were analyzed in Sect. 4.1 applying decomposition methods; see Examples 4.2 and 4.4).

By means of EViews (see Table 6.15) one has constructed the multiplicative seasonal model SARIMA (1, 0, 0) × (0, 1, 0)$_{12}$ (with deterministic trend) of the form

$$(1 - 0.93B)(1 - B^{12})y_t = -56\ \ 506.21 + 51.11 \cdot t + \varepsilon_t, \widehat{\sigma} = 9\ \ 658.47$$

or equivalently

$$y_t - 0.93y_{t-1} - y_{t-12} + 0.93y_{t-13} = -56\,506.21 + 51.11 \cdot t + \varepsilon_t,\ \widehat{\sigma} = 9\,658.47.$$

◇

Remark 6.21 The seasonal time series in financial practice are frequently modeled using so-called *airline model*, which is the process SARIMA (0, 1, 1) × (0, 1, 1)$_s$

$$(1 - B)(1 - B^s)y_t = (1 + \theta_1 B)(1 + \Theta_1 B^s)\varepsilon_t. \tag{6.107}$$

◇

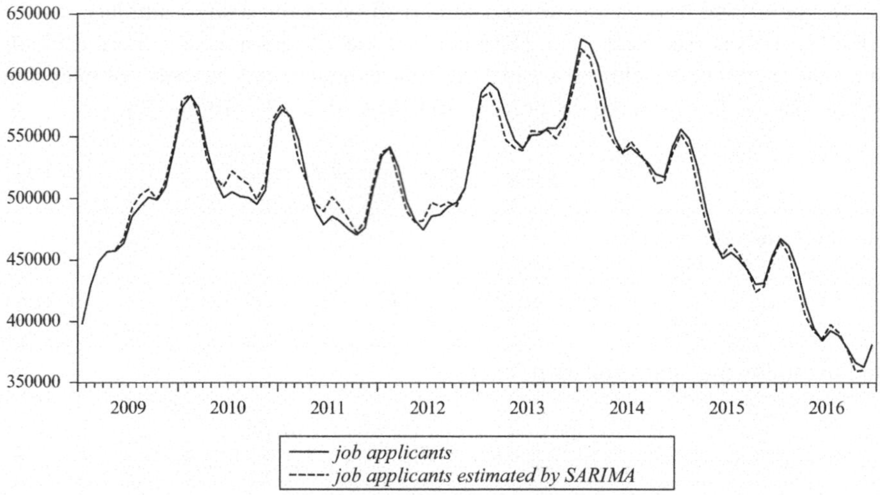

Fig. 6.10 Monthly data 2009M1-2016M12 and the values estimated by model SARIMA in Example 6.7 (*job applicants kept in Czech labor office register*); see Table 4.4. Source: Czech Statistical Office

Table 6.15 Estimation of the process SARIMA $(1, 0, 0) \times (0, 1, 0)_{12}$ (with deterministic trend) from Example 6.7 (*job applicants kept in Czech labor office register*) calculated by means of EViews

Dependent Variable: JOBAPPLICANTS—JOBAPPLICANTS(−12)				
Method: Least Squares				
Variable	Coefficient	Error	t-Statistic	Prob.
C	−56,506.21	70,358.24	−0.803121	0.4243
T	51.11415	845.0047	0.060490	0.9519
AR(1)	0.931628	0.023134	40.27022	0.0000
R-squared	0.971096	Mean dependent var		−12,300.01
Adjusted *R*-squared	0.970373	S.D. dependent var		56,113.27
S.E. of regression	9658.470	Akaike info criterion		21.22453
Sum squared resid	7.46E+09	Schwarz criterion		21.31196
Log likelihood	−877.8181	Hannan–Quinn criter.		21.25966
F-statistic	1343.879	Durbin–Watson stat		0.865019
Prob(*F*-statistic)	0.000000			

Source: calculated by EViews

6.6 Predictions in Box–Jenkins Methodology

One of convenient features of Box–Jenkins methodology consists in easy construction of predictions. We shall demonstrate basic principles of corresponding forecasting philosophy by means of stationary and invertible process ARMA(p, q) with zero mean value

$$y_t = \varphi_1 y_{t-1} + \ldots + \varphi_p y_{t-p} + \varepsilon_t + \theta_1 \varepsilon_{t-1} + \ldots + \theta_q \varepsilon_{t-q}. \tag{6.108}$$

Similarly as in the previous chapters, the symbol $\widehat{y}_{t+k}(t)$ will denote the prediction of value y_{t+k} constructed in time t, i.e., the prediction for time $t + k$ in time t (*k-step-ahead prediction*).

For simplicity, we shall construct the *linear* prediction, i.e., the prediction which is a linear function of values y_t, y_{t-1}, ... or equivalently a linear function of ε_t, ε_{t-1}, ... (since we assume the stationarity and invertibility). In addition, the mean square error of constructed prediction

$$MSE = \mathrm{E}\left(y_{t+k} - \widehat{y}_{t+k}(t)\right)^2 \tag{6.109}$$

should be minimal over all linear predictions. If one takes such a prediction in the form

$$\widehat{y}_{t+k}(t) = \psi_k^* \varepsilon_t + \psi_{k+1}^* \varepsilon_{t-1} + \ldots, \tag{6.110}$$

and keeps the usual form of linear processes

$$y_{t+k} = \varepsilon_{t+k} + \psi_1 \varepsilon_{t+k-1} + \psi_2 \varepsilon_{t+k-2} + \ldots + \psi_{k-1} \varepsilon_{t+1} + \psi_k \varepsilon_t + \ldots \tag{6.111}$$

(see (6.17)), then obviously one should look for such coefficients $\psi_k^*, \psi_{k+1}^*, \ldots,$ which minimize the expression

$$\left(1 + \psi_1^2 + \ldots + \psi_{k-1}^2 + \sum_{j=k}^{\infty} \left(\psi_j - \psi_j^*\right)^2\right) \cdot \sigma^2. \tag{6.112}$$

Evidently, the expression (6.112) attains its minimum for

$$\psi_j^* = \psi_j, \qquad j = k,\ k+1, \ldots. \tag{6.113}$$

Hence we have derived that

$$\widehat{y}_{t+k}(t) = \psi_k \varepsilon_t + \psi_{k+1} \varepsilon_{t-1} + \ldots\ . \tag{6.114}$$

Moreover, the error of prediction $\widehat{y}_{t+k}(t)$ defined as

$$e_{t+k}(t) = y_{t+k} - \widehat{y}_{t+k}(t) \tag{6.115}$$

can be obviously expressed as

$$e_{t+k}(t) = \varepsilon_{t+k} + \psi_1 \varepsilon_{t+k-1} + \ldots + \psi_{k-1} \varepsilon_{t+1} \tag{6.116}$$

with zero mean value and variance

$$\mathrm{var}(e_{t+k}(t)) = \left(1 + \psi_1^2 + \ldots + \psi_{k-1}^2\right) \cdot \sigma^2. \tag{6.117}$$

Particularly, it holds

$$e_t(t-1) = y_t - \widehat{y}_t(t-1) = \varepsilon_t, \tag{6.118}$$

i.e., the white noise can be looked upon as the one-step-ahead prediction errors (this fact also justifies why one sometimes denotes the white noise as *innovation*; see also Remark 6.4).

So far we have dealt with theoretical features of predictions. Now we will show how to construct predictions according to Box–Jenkins methodology in reality. As it holds

$$y_{t+k} = \varphi_1 y_{t+k-1} + \ldots + \varphi_p y_{t+k-p} + \varepsilon_{t+k} + \theta_1 \varepsilon_{t+k-1} + \ldots + \theta_q \varepsilon_{t+k-q}, \tag{6.119}$$

one can write (due to linearity of predictions)

$$\widehat{y}_{t+k}(t) = \varphi_1 \widehat{y}_{t+k-1}(t) + \ldots + \varphi_p \widehat{y}_{t+k-p}(t) + \widehat{\varepsilon}_{t+k}(t) + \theta_1 \widehat{\varepsilon}_{t+k-1}(t) + \ldots \\ + \widehat{\theta}_q \varepsilon_{t+k-q}(t). \tag{6.120}$$

The relation (6.120) is basic one for real calculations of predictions since one can substitute

$$\widehat{y}_{t+j}(t) = y_{t+j} \quad \text{for } j \leq 0, \tag{6.121}$$

$$\widehat{\varepsilon}_{t+j}(t) = \begin{cases} 0 & \text{for } j > 0\ , \\ \varepsilon_{t+j} = y_{t+j} - \widehat{y}_{t+j}(t+j-1) & \text{for } j \leq 0\ . \end{cases} \tag{6.122}$$

Considering the previous relations, it is possible to formulate the following algorithm for real calculations of predictions (this algorithm can be also used in the case of nonstationary models of the type ARIMA and SARIMA):

(a) One proceeds recursively from a suitable time t, i.e., at first we construct one-step-ahead predictions

$$\widehat{y}_{t+1}(t),\ \widehat{y}_{t+2}(t+1), \ldots,$$

then two-step-ahead predictions (by means of one-step-ahead predictions)

$$\widehat{y}_{t+2}(t),\ \widehat{y}_{t+3}(t+1),\ldots$$

etc., until we reach the prediction horizon and the prediction time (mostly the end $t=n$ of observed time series), which correspond to the our real prediction problem.

(b) To realize (a) one makes use of the formula (6.120) with estimated parameters and substituting relations (6.121) and (6.122) (in order to start the recursive calculations, one must choose initial values, e.g., in the model MA(q) one can start with $\varepsilon_1 = \varepsilon_2 = \ldots = \varepsilon_q = 0$).

(c) One can also construct the interval predictions. For example assuming the normality, the 95 % prediction interval can be approximated by means of (6.117) as

$$\left(\widehat{y}_{t+k}(t) - 2\widehat{\sigma}\left(1+\sum_{j=1}^{k-1}\widehat{\psi}_j^2\right)^{1/2},\quad \widehat{y}_{t+k}(t) + 2\widehat{\sigma}\left(1+\sum_{j=1}^{k-1}\widehat{\psi}_j^2\right)^{1/2}\right). \tag{6.123}$$

Remark 6.22 Let us stress once more that in practice one substitutes to the prediction formulas the estimated parameter (see, e.g., (6.123)). Fortunately in routine situations, the predictions remain after such a substitution acceptable (particularly for longer time series).

◇

Example 6.8 This example demonstrates how to construct predictions in three estimated models of different types:

1. *Stationary* AR(1) *process with deterministic linear trend*:

$$y_t - 9.58 - 1.75\cdot t = 0.48(y_{t-1} - 9.58 - 1.75\cdot(t-1)) + \varepsilon_t$$

(see ((6.79) and ((6.80)); one calculates step by step:

$$\begin{aligned}
\widehat{y}_2(1) &= 9.58 + 1.75\cdot 2 + 0.48(y_1 - 9.58 - 1.75\cdot 1) = 7.64 + 0.48y_1\\
\widehat{y}_3(2) &= 9.58 + 1.75\cdot 3 + 0.48(y_2 - 9.58 - 1.75\cdot 2) = 8.55 + 0.48y_2\\
&\vdots\\
\widehat{y}_3(1) &= 9.58 + 1.75\cdot 3 + 0.48(\widehat{y}_2(1) - 9.58 - 1.75\cdot 2) = 8.55 + 0.48\widehat{y}_2(1)\\
\widehat{y}_4(2) &= 9.58 + 1.75\cdot 4 + 0.48(\widehat{y}_3(2) - 9.58 - 1.75\cdot 3) = 9.46 + 0.48\widehat{y}_3(2)\\
&\vdots
\end{aligned}$$

2. *Process* ARIMA(0, 1, 1):

$$y_t = y_{t-1} + \varepsilon_t + 0.39\varepsilon_{t-1};$$

one again calculates step by step (for $\varepsilon_1 = 0$):

$$\begin{aligned}
&\widehat{y}_2(1) = y_1 + 0.39\varepsilon_1 = y_1\\
&\widehat{y}_3(2) = y_2 + 0.39(y_2 - \widehat{y}_2(1))\\
&\vdots\\
&\widehat{y}_3(1) = \widehat{y}_2(1)\\
&\widehat{y}_4(2) = \widehat{y}_3(2)
\end{aligned}$$

(obviously, it holds in general $\widehat{y}_{t+k}(t) = \widehat{y}_{t+1}(t)$ for $k \geq 1$).

3. *Process* SARIMA $(0, 0, 1) \times (0, 0, 1)_4$:

$$y_t = \varepsilon_t - 0.4\varepsilon_{t-1} + 0.5\varepsilon_{t-4} - 0.2\varepsilon_{t-5};$$

one calculates step by step (for $\varepsilon_1 = \ldots = \varepsilon_5 = 0$):

$$\begin{aligned}
&\widehat{y}_6(5) = 0\\
&\widehat{y}_7(6) = -0.4\big(y_6 - \widehat{y}_6(5)\big) = -0.4y_6\\
&\widehat{y}_8(7) = -0.4\big(y_7 - \widehat{y}_7(6)\big) = -0.4y_7 - 0.16y_6\\
&\vdots\\
&\widehat{y}_7(5) = \widehat{y}_8(6) = \widehat{y}_9(7) = 0\\
&\widehat{y}_{10}(8) = 0.5(y_6 - \widehat{y}_6(5)) = 0.5y_6\\
&\vdots
\end{aligned}$$

As the interval predictions are concerned, e.g., in the second example (i.e., for ARIMA) one can write

$$y_{t+k} - y_t = \varepsilon_{t+k} + (1 + 0.39)(\varepsilon_{t+k-1} + \ldots + \varepsilon_{t+1}) + 0.39\varepsilon_t,$$

so that

$$\widehat{\sigma}(e_{t+k}(t)) = \left(1 + (k-1)(1 + 0.39)^2 + 0.39^2\right)^{1/2} \widehat{\sigma},$$

and hence the 95 % prediction interval can be approximated by

$$\big(\widehat{y}_{t+k}(t) - 2\widehat{\sigma}(e_{t+k}(t)), \widehat{y}_{t+k}(t) + 2\,\widehat{\sigma}(e_{t+k}(t))\big).$$

◇

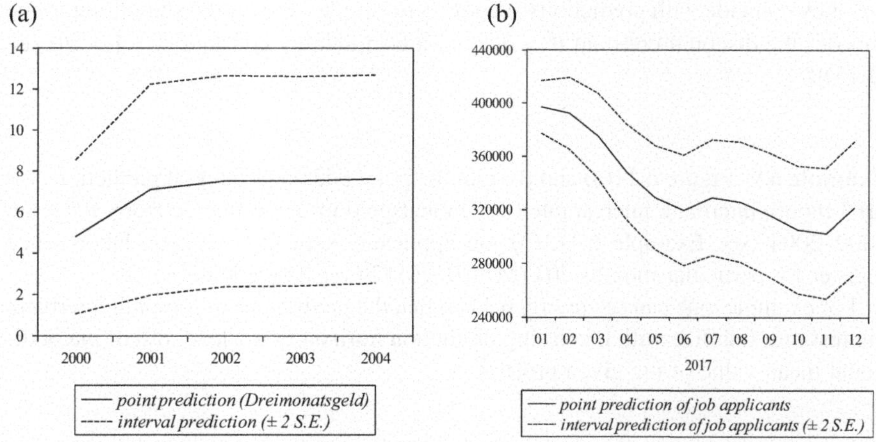

Fig. 6.11 Point and 95 % interval predictions from Example 6.9 for (**a**) 3-month interbank interest rate (Dreimonatsgeld in % p.a.) in Germany for years 2000–2004 (see also Example 6.1); (**b**) job applicants kept in the Czech labor office register for particular months 2017M1-2017M12 (see also Example 6.7) calculated by means of EViews

Remark 6.23 The behavior of predictions is different for stationary and nonstationary processes. If the prediction horizon increases in a stationary process, then the prediction converges to the mean value of the process (*mean reversion*) and the variance of prediction error converges to the variance of process: $\widehat{y}_{t+k}(t) \rightarrow \mathrm{E}(y_t) = \mu$ in the sense of convergence in mean square (it follows from (6.114) for $k \rightarrow \infty$ where $\mu = 0$), $\mathrm{var}(e_{t+k}(t)) \rightarrow \mathrm{var}(y_t) = \sigma^2$ in nondecreasing way (it follows from (6.116) again for $k \rightarrow \infty$). On the contrary in nonstationary processes (e.g., ARIMA), if the prediction horizon increases then the width of prediction horizon grows to infinity: $\mathrm{var}(e_{t+k}(t)) \rightarrow \infty$. Therefore, the applicability of these predictions is more and more dubious for $k \rightarrow \infty$) and also the (unconditional) variance of such processes is infinite (i.e., y_t can attain any real value for sufficiently large t). In any case, the prediction band composed from particular prediction intervals has a "funnel" shape (see, e.g., Fig. 6.11).

◇

Remark 6.24 If we write an estimated model ARIMA(0,1,1) as

$$\Delta y_t = \varepsilon_t - \widehat{\theta}\varepsilon_{t-1}, \tag{6.124}$$

then the predictions have the form

$$\widehat{y}_{t+1}(t) = \left(1 - \widehat{\theta}\right) \sum_{j=0}^{\infty} \widehat{\theta}^j y_{t-j}, \tag{6.125}$$

i.e., they coincide with predictions according to simple exponential smoothing if one chooses the discount constant $\beta = \widehat{\theta}_1$ (i.e., the smoothing constant $\alpha = 1 - \widehat{\theta}_1$; see (3.75)).

◇

Example 6.9 Figure 6.11(a) and (b) plot the point and 95 % interval predictions for (a) 3-month interbank interest rate (Dreimonatsgeld in % p.a.) in Germany for years 2000–2004 (see Example 6.1); (b) job applicants kept in the Czech labor office register for particular months 2017M1-2017M12 (see Example 6.7).

For example, one can see in Fig. 6.11(a) that the predictions of 3-month interbank interest rate stabilize with increasing prediction horizon to the level 8% of unconditional mean value of the given process.

◇

6.7 Long Memory Process

Some financial (but also economic or hydrologic) data remain autocorrelated even over very long time distances. Strictly speaking, their estimated correlogram and partial correlogram decrease hyperbolically (by a polynomial rate; see (6.132)). Such a rate lies between a very slow linear decrease for the processes ARIMA with some characteristic roots nearly on the border of unit circle and a very fast exponential decrease for the stationary processes ARMA. The time series of this type are usually called *long-memory process* (or *persistent process*). A successful method how to model it consists in so-called *fractional differencing* (see, e.g., Hurst (1951) in hydrology and Granger (1980) in economy).

The simplest example of a long-memory process is the *fractionally integrated process of order d* (*d* is a non-integer) denoted by acronym FI(d) of the form

$$(1 - B)^d y_t = \varepsilon_t \quad \text{or} \quad \Delta^d y_t = \varepsilon_t. \tag{6.126}$$

For this process, one should distinguish the following cases:

1. $d < 0.5$: The process is (weakly) stationary with representation in the form of linear process

$$y_t = \varepsilon_t + \sum_{i=1}^{\infty} \psi_i \varepsilon_{t-i}, \tag{6.127}$$

where

$$\psi_i = \frac{d(d+1)\cdot \ldots \cdot (d+i-1)}{i\ !} \tag{6.128}$$

(it follows from the extension of $(1-z)^{-d}$ to power series). The process is nonstationary for $d \geq 0.5$.

2. $d > -0.5$: The process is invertible with representation in the form of (infinite) autoregressive process

$$y_t = \sum_{i=1}^{\infty} \pi_i y_{t-i} + \varepsilon_t, \tag{6.129}$$

where

$$\pi_i = (-1)^i \frac{d(d-1)\cdot \ldots \cdot (d-i+1)}{i\ !} \tag{6.130}$$

(it follows from the extension of $(1-z)^{d}$ to power series). The process is noninvertible for $d \leq -0.5$.

3. $-0.5 < d < 0.5$: The process is stationary and invertible with autocorrelation and partial autocorrelation function of the form

$$\rho_k = \frac{d(d+1)\cdot \ldots \cdot (d+k-1)}{(1-d)(2-d)\cdot \ldots \cdot (k-d)},\ \rho_{kk} = \frac{d}{k-d},\ k = 1, 2, \ldots . \tag{6.131}$$

Particularly for large lags k it holds

$$\psi_k = O\left(k^{d-1}\right),\ \pi_k = O\left(k^{-d-1}\right),\ \rho_k = O\left(k^{2d-1}\right). \tag{6.132}$$

4. $0 < d < 0.5$: The autocorrelation function is positive and fulfills

$$\sum_{k=1}^{\infty} \rho_k = \infty. \tag{6.133}$$

Particularly in this case one uses explicitly the attribution *long-memory process* (or *persistent process*). The partial sums $y_1 + \ldots + y_t$ grow with a quicker rate than the linear one.

5. –0.5 < *d* < 0: The autocorrelation function fulfills

$$\sum_{k=1}^{\infty} |\rho_k| < \infty. \tag{6.134}$$

In this case, one uses the attribution *intermediate-memory process* or *antipersistent process*. The partial sums $y_1 + \ldots + y_t$ grow with a slower rate than the linear one.

In general, one can apply model ARFIMA(p, d, q) (*autoregressive fractionally integrated moving average process of order p, d, q*) of the form

$$\varphi(B)w_t = \theta(B)\varepsilon_t \tag{6.135}$$

(i.e., a stationary and invertible model ARMA(p, q) for w_t), where w_t arises from the original process y_t as a fractionally integrated process of order d

$$w_t = (1 - B)^d y_t. \tag{6.136}$$

Remark 6.25 Sometimes the processes of the type or random walk

$$y_t = y_1 + \sum_{\tau=2}^{t} \varepsilon_\tau \tag{6.137}$$

(see (6.90)) are called *strong-memory processes* since they “remember” all last shocks ε_t, ε_{t-1}, ...

◇

Remark 6.26 A sudden structural break (i.e., an abrupt change of the model) within an observed time series may lead, when one fits a model, to a pseudo-long-memory behavior. Even a mean change of this type already pretends a long-memory behavior in sample (partial) correlograms. Particularly, it is typical for some financial time series which are modeled as long-memory processes since a structural break has been not taken into account.

◇

6.8 Exercises

Exercise 6.1 Repeat the analysis from Examples 6.1–6.4 and 6.9(a) (the stationary process of “Dreimonatsgeld”), but only for data since 1965 (*hint:* $y_t = 6.466 + 0.924y_{t-1} - 0.735y_{t-2} + 0.508y_{t-3} - 0.510y_{t-4} + \varepsilon_t$, $\widehat{\sigma} = 1.785$).

Exercise 6.2 Repeat the analysis from Examples 6.5 and 6.6 (the nonstationary index PX), but only for last 150 observations (*hint:* $\widehat{\sigma} = 8.33$).

Exercise 6.3 Repeat the analysis from Examples 6.7 and 6.9(b) (the seasonal process of job applicants), but only for data since 2010 (*hint:* $y_t - 0.98y_{t-1} - y_{t-12} + 0.98y_{t-13} = 193\ 593.5 - 2\ 684.2 \cdot t + \varepsilon_t, \widehat{\sigma} = 7\ 267.96$).

Chapter 7
Autocorrelation Methods in Regression Models

7.1 Dynamic Regression Model

Box–Jenkins methodology is often applied in the context of so-called *dynamic regression models* with dynamics of explanatory variables on the right-hand side of model (including lagged values of the explained variable y from the left-hand side of model) and the residual component u with a dynamic correlation structure (including ARMA structure), e.g.,

$$\Delta y_t = \beta_1 + \beta_2 \Delta x_{t2} + \beta_3 x_{t3} + \beta_4 x_{t-1,2} + \beta_5 y_{t-1} + u_t. \tag{7.1}$$

Formally, one can write the (linear) dynamic regression model for *explained variable* y_t as

$$y_t = \beta_1 + \beta_2 x_{t2} + \ \dots \ + \beta_k x_{tk} + u_t, \tag{7.2}$$

where all *explanatory variables* x_{ti} are orthogonal to the residual u_t in the same time t, i.e., $\text{cov}(x_{ti}, u_t) = 0$ (so-called *simultaneous uncorrelatedness*). Such a condition of orthogonality guarantees some useful statistical properties of the model, e.g., OLS estimates of regression parameters in (7.2) are consistent under this condition. Dynamic regression models are broadly used in econometric modeling, where the orthogonal explanatory variables x_{ti} may be either (*strictly*) *exogenous* (i.e., originating outside the model (7.2)) or *predetermined* (i.e., originating within the model, but in a past time viewed from the perspective of present time t, e.g., originating in time $t - 1$).

More specifically, one usually assumes that the residual component u_t is modeled as an ARMA process

T. Cipra, *Time Series in Economics and Finance*,
https://doi.org/10.1007/978-3-030-46347-2_7

$$u_t = \varphi_1 u_{t-1} + \ldots + \varphi_p u_{t-p} + \varepsilon_t + \theta_1 \varepsilon_{t-1} + \ldots + \theta_q \varepsilon_{t-q}, \tag{7.3}$$

where ε_t is white noise with variance σ^2 and $\beta_1, \ldots, \beta_k, \varphi_1, \ldots, \varphi_p, \theta_1, \ldots, \theta_q, \sigma^2$ are (unknown) parameters.

One should stress that the typical feature of dynamic regression models is the exploitation of time lagged (delayed) variables. It has practical reasons, e.g.:

- *Decelerated responses to changes*: Some economic and financial variables change slowly so that a response to the changes of this type (e.g., changes in the structure of financial markets, government politics, bank strategy) is measurable with a substantial time delay and not within the same time period. In the economic and financial context, the reasons of such delay can be manifold:
  - Psychological: e.g., market subjects disbelieve at first new messages or underestimate their consequences.
  - Technological: e.g., the speed of transactions depends on technical facilities of financial exchanges.
  - Due to liquidity: e.g., new investment positions cannot be open until the old ones are closed (or sold), or until a necessary capital is available.

  Moreover, the speed and intensity of responses depend on the character of changes, e.g., whether the changes are permanent or transient. In any case, a complex dynamic structure complicates the model interpretation.
- *Overreaction to changes*: Sometimes pessimistic economic prognoses cause immediate decreases of prices which are stabilized later (i.e., with a time gap) as soon as the real results are announced.
- *Modeling autocorrelated residuals*: The application of dynamic models instead of static ones can sometimes remove the problem of autocorrelated residuals (see Sect. 7.2).

In this chapter, we deal with several special cases of dynamic regression models which are important for economic and financial time series, namely:

- *Linear regression model with autocorrelated residuals*: does not contain any lagged variables (neither explanatory nor explained), but the delay is comprised in the residual component (see Sect. 7.2).
- *Distributed lag model*: contains the lagged explanatory variables but no lagged explained variable (see Sect. 7.3).
- *Autoregressive distributed lag model*: contains the lagged explained variable (and possibly also lagged explanatory variables (see Sect. 7.4)).

7.2 Linear Regression Model with Autocorrelated Residuals

Linear regression model with autocorrelated residuals is usually the classical regression

$$y_t = \beta_1 + \beta_2 x_{t2} + \ \ldots \ + \beta_k x_{tk} + u_t \tag{7.4}$$

with the ARMA structure (7.3) of residuals u_t, but in contrast to (7.2) the explanatory variables x_{ti} are looked upon as *deterministic regressors*. It is a popular generalization of linear regression where the residual component has the form of uncorrelated white noise to the case with correlated observations. Such a correlatedness must be taken into account since it is usual in practice (e.g., the delayed values of some variables, which should be included among regressors of (7.4), are present only in the residuals u_t causing correlatedness in time).

The simplest type of correlatedness covering majority of routine situations consists in modeling the residual component u_t by means of the stationary autoregressive model of the first order (see the process AR(1) in Remark 6.9) written as

$$u_t = \rho u_{t-1} + \varepsilon_t, \tag{7.5}$$

where the autoregressive parameter ρ ($-1 < \rho < 1$) equals the first autocorrelation of the process u_t (it is denoted ρ instead of ρ_1 for simplicity) and ε_t is white noise. The sign of ρ plays an important role here: the positive $\rho > 0$ (so-called *positive correlatedness* plotted for a trajectory of residuals u_t in the scatterplot on the left-hand side of Fig. 7.1) induces the inertia for the signs of neighboring values u_t (see the right-hand side of Fig. 7.1 with a relatively rare crossing of time axis), while on the contrary the negative $\rho < 0$ (so-called *negative correlatedness* plotted in the scatterplot on the left-hand side of Fig. 7.2) induces frequent changes of the signs of neighboring values u_t (see the right-hand side of Fig. 7.2 with a relatively dense crossing of time axis).

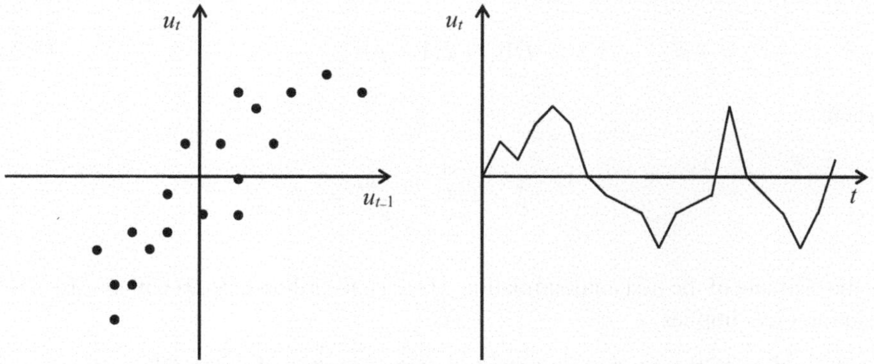

Fig. 7.1 Positive autocorrelatednes ($\rho > 0$)

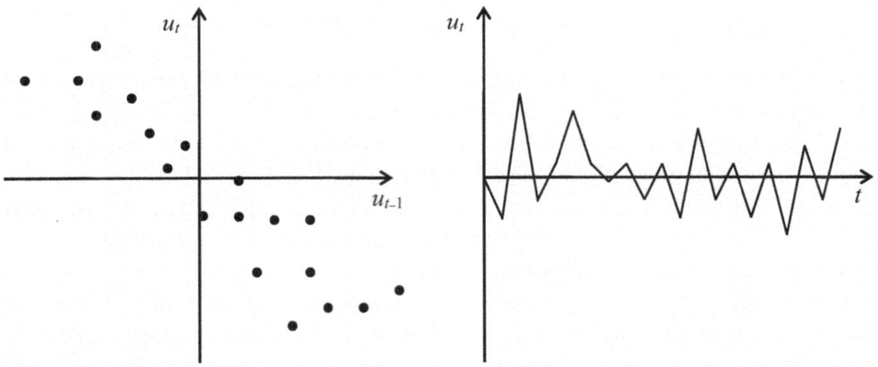

Fig. 7.2 Negative autocorrelatedness ($\rho < 0$)

7.2.1 Durbin–Watson Test

Durbin–Watson test (also called the *test of autocorrelatedness of residuals*) is one of the most frequent tests in regression analysis. In contrast to the subjective graphical instruments in Figures 7.1 and 7.2 (with the estimated OLS-residuals $\widehat{u}_t$, i.e., the residuals calculated by the method of ordinary least squares in the linear regression model (7.4)), it is the statistical test with null hypothesis H_0: $\rho = 0$ in (7.5). Its test statistics has the form

$$DW = \frac{\sum_{t=2}^{n}(\widehat{u}_t - \widehat{u}_{t-1})^2}{\sum_{t=1}^{T}\widehat{u}_t^2}. \tag{7.6}$$

Obviously, it can be approximated as

$$DW \approx 2(1 - \widehat{\rho}), \tag{7.7}$$

where

$$\widehat{\rho} = \frac{\sum_{t=2}^{n}\widehat{u}_{t-1}\widehat{u}_t}{\sum_{t=1}^{T}\widehat{u}_t^2} \tag{7.8}$$

is the estimate of the first autocorrelation ρ (see r_1 according to (6.9) with $\overline{u} = 0$). The relation (7.7) implies:

- If $\widehat{\rho} \approx 0$ (i.e., the neighboring residuals are uncorrelated), then $DW \approx 2$.
- If $\widehat{\rho} \approx 1$ (i.e., the neighboring residuals are extremely positively correlated), then $DW \approx 0$.
- If $\widehat{\rho} \approx -1$ (i.e., the neighboring residuals are extremely negatively correlated), then $DW \approx 4$.

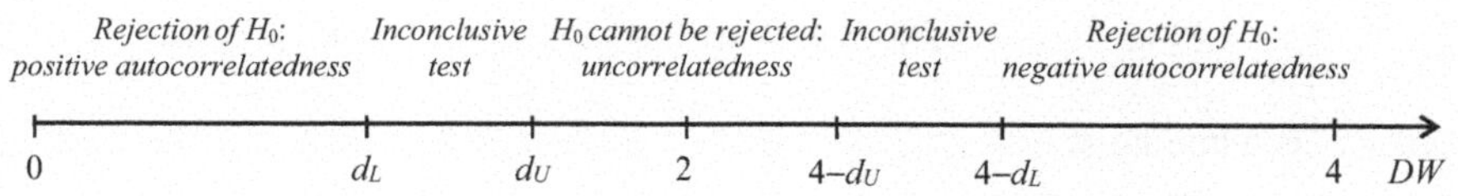

Fig. 7.3 Conclusions of Durbin–Watson test corresponding to particular values of statistics *DW*

The statistics *DW* does not have any standard probability distribution. However, if one assumes the normality of white noise ε_t, then *DW* has two critical values d_L (lower) and d_U (upper), which depend only on the number of observations n and regressors k (but not on the values of these regressors). Nowadays, DW test is mainly used as an informal instrument indicating possibility of existence of autocorrelated residuals. Its conclusions on the null hypothesis H_0: $\rho = 0$ are summarized in Fig. 7.3 (however, the test is inconclusive for some values of the statistics *DW*, so that no conclusions are possible in such a case):

The critical values d_L and d_U can be found in statistical tables or they are calculated by means of simulations directly in software systems (in the form of *p*-values). Moreover, in practice one can apply simplified rules ("rules of thumb"): e.g., if one has more observations n (more than fifty) and the number of regressors k is not too high, then the value of *DW* lower than 1.5 usually implies the positive autocorrelatedness (see Example 7.1).

7.2.2 *Breusch–Godfrey Test*

Later some more general tests have been also suggested that enable to detect even higher order autocorrelations than the order one in DW test (of course, it would be possible to try sequentially in the numerator of *DW* statistics (7.6) various differences of non-neighboring OLS residuals, but such an approach is too tedious). The other alternative is to apply procedures suggested originally for verification of Box–Jenkins models, mainly *Q*-tests (e.g., Box–Pierce test or Ljung–Box test; see (6.66) and (6.67)).

Nowadays in the context of regression models of the type (7.4), econometric software systems offer particularly *Breusch–Godfrey test* of autocorrelated residuals. It was suggested for the alternative hypothesis that the residual component u_t is the autoregressive model AR(p) of a higher order $p \geq 1$. The BG test proceeds in the following way:

1. One calculates OLS residuals $\widehat{u}_t$ in the model (7.4) (i.e., by the classical method of least squares in the same way as for DW test).
2. One estimates an auxiliary model

$$\widehat{u}_t = \gamma_1 + \gamma_2 x_{t2} + \ldots + \gamma_k x_{tk} + \varphi_1 \widehat{u}_{t-1} + \varphi_2 \widehat{u}_{t-2} + \ldots + \varphi_p \widehat{u}_{t-p} + \varepsilon_t. \tag{7.9}$$

where ε_t is white noise.

3. One tests

$$\begin{aligned} \mathrm{H}_0 : \varphi_1 = \varphi_2 = \ldots = \varphi_p = 0 \ \text{ against } \ \mathrm{H}_1 : \varphi_1 \neq 0 \ \text{or} \\ \varphi_2 \neq 0 \ \text{or} \ \ldots \ \text{or} \ \varphi_p \neq 0 \end{aligned} \tag{7.10}$$

applying the classical *F*-test in the model (7.10). In particular, the critical value of this test with significance level α is the quantile $F_{1-\alpha}(p, n{-}k{-}p)$ of *F*-distribution. Other tests instead of *F*-test, e.g., LM test (Lagrange multiplier), are also possible (see Example 7.1).

The problematic point of BG test is how to choose the autoregressive order *p*. A simple recommendation is the choice corresponding to the frequency of data (e.g., $p = 4$ for quarterly observations, $p = 12$ for monthly observations, etc.), since the residual component is usually correlated mainly with the residual component for the same seasonal period of previous year. On the other hand, if the model is adequate from the statistical point of view, then no autocorrelations should be significant for arbitrary choice of *p*.

7.2.3 Construction of Linear Regression Model with ARMA Residuals

Let us return to the construction of model (7.4) with autocorrelated residuals. If DW test confirmed the correlation structure in the form of process AR(1) (i.e., the simplest case (7.5) of correlatedness), then one can estimate the linear regression (7.4) by means of *Cochrane–Orcutt method*, which has been very popular in practice. This method makes use of so-called *Koyck transformation*, when one subtracts from the regression equation (7.4) in time *t* the same equation in time $t - 1$ multiplied by the constant ρ:

$$y_t - \rho y_{t-1} = (1-\rho)\beta_1 + \beta_2(x_{t2} - \rho x_{t-1,2}) + \ \ldots \ + \beta_k(x_{tk} - \rho x_{t-1,k}) + \varepsilon_t \tag{7.11}$$

(obviously, $\varepsilon_t = u_t - \rho u_{t-1}$; see (7.5)). Should the value of ρ be known, then one would obtain by means of Koyck transformation the classical linear regression

$$y_t^* = \beta_1^* + \beta_2 x_{t2}^* + \ \ldots \ + \beta_k x_{tk}^* + \varepsilon_t, \tag{7.12}$$

where $y_t^* = y_t - \rho y_{t-1}$, $\beta_1^* = (1-\rho)\beta_1$, $x_{t2}^* = x_{t2} - \rho x_{t-1,2}$, $\ldots$, $x_{tk}^* = x_{tk} - \rho x_{t-1,k}$. This fact is the principle of Cochrane–Orcutt method that is phased in the following steps:

1. One calculates the OLS residuals $\widehat{u}_t$ in the model (7.4).
2. One constructs the estimate $\widehat{\rho}$ of the parameter ρ according to (7.8).
3. One constructs the OLS estimate in the model (7.12) replacing the parameter ρ by the estimate $\widehat{\rho}$.
4. The procedure goes on iteratively by repeating the steps 1 to 3, where in the step 1 one applies the OLS residuals calculated by means of the OLS estimate from the previous step 3. The procedure stops finally using a suitable stopping rule (e.g., if the change in the estimated value of ρ between neighboring iteration cycles drops under a limit fixed in advance).

The disadvantage of Cochrane–Orcutt method consists mainly in the fact that it delivers estimated parameters of the transformed model (7.12). Some software systems (e.g., EViews in Example 7.1) enable to return to the estimated parameters in the original model (7.4) (i.e., before Koyck transformation) by testing the parametric constraints following from this transformation.

If we consider the linear regression model (7.4) with general ARMA structure (7.3) of residuals u_t (and not specifically AR(1)), then there are various sophisticated method for its construction (e.g., the two-stage estimation procedures using the concept of instrumental variables or other methods; see EViews).

On the other hand, an opposite approach ignoring the residual correlations is also possible. Namely, if we apply the classical OLS methodology directly to the model (7.4) (i.e., ignoring the fact that u_t need not be white noise), then the corresponding OLS estimates of parameters $\beta_1, \ldots, \beta_k$ are not the best linear unbiased estimates, but remain consistent (i.e., for large n they are near to the true theoretical values of these parameters with a high probability). The only weak point of OLS estimates if they are used in the models with autocorrelated residuals consists in underestimating their errors (it can, e.g., impair t-tests of parameter significance). The remedy of this weakness is the application of *Newey–West estimate* of the error matrix of OLS estimates of parameters $\beta_1, \ldots, \beta_k$ denoted in software as HAC (*heteroscedasticity and autoregression consistent covariances*; see, e.g., EViews in Example 7.1).

Example 7.1 Table 7.1 presents the values AAA_t and $TBILL_t$ of (average) yields to maturity (YTM in % p.a.) of corporate bonds of the highest quality AAA and 3-month T-bills according to S&P in the USA for particular quarters 1990–1994 ($t = 1$, ..., 20). These are two alternative ways of risk-free investing so that one should expect that yields AAA_t depend positively on short-term interest rates $TBILL_t$ in time. This expectation is confirmed in Table 7.2 for model

$$AAA_t = \beta_1 + \beta_2 TBILL_t + \varepsilon_t, \qquad t = 1, \ \ldots, \ 20 \tag{7.13}$$

(the estimated model is highly significant according to t-ratios and F-test for the coefficient of determination R^2 with significantly positive estimate $b_2 = 0.426$ of the

Table 7.1 Quarterly data 1990–1994 in Example 7.1 (*yields to maturity of corporate bonds of the highest quality AAA and three-month T-bills in the USA in % p.a.*)

t	Year	Quarter	AAA_t	$TBILL_t$
1	1990	1	9.37	7.76
2	1990	2	9.26	7.77
3	1990	3	9.56	7.49
4	1990	4	9.05	7.02
5	1991	1	8.93	6.05
6	1991	2	9.01	5.59
7	1991	3	8.61	5.41
8	1991	4	8.31	4.58
9	1992	1	8.20	3.91
10	1992	2	8.22	3.72
11	1992	3	7.92	3.13
12	1992	4	7.98	3.08
13	1993	1	7.58	2.99
14	1993	2	7.33	2.98
15	1993	3	6.66	3.02
16	1993	4	6.93	3.08
17	1994	1	7.48	3.25
18	1994	2	7.97	4.04
19	1994	3	8.34	4.51
20	1994	4	8.46	5.28

Source: FRED (Federal Reserve Bank of St. Louis) (https://fred.stlouisfed.org/graph/?id=AAA, https://fred.stlouisfed.org/graph/?id=TB3MS,TB3MA)

Table 7.2 Estimation of the model (7.13) from Example 7.1 (*yields to maturity of corporate bonds AAA*) calculated by means of EViews

Dependent Variable: AAA				
Method: Least Squares				
Sample: 1 20				
Included observations: 20				
Variable	Coefficient	Std. Error	t-Statistic	Prob.
C	6.242529	0.230299	27.10622	0.0000
TBILL	0.425939	0.045876	9.284506	0.0000
R-squared	0.827259	F-statistic		86.20205
S.E. of regression	0.343246	Prob (F-statistic)		0.000000
Durbin–Watson stat	0.778482			

Source: Calculated by EViews

parameter β_2; see Table 7.2). The strong positive autocorrelatedness between ε_{t-1} and ε_t is also demonstrated by means of the scatterplot in Fig. 7.4 (the corresponding correlation coefficient is 0.602).

As statistical tests are concerned, the rule of thumb in the framework of Durbin–Watson test confirms the positive autocorrelatedness, since $DW = 0.778$ is much lower than 1.5 (see Table 7.2). The results of Breusch–Godfrey test with the residual

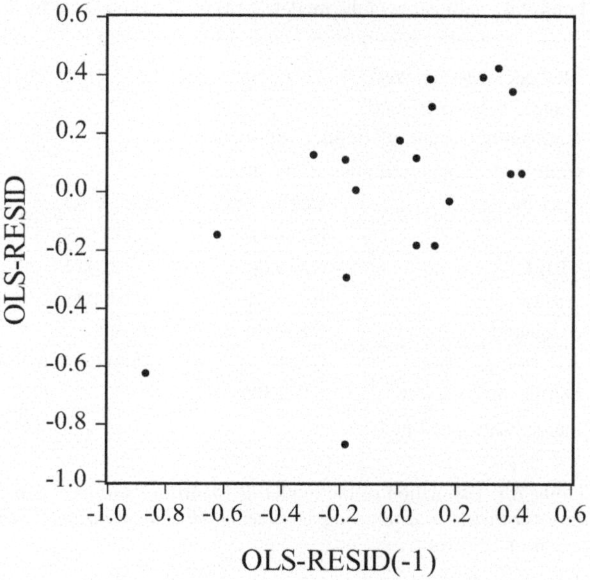

Fig. 7.4 Scatterplot demonstrating positive autocorrelatedness of residuals in Example 7.1 (*yields to maturity of corporate bonds AAA*)

Table 7.3 Breusch–Godfrey test of the model (7.13) with $p = 1$ from Example 7.1 (*yields to maturity of corporate bonds AAA*) calculated in EViews

Breusch–Godfrey Serial Correlation F- and LM Test:			
F-statistic	9.759497	Prob. F(1,17)	0.006178
Obs$*R$-squared	7.294230	Prob. Chi-Square(1)	0.006918

Source: Calculated by EViews

autoregressive model of order $p = 1$ in Table 7.3 (both in the form of F-test, and in the form of LM test) give the same conclusion.

The final estimate of the identified model

$$AAA_t = \beta_1 + \beta_2 TBILL_t + u_t, \qquad u_t = \rho u_{t-1} + \varepsilon_t \tag{7.14}$$

has been obtained by Cochrane–Orcutt method (see Table 7.4).

Table 7.5 presents the estimate of the same model obtained applying HAC approach of Newey–West. In comparison with Table 7.2, the estimated standard deviations of OLS estimates $b_1 = 6.243$ and $b_2 = 0.426$ are much higher (compare 0.377 with 0.230 for b_1 and 0.061 with 0.046 for b_2) so that the corresponding t-ratios are lower according to Newey–West.

◊

Table 7.4 Estimation of the model (7.14) from Example 7.1 by means of Cochrane–Orcutt method (*yields to maturity of corporate bonds AAA*) calculated in EViews

Method: Least Squares				
Sample (adjusted): 2 20				
Included observations: 19 after adjustments				
Convergence achieved after 10 iterations				
Variable	Coefficient	Std. Error	*t*-Statistic	Prob.
C	6.246693	0.487125	12.82358	0.0000
TBILL	0.429785	0.105027	4.092135	0.0009
AR(1)	0.601729	0.199094	3.022333	0.0081
R-squared	0.880105	*F*-statistic		58.72528
S.E. of regression	0.286794	Prob (*F*-statistic)		0.000000
Durbin–Watson stat	1.722505			

Source: Calculated by EViews

Table 7.5 Estimation of the model (7.14) from Example 7.1 including Newey–West estimate of the error matrix of estimated parameters (*yields to maturity of corporate bonds AAA*) calculated by means of EViews

Dependent Variable: AAA				
Method: Least Squares				
Sample: 1 20				
Included observations: 20				
Newey-West HAC Standard Errors and Covariance (lag truncation = 2)				
Variable	Coefficient	Std. Error	*t*-Statistic	Prob.
C	6.242529	0.376704	16.57143	0.0000
TBILL	0.425939	0.061072	6.974403	0.0000
R-squared	0.827259	*F*-statistic		86.20205
S.E. of regression	0.343246	Prob (*F*-statistic)		0.000000
Durbin–Watson stat	0.778482			

Source: Calculated by EViews

7.3 Distributed Lag Model

Distributed lag model (or *DL model*) contains lagged explanatory variables but no lagged explained variable (obviously, it fulfills in this way the condition that explanatory variables are orthogonal to the residual in the same time, the lagged regressors being predetermined in the previous times). For simplicity, we confine ourselves to the case with a single explanatory variable x and the residual component in the form of a white noise (the generalization for more lagged explanatory variables does not cause a serious complication):

$$y_t = \alpha + \sum\nolimits_{i=0}^{\infty} \beta_i x_{t-i} + \varepsilon_t. \tag{7.15}$$

In such a model, the influence of the explanatory variable is distributed to a large number of past time moments. In this context, one can distinguish:

- *Short-run effect* in consequence of an immediate change of the explanatory variable: it is described by the parameter β_0 (so-called *impact multiplier* or *short-run multiplier*).
- *Cumulative effect* in consequence of changes of the explanatory variable till a lag τ: it is described by the finite sum of parameters

$$\beta(\tau) = \sum_{i=0}^{\tau} \beta_i, \tag{7.16}$$

- *Long-run effect* in consequence of changes of the explanatory variable: it is described by the infinite sum of parameters (so-called *equilibrium multiplier* or *long-run multiplier* reflecting the influence of the explanatory variable after its transition to an equilibrium, i.e., to a balanced state)

$$\beta = \sum_{i=0}^{\infty} \beta_i. \tag{7.17}$$

Moreover, one usually calculates also other characteristics of lagged effects, e.g.:

$$\textit{median lag} = \text{smallest } q^* \text{ such that } \frac{\sum_{i=0}^{q*} \beta_i}{\sum_{i=0}^{\infty} \beta_i} \geq 0.5 \tag{7.18}$$

and

$$\textit{mean lag} = \frac{\sum_{i=0}^{\infty} i \cdot \beta_i}{\sum_{i=0}^{\infty} \beta_i}. \tag{7.19}$$

7.3.1 Geometric Distributed Lag Model

The DL model of the form (7.15) is too general to be applied in practice. Therefore, various modifications have been suggested. The *geometric distributed lag model* (or *GDL model*) is a very pragmatic solution since it uses only a finite number of parameters α, β, and λ:

$$y_t = \alpha + \beta \sum_{i=0}^{\infty} (1-\lambda)\lambda^i x_{t-i} \; + \; \varepsilon_t, \qquad 0 < \lambda < 1. \tag{7.20}$$

In this case, the long-run effect (7.17) equals directly the parameter β since

$$\beta\sum_{i=0}^{\infty}(1-\lambda)\lambda^i = \beta. \tag{7.21}$$

Applying Koyck transformation, when one subtracts from the regression equation (7.20) in time t the same equation in time $t-1$ multiplied by the constant λ (compare with (7.11)), one obtains

$$y_t = \alpha(1-\lambda) + \beta(1-\lambda)x_t + \lambda y_{t-1} + \eta_t, \quad \text{where} \quad \eta_t = \varepsilon_t - \lambda\varepsilon_{t-1}. \tag{7.22}$$

Even though the model (7.22) presents a substantial simplification in comparison with (7.20), the MA(1) residual structure of η_t (see (7.22)) can complicate the estimation procedures (the OLS estimates need not be consistent).

7.3.2 Polynomial Distributed Lag Model

Polynomial distributed lag model (or *PDL model*) was suggested by Almon (1965) as a special case of DL model with simplified expression of coefficients by means of polynomials. This approach can reduce substantially the number of parameters which must be estimated when constructing "trimmed" distributed lag model

$$y_t = \alpha + \sum_{i=0}^{k}\beta_i x_{t-i} + \varepsilon_t \tag{7.23}$$

(obviously, one must choose a priori an adequate length k of trimming). The PDL models suppose the possibility of approximation

$$\beta_i = \alpha_0 + \alpha_1 i + \alpha_2 i^2 + \ldots + \alpha_r i^r, \qquad i = 0, 1, \ldots, k, \tag{7.24}$$

where the order r of approximative polynomial is much lower than the maximum lag k ($r << k$). After substituting (7.24) into (7.23), one obtains

$$\begin{aligned} y_t &= \alpha + \alpha_0\sum_{i=0}^{k}x_{t-i} + \alpha_1\sum_{i=0}^{k}i\cdot x_{t-i} + \ldots + \alpha_r\sum_{i=0}^{k}i^r\cdot x_{t-i} + \varepsilon_t = \\ &= \alpha + \alpha_0 z_{0t} + \alpha_1 z_{1t} + \ldots + \alpha_r z_{rt} + \varepsilon_t, \end{aligned} \tag{7.25}$$

i.e., each z_{jt} is a linear combination of actual value and k lagged values x_t, x_{t-1}, ..., x_{t-k}. In the model (7.25), one can use mostly without problems the classical OLS methodology and then according to (7.24) find the corresponding estimates of original parameters β_i. Moreover, one can also include the constraint

$$\beta_{-1} = \alpha_0 + \alpha_1(-1) + \alpha_2(-1)^2 + \ldots + \alpha_r(-1)^r = 0 \tag{7.26}$$

ensuring the null influence of x_{t+1} on y_t (i.e., the null influence from the future time), or the constraint

$$\beta_{k+1} = \alpha_0 + \alpha_1(k+1) + \alpha_2(k+1)^2 + \ldots + \alpha_r(k+1)^r = 0 \tag{7.27}$$

ensuring the null influence of x_{t-k-1} on y_t (i.e., the null influence from the past time beyond the used trimming).

Example 7.2 Table 7.6 presents the values of money supply $M1_t$ and gross domestic product GDP_t (in billions of USD) in the USA for particular quarters 1950–2000 ($t = 1, \ldots, 204$). For these data, we shall estimate the DL model explaining the gross domestic product by means of lagged money supplies (since there is usually an inertia in the effect of M1):

$$\ln GDP_t = \alpha + \sum\nolimits_{i=0}^{4} \beta_i \ln M1_{t-i} + \varepsilon_t, \qquad t = 5, \ \ldots, \ 204. \tag{7.28}$$

The model (7.28) trimmed beyond the lag of four quarters is estimated in Table 7.7 (the coefficient of determination R^2 is relatively high, but the statistics DW near to zero indicates the strong positive autocorrelatedness). Hence the long-run effect of money supply on gross domestic product is

$$\beta_{DL} = \sum\nolimits_{i=0}^{4} \beta_i = 1.314 + \ldots + (-0.021) = 0.579.$$

Alternatively the corresponding geometric distributed lag model

$$\ln GDP_t = \alpha + \beta \sum\nolimits_{i=0}^{\infty} (1-\lambda)\lambda^i \ln M1_{t-i} + \varepsilon_t \tag{7.29}$$

rewritten by means of Koyck transformation to the form

$$\ln GDP_t = \alpha(1-\lambda) + \beta(1-\lambda)\ln M1_t + \lambda \ln GDP_{t-1} + \eta_t, \text{ where } \eta_t = \varepsilon_t - \lambda\varepsilon_{t-1} \tag{7.30}$$

is estimated in Table 7.8. In this case, the long-run effect of M1 on GDP works out

$$\beta_{GDL} = \frac{\beta_{GDL}(1-\lambda)}{1-\lambda} = \frac{0.004\ 530}{1-0.990\ 601} = 0.482.$$

Finally, Table 7.9 presents the results when applying PDL approach (7.25) to model (7.28) with a higher trimming lag $k = 12$ and lower order $r = 3$ of approximative polynomial. In the first part of Table 7.9, one estimates the model (7.25) (e.g., PDL01 is the regressor z_{0t}, etc.). In the second part of Table 7.9 one calculates according to (7.24) the estimates of original parameters β_i (till the lag of 12) including their graphical plot.

Table 7.6 Quarterly data 1950–2004 in Example 7.2 (*money supply M1 and gross domestic product GDP in the USA in billions of USD*)

Quarter	t	Year	GDP	M1	t	Year	GDP	M1	t	Year	GDP	M1
1	1	1950	1610.5	110.20	69	1967	3291.8	174.80	137	1984	5402.3	530.80
2	2	1950	1658.8	111.75	70	1967	3289.7	177.00	138	1984	5493.8	540.50
3	3	1950	1723.0	112.95	71	1967	3313.5	180.70	139	1984	5541.3	543.90
4	4	1950	1753.9	113.93	72	1967	3338.3	183.30	140	1984	5583.1	551.20
1	5	1951	1773.5	115.08	73	1968	3406.2	185.50	141	1985	5629.7	565.70
2	6	1951	1803.7	116.19	74	1968	3464.8	189.40	142	1985	5673.8	582.90
3	7	1951	1839.8	117.76	75	1968	3489.2	192.70	143	1985	5758.6	604.40
4	8	1951	1843.3	119.89	76	1968	3504.1	197.40	144	1985	5806.0	619.10
1	9	1952	1864.7	121.31	77	1969	3558.3	200.00	145	1986	5858.9	632.60
2	10	1952	1866.2	122.37	78	1969	3567.6	201.30	146	1986	5883.3	661.20
3	11	1952	1878.0	123.64	79	1969	3588.3	202.10	147	1986	5937.9	688.40
4	12	1952	1940.2	124.72	80	1969	3571.4	203.90	148	1986	5969.5	724.00
1	13	1953	1976.0	125.33	81	1970	3566.5	205.70	149	1987	6013.3	732.80
2	14	1953	1992.2	126.05	82	1970	3573.9	207.60	150	1987	6077.2	743.50
3	15	1953	1979.5	126.22	83	1970	3605.2	211.90	151	1987	6128.1	748.50
4	16	1953	1947.8	126.37	84	1970	3566.5	214.30	152	1987	6234.4	749.40
1	17	1954	1938.1	126.54	85	1971	3666.1	218.70	153	1988	6275.9	761.10
2	18	1954	1941.0	127.18	86	1971	3686.2	223.60	154	1988	6349.8	778.80
3	19	1954	1962.0	128.38	87	1971	3714.5	226.60	155	1988	6382.3	784.60
4	20	1954	2000.9	129.72	88	1971	3723.8	228.20	156	1988	6465.2	786.10
1	21	1955	2058.1	131.07	89	1972	3796.9	234.20	157	1989	6543.8	782.70
2	22	1955	2091.0	131.88	90	1972	3883.8	236.80	158	1989	6579.4	773.90
3	23	1955	2118.9	132.40	91	1972	3922.3	243.30	159	1989	6610.6	782.00
4	24	1955	2130.1	132.64	92	1972	3990.5	249.10	160	1989	6633.5	792.10
1	25	1956	2121.0	133.11	93	1973	4092.3	251.50	161	1990	6716.3	800.80
2	26	1956	2137.7	133.38	94	1973	4133.3	256.90	162	1990	6731.7	809.70

3	27	1956	2135.3	133.48	95	1973	4117.0	258.00	163	1990	6719.4	821.10
4	28	1956	2170.4	134.09	96	1973	4151.1	262.70	164	1990	6664.2	823.90
1	29	1957	2182.7	134.29	97	1974	4119.3	266.50	165	1991	6631.4	838.00
2	30	1957	2177.7	134.36	98	1974	4130.4	268.60	166	1991	6668.5	857.40
3	31	1957	2198.9	134.26	99	1974	4084.5	271.30	167	1991	6684.9	871.20
4	32	1957	2176.0	133.48	100	1974	4062.0	274.00	168	1991	6720.9	895.90
1	33	1958	2117.4	133.72	101	1975	4010.0	276.20	169	1992	6783.3	935.80
2	34	1958	2129.7	135.22	102	1975	4045.2	282.70	170	1992	6846.8	954.50
3	35	1958	2177.5	136.64	103	1975	4115.4	286.00	171	1992	6899.7	988.70
4	36	1958	2226.5	138.48	104	1975	4167.2	286.80	172	1992	6990.6	1024.00
1	37	1959	2273.0	139.70	105	1976	4266.1	292.40	173	1993	6988.7	1038.10
2	38	1959	2332.4	141.20	106	1976	4301.5	296.40	174	1993	7031.2	1075.30
3	39	1959	2331.4	141.00	107	1976	4321.9	300.00	175	1993	7062.0	1105.20
4	40	1959	2339.1	140.00	108	1976	4357.4	305.90	176	1993	7168.7	1129.20
1	41	1960	2391.0	139.80	109	1977	4410.5	313.60	177	1994	7229.4	1140.00
2	42	1960	2379.2	139.60	110	1977	4489.8	319.00	178	1994	7330.2	1145.60
3	43	1960	2383.6	141.20	111	1977	4570.6	324.90	179	1994	7370.2	1152.10
4	44	1960	2352.9	140.70	112	1977	4576.1	330.50	180	1994	7461.1	1149.80
1	45	1961	2366.5	141.90	113	1978	4588.9	336.60	181	1995	7488.7	1146.50
2	46	1961	2410.8	142.90	114	1978	4765.7	347.10	182	1995	7503.3	1144.10
3	47	1961	2450.4	143.80	115	1978	4811.7	352.70	183	1995	7561.4	1141.90
4	48	1961	2500.4	145.20	116	1978	4876.0	356.90	184	1995	7621.9	1126.20
1	49	1962	2544.0	146.00	117	1979	4888.3	362.10	185	1996	7676.4	1122.00
2	50	1962	2571.5	146.60	118	1979	4891.4	373.60	186	1996	7802.9	1115.00
3	51	1962	2596.8	146.30	119	1979	4926.2	379.70	187	1996	7841.9	1095.80
4	52	1962	2603.3	147.80	120	1979	4942.6	381.40	188	1996	7931.3	1080.50

(continued)

Table 7.6 (continued)

Quarter	t	Year	GDP	M1	t	Year	GDP	M1	t	Year	GDP	M1
1	53	1963	2634.1	149.20	121	1980	4958.9	388.10	189	1997	8016.4	1072.00
2	54	1963	2668.4	150.40	122	1980	4857.8	389.40	190	1997	8131.9	1066.20
3	55	1963	2719.6	152.00	123	1980	4850.3	405.40	191	1997	8216.6	1065.30
4	56	1963	2739.4	153.30	124	1980	4936.6	408.10	192	1997	8272.9	1073.40
1	57	1964	2800.5	154.50	125	1981	5032.5	418.70	193	1998	8396.3	1080.30
2	58	1964	2833.8	155.60	126	1981	4997.3	425.50	194	1998	8442.9	1077.60
3	59	1964	2872.0	158.70	127	1981	5056.8	427.50	195	1998	8528.5	1076.20
4	60	1964	2879.5	160.30	128	1981	4997.1	436.20	196	1998	8667.9	1097.00
1	61	1965	2950.1	161.50	129	1982	4914.3	442.40	197	1999	8733.5	1102.20
2	62	1965	2989.9	162.20	130	1982	4935.5	447.90	198	1999	8771.2	1099.80
3	63	1965	3050.7	164.90	131	1982	4912.1	457.50	199	1999	8871.5	1093.40
4	64	1965	3123.6	167.80	132	1982	4915.6	474.30	200	1999	9049.9	1124.80
1	65	1966	3201.1	170.50	133	1983	4972.4	490.20	201	2000	9102.5	1113.70
2	66	1966	3213.2	171.60	134	1983	5089.8	504.40	202	2000	9229.4	1105.30
3	67	1966	3233.6	172.00	135	1983	5180.4	513.40	203	2000	9260.1	1096.00
4	68	1966	3261.8	172.00	136	1983	5286.8	520.80	204	2000	9303.9	1088.10

Source: FRED (Federal Reserve Bank of St. Louis) (http://www.economagic.com/fedstl.htm#GDP, http://www.economagic.com/em-cgi/data.exe/frbH6/m1)

Table 7.7 Estimation of distributed lag model (7.28) from Example 7.2 (*gross domestic product GDP explained by lagged money supply M1*) calculated by means of EViews

Dependent Variable: LOG(GDP)				
Method: Least Squares				
Sample (adjusted): 5 204				
Included observations: 200 after adjustments				
Variable	Coefficient	Std. Error	*t*-Statistic	Prob.
C	4.946307	0.056813	87.06268	0.0000
LOG(M1)	1.313582	0.867809	1.513677	0.1317
LOG(M1(-1))	–0.406028	1.489386	–0.272614	0.7854
LOG(M1(-2))	–0.055179	1.493922	–0.036936	0.9706
LOG(M1(-3))	–0.252407	1.490153	–0.169383	0.8657
LOG(M1(-4))	–0.021173	0.878340	–0.024106	0.9808
R-squared	0.949165	*F*-statistic		724.4522
S.E. of regression	0.108728	Prob (*F*-statistic)		0.000000
Durbin–Watson stat	0.018483			

Source: Calculated by EViews

Table 7.8 Estimation of geometric distributed lag model (7.29) after Koyck transformation (7.30) from Example 7.2 (*gross domestic product GDP explained by lagged money supply M1*) calculated by means of EViews

Dependent Variable: LOG(GDP)				
Method: Least Squares				
Sample (adjusted): 2 204				
Included observations: 203 after adjustments				
Variable	Coefficient	Std. Error	*t*-Statistic	Prob.
C	0.060444	0.030959	1.952384	0.0523
LOG(M1)	0.004530	0.003765	1.203104	0.2304
LOG(GDP(-1))	0.990601	0.006235	158.8895	0.0000
R-squared	0.999584	*F*-statistic		240302.5
S.E. of regression	0.009932	Prob (*F*-statistic)		0.000000
Durbin–Watson stat	1.286477			

Source: Calculated by EViews

The sum of estimated parameters corresponding to the actual value and lagged values of ln*M1* in Table 7.9 (i.e., the long-run effect of money supply on gross domestic product) is $\beta_{PDL} = 0.565$.

◇

Table 7.9 Estimation of polynomial distributed lag model from Example 7.2 (*gross domestic product GDP explained by lagged money supply M1*) calculated by means of EViews

<table>
<tr><td colspan="6">Dependent Variable: LOG(GDP)</td></tr>
<tr><td colspan="6">Method: Least Squares</td></tr>
<tr><td colspan="6">Sample (adjusted): 13 204</td></tr>
<tr><td colspan="6">Included observations: 192 after adjustments</td></tr>
<tr><td>Variable</td><td colspan="2">Coefficient</td><td>Std. Error</td><td>t-Statistic</td><td>Prob.</td></tr>
<tr><td>C</td><td colspan="2">5.041623</td><td>0.054881</td><td>91.86544</td><td>0.0000</td></tr>
<tr><td>PDL01</td><td colspan="2">−0.193079</td><td>0.073166</td><td>−2.638932</td><td>0.0090</td></tr>
<tr><td>PDL02</td><td colspan="2">0.058884</td><td>0.090832</td><td>0.648271</td><td>0.5176</td></tr>
<tr><td>PDL03</td><td colspan="2">0.016896</td><td>0.005232</td><td>3.229455</td><td>0.0015</td></tr>
<tr><td>PDL04</td><td colspan="2">−0.003285</td><td>0.003699</td><td>−0.887938</td><td>0.3757</td></tr>
<tr><td>R-squared</td><td colspan="2">0.952468</td><td colspan="2">F-statistic</td><td>936.7900</td></tr>
<tr><td>Durbin–Watson stat</td><td colspan="2">0.016460</td><td colspan="2">Prob(F-statistic)</td><td>0.000000</td></tr>
<tr><td colspan="2">Lag Distribution of LOG(M1)</td><td>i</td><td>Coefficient</td><td>Std. Error</td><td>t-Statistic</td></tr>
<tr><td colspan="2">| *</td><td>0</td><td>0.77136</td><td>0.27859</td><td>2.76880</td></tr>
<tr><td colspan="2">| *</td><td>1</td><td>0.34548</td><td>0.06241</td><td>5.53531</td></tr>
<tr><td colspan="2">|*</td><td>2</td><td>0.05194</td><td>0.12823</td><td>0.40505</td></tr>
<tr><td colspan="2">* |</td><td>3</td><td>−0.12898</td><td>0.17439</td><td>−0.73960</td></tr>
<tr><td colspan="2">* |</td><td>4</td><td>−0.21699</td><td>0.15996</td><td>−1.35647</td></tr>
<tr><td colspan="2">* |</td><td>5</td><td>−0.23178</td><td>0.10951</td><td>−2.11655</td></tr>
<tr><td colspan="2">* |</td><td>6</td><td>−0.19308</td><td>0.07317</td><td>−2.63893</td></tr>
<tr><td colspan="2">* |</td><td>7</td><td>−0.12058</td><td>0.11146</td><td>−1.08182</td></tr>
<tr><td colspan="2">*|</td><td>8</td><td>−0.03400</td><td>0.16172</td><td>−0.21026</td></tr>
<tr><td colspan="2">|*</td><td>9</td><td>0.04695</td><td>0.17525</td><td>0.26793</td></tr>
<tr><td colspan="2">|*</td><td>10</td><td>0.10258</td><td>0.12791</td><td>0.80200</td></tr>
<tr><td colspan="2">| *</td><td>11</td><td>0.11317</td><td>0.06468</td><td>1.74959</td></tr>
<tr><td colspan="2">|*</td><td>12</td><td>0.05902</td><td>0.28403</td><td>0.20778</td></tr>
<tr><td colspan="3">Sum of Lags</td><td>0.56508</td><td>0.00954</td><td>59.2126</td></tr>
</table>

Source: Calculated by EViews

7.4 Autoregressive Distributed Lag Model

Autoregressive distributed lag model (or *ADL model*) contains both lagged explanatory variables and lagged explained variable. It can be looked upon as a special (linear) filtering scheme (therefore, it is sometimes also called *transfer function model*, since transfer functions are typical concepts in filtering theories). This model can be formally written by means of operators used in Box–Jenkins methodology (see Sect. 6.2) as

$$\varphi(B)y_t = \alpha + \beta(B)x_t + \varepsilon_t, \tag{7.31}$$

where $\varphi(B) = 1 - \varphi_1 B - \ldots - \varphi_p B^p$ is the autoregressive operator, $\beta(B) = \beta_0 + \beta_1 B + \ldots + \beta_k B^k$ is the operator of distributed lags of explanatory variable x, and ε_t is the residual in the form of stationary process ARMA(r, s). In addition, more explanatory variables can be included, but then the model (7.31) must be extended, e.g., for two explanatory variables x_1 and x_2 to the form

$$\varphi(B)y_t = \alpha + \beta_1(B)x_{t1} + \beta_2(B)x_{t2} + u_t. \tag{7.32}$$

If the autoregressive operator $\varphi(B)$ is stationary (i.e., its roots lie outside the unit circle in complex plane), then (7.31) can be rewritten as

$$y_t = \mu + \frac{\beta(B)}{\varphi(B)} x_t + \eta_t, \tag{7.33}$$

where $\mu = \alpha/(1 - \varphi_1 - \ldots - \varphi_p)$ and the residual component $\eta_t = \varphi(B)^{-1}u_t$ is a stationary process ARMA($p + r$, s). According to (7.33), the process $\{y_t\}$ originates by filtering the process $\{x_t\}$ (i.e., one can indeed look upon the ADL model as the "model based on the principle of filters"; see above).

Remark 7.1 Obviously, also other models described in this chapter can be presented as ADL models:

- Linear regression models with ARMA residuals (7.4) from Sect. 7.2 (see ADL model written as (7.32) with $\varphi(B) = \beta_1(B) = \beta_2(B) = \ldots = 1$).
- DL models (7.15) from Sect. 7.3 (see ADL model written as (7.33)).
- GDL models after Koyck transformation (7.22) from Sect. 7.3.1 (see ADL model written as (7.31) with $\varphi(B) = 1 - \lambda B$, $\beta(B) = \beta(1 - \lambda)$ and $u_t = \varepsilon_t - \lambda\varepsilon_{t-1}$).

◇

The construction of ADL model is analogous to the classical procedures in Box–Jenkins methodology and supposes the application of a suitable software (see, e.g., EViews). Moreover, the models have usually specific forms since they are constructed for specific situations. We describe here two specific cases, for which the ADL models seem to be useful:

7.4.1 Intervention Analysis

An important application of ADL models is so-called *intervention analysis* that is suitable for situations when the course of a time series is evidently impaired in a time period t_0 by a one-shot incidence from outside which changes in a significant way the course of this time series. For instance, the sale of a product suddenly jumps up due to successful advertisement, the financial market is influenced by a change in legislative, and the like. Such interferences are denoted in time series analysis as *interventions*.

A recommended method how to construct a model of intervention analysis is to apply the scheme of ADL models (7.31), in which the explanatory variable x_t has the form of *jumps* S_t or *pulses* P_t defined as

$$S_t = \begin{cases} 0 & \text{for } t < t_0, \\ 1 & \text{for } t \geq t_0, \end{cases} \qquad P_t = \begin{cases} 0 & \text{for } t \neq t_0 \\ 1 & \text{for } t = t_0 \end{cases} \tag{7.34}$$

with the moment of intervention t_0 (jumps S_t and pulses P_t are obviously special examples of so-called *dummy variables* (or *dummies*) which are popular in the econometric modeling). Choosing a jump or a pulse and a suitable form of ADL scheme (7.31), one can model various modes how the intervention fades away (such an analysis is sometimes also denoted as the *impulse response*). For example, an immediate dynamic change in a given time series can be modeled using (7.33) in the form

$$y_t = \mu + \frac{\beta_0}{1 - \varphi_1 B} S_t + \eta_t = \mu + \beta_0 \left(S_t + \varphi_1 S_{t-1} + \varphi_1^2 S_{t-2} + \ldots \right) + \eta_t. \tag{7.35}$$

The response to such an intervention mode corresponds to shifting the time series by the value $\beta_0(1 + \varphi_1 + _{...} + \varphi_1{}^h)$ in each time $t_0 + h$ $(h = 0, 1, ...)$. If $|\varphi_1| < 1$, then this shift achieves asymptotically the value $\beta_0/(1 - \varphi_1)$. If $\varphi_1 = 1$, then the level of time series changes linearly with the accrual of β_0 during each time unit. Models for other modes of intervention changes including practical applications are described, e.g., in Box and Tiao (1975).

7.4.2 Outliers

Another possible application of ADL schemes consists in modeling outliers (on the other hand, the outliers can be also handled by applying other approaches, e.g., by robustifying statistical methods to be insensitive to the outlying values; see Sect. 2.2.1.2). The approach based on ADL modeling may be convenient (especially, if the aim is predicting time series). In general, two types of outliers should be distinguished:

1. *Additive outlier* (abbreviated as *AO*) is linked additively to the basic (e.g., stationary) process in time t_0, i.e.:

$$y_t = z_t + \delta \cdot P_t, \tag{7.36}$$

where z_t is a stationary process, P_t is the pulse according to (7.34), and δ is the size of modeled outlier. Particularly in the case of stationary autoregressive process of the form $\varphi(B)z_t = \alpha + \varepsilon_t$, it holds for the observed (contaminated) time series y_t (simply substituting $z_t = y_t - \delta \cdot P_t$ to this autoregressive model)

$$y_t = \alpha + \sum_{j=1}^{p} \varphi_j y_{t-j} + \delta \cdot P_t - \sum_{j=1}^{p} \delta \varphi_j P_{t-j} + \varepsilon_t, \tag{7.37}$$

where the dummy variable P_{t-j} is unit in time $t_0 + j$ and otherwise zero. Neglecting the outlier (i.e., applying the classical autoregressive model directly for y_t) is incorrect and can cause substantial estimation and prediction errors. In addition under suspicion on an outlier in time t_0, one can test the significance of parameters δ in (7.37) by means of the classical t-test.

2. *Innovation outlier* (abbreviated as *IO*) is generated in the innovation process so that, e.g., in the case of stationary autoregressive structure of observed process y_t one should write

$$y_t = \sum_{j=1}^{p} \varphi_j y_{t-j} + \delta \cdot P_t + \varepsilon_t. \tag{7.38}$$

Such an innovation irregularity has the main impact only in time t_0 and then its influence decays so that its ignoring is not so dangerous for estimation or prediction as in the case of AO. The test for IO is analogical as for AO, i.e., by means of the classical t-test of significance of the parameter δ in (7.38). However, if the observed process $\{y_t\}$ has the nonstationary ARIMA structure, then the influence of innovation outlier persists over long time horizons.

7.5 Exercises

Exercise 7.1 Repeat the analysis from Example 7.1 (the yields to maturity of corporate bonds of the highest quality AAA), but only for data since 1991 (*hint:* $AAA_t = 5.84 + 0.535TBILL_t + \varepsilon_t$, $DW = 0.770$).

Exercise 7.2 Repeat the analysis from Example 7.2 (the gross domestic product GDP in the USA), but only for data since 1980 (*hint:* $\beta_{DL} = 0.469$, $\beta_{GDL} = 0.907$, $\beta_{PDL} = 0.421$).

Part IV
Financial Time Series

Chapter 8
Volatility of Financial Time Series

8.1 Characteristic Features of Financial Time Series

The models introduced in previous chapters can be mostly considered as linear models (e.g., the linear process from Sect. 6.2 is linear function of white noise values) or can be linearized by a simple transformation (e.g., the logarithmic transformation). However, many relations in economy and particularly in finance are principally nonlinear (e.g., dependence of volatility of financial time series on previous time series values). Therefore, various nonlinear models are preferred in finance, since they fit better the substance of financial data.

The financial time series are usually derived from *prices* P_t of financial assets at time (e.g., stocks or commodities, but also indices) and *price variations* $\Delta P_t = P_t - P_{t-1}$. However, one prefers to analyze returns (*relative prices*) since in contrast to the prices they do not depend on monetary units which facilitates comparisons among assets. The most usual relative prices are so-called *log returns* (namely continuously compounded returns) defined as

$$r_t = \ln\left(\frac{P_t}{P_{t-1}}\right) = \ln P_t - \ln P_{t-1} = p_t - p_{t-1} \tag{8.1}$$

(see also (6.91)), where $p_t = \ln P_t$ are *logarithmic prices* at time t. Sometimes also *relative price variations* or simply *returns* are used (even if sometimes the term "return" denotes the log return (8.1))

$$return_t = \frac{P_t - P_{t-1}}{P_{t-1}}. \tag{8.2}$$

Obviously, for smaller $| r_t |$ one can approximate

T. Cipra, *Time Series in Economics and Finance*,
https://doi.org/10.1007/978-3-030-46347-2_8

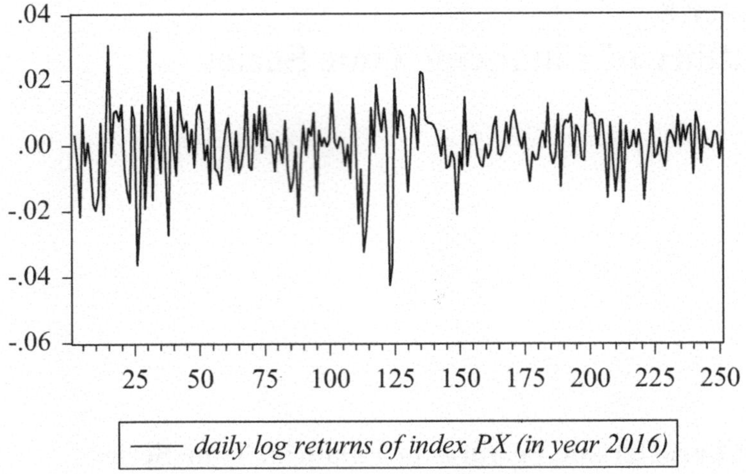

Fig. 8.1 Daily log returns of index PX in 2016 (251 trading days)

$$r_t = \ln(1 + return_t) \approx return_t \tag{8.3}$$

(more exactly, it holds $r_t = return_t + O\left(return_t^2\right)$).

In particular, the linear models from Chaps. 6 and 7 are not capable of covering some typical properties of financial time series:

- *Nonstationarity of prices and stationarity of (log) returns*: The time series $\{P_t\}$ are usually nonstationary in the form of random walk without intercept (see, e.g., the daily values of index PX during the year 2016 in Example 6.6 including Fig. 6.8(a)). On the contrary, the log returns $\{r_t\}$ (or price variations ΔP_t) can be mostly considered as (weakly) stationary (i.e., with the first and second moments invariant in time): (1) $\{r_t\}$ oscillate around zero; (2) even if the oscillations vary in magnitude, their averages over longer periods are almost constant (see daily log returns of index PX during 2016 in Fig. 8.1); (3) $\{r_t\}$ are almost uncorrelated in time (see below).
- *Uncorrelated log returns*: The time series $\{r_t\}$ of log returns generally display relatively small autocorrelations so that they are close to white noise (due to their weak stationarity; see above). It is demonstrated in Fig. 8.2(a) with estimated autocorrelations of log returns of index PX during 2016 (there are no significance bands in Fig. 8.2, since the classical Bartlett's approximation (6.12) cannot be applied here). On the other hand, for intraday series with very small time intervals between observations (measured in minutes or seconds) significant autocorrelations can be observed due to so-called *microstructure effect*.
- *Correlated log returns squared*: In contrast to the previous property, the time series $\{r_t^2\}$ of log returns squared (or $\{|r_t|\}$ of log returns in absolute values) are usually strongly correlated (see Fig. 8.2(b)). This property is important for the

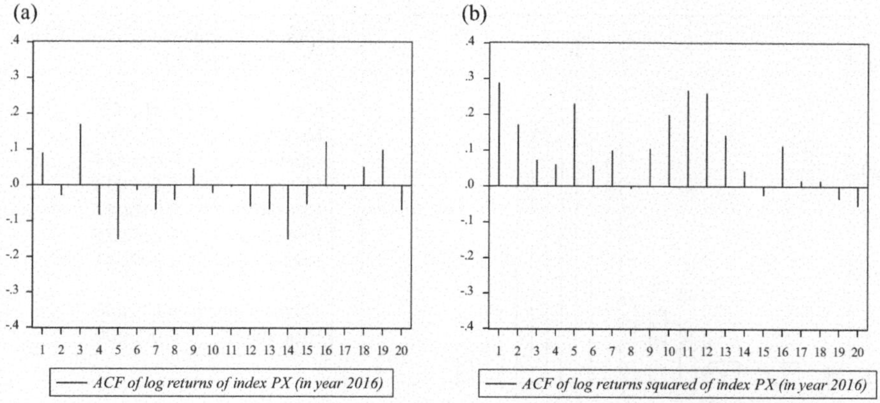

Fig. 8.2 Estimated autocorrelations of (**a**) log returns and (**b**) log returns squared of index PX in 2016 (251 trading days)

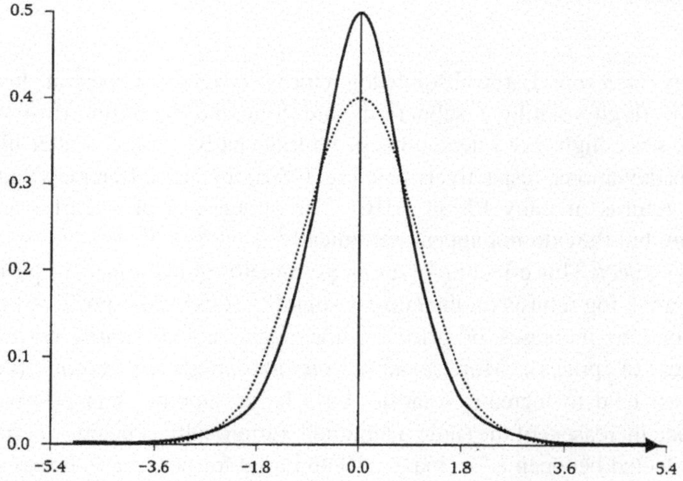

Fig. 8.3 Probability density of $N(0, 1)$ (dotted line) versus probability density of a leptokurtic distribution with zero mean value, unit variance, and kurtosis $\gamma_2 > 0$ (solid line)

construction of GARCH models based on the principle of conditional heteroscedasticity (see Sect. 8.3).

- *Leptokurtic* (or *heavy-tailed* or *fat-tailed*) *distributions*: Log returns of financial assets have mostly probability distributions which are sharply peaked at zero, have fat tails (decreasing to zero more slowly than $\exp(-x^2/2)$), and have narrow shoulders (one can identify "narrower waist and heavier tails"; see Figs. 8.3 and 8.4). The typical characteristic of such distributions is significantly positive kurtosis coefficient (it is a measure for tail thickness).

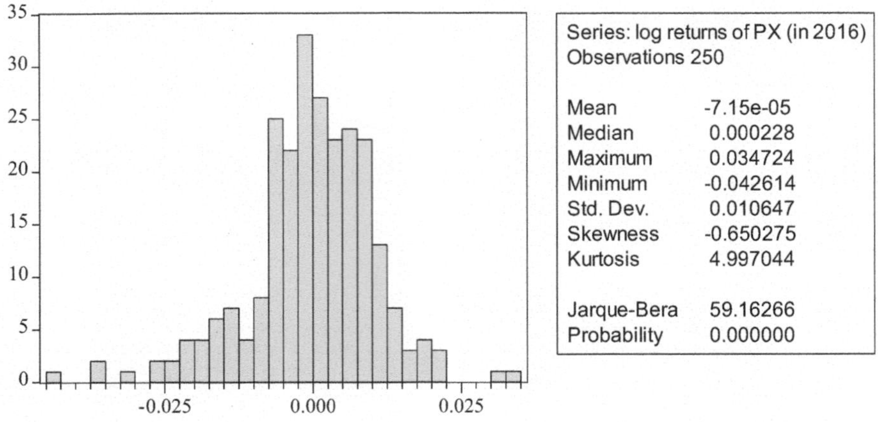

Fig. 8.4 Leptokurtic distribution (particularly with higher kurtosis coefficient compared with the corresponding normal distribution) for daily log returns of index PX in 2016 from Example 8.1. Source: calculated by EViews

- *Volatility clustering*: Large absolute log returns $|r_t|$ tend to appear in clusters, i.e., turbulent (high-volatility) subperiods are followed by quiet (low-volatility) periods, since high (low) deviations of returns can be expected after high (low) previous deviations, respectively (see Fig. 6.8(b) for first differences and Fig. 8.1 for log returns of daily PX in 2016). The subperiods of volatility bursts are recurrent, but they do not appear periodically.
- *Leverage effect*: This effect involves an asymmetry of the impact of past positive and negative log returns on the current volatility (obviously, positive log returns correspond to increases of prices, while negative log returns correspond to decreases of prices). More exactly, previous negative returns (i.e., price decreases) tend to increase volatility by a larger amount than positive returns (i.e., price increases) of the same magnitude. Empirically, a positive correlation is often detected between $r_t^+ = \max(r_t, 0)$ and $|r_{t+h}|$ for positive h, but this correlation is generally less than between $-r_t^- = \max(-r_t, 0)$ and $|r_{t+h}|$ (e.g., for log returns of daily PX in 2016 one has $\text{corr}(r_t^+, |r_{t+1}|) = 0.377$, while $\text{corr}(-r_t^-, |r_{t+1}|) = 0.527$).
- *Calendar effects*: Various calendar effects should be also mentioned in the context of financial time series: the day of week, the proximity of holidays, seasonality, and other factors may have significant effects on returns. Following a period of market closure, volatility tends to increase, reflecting the information cumulated during this break. Similar effects appear in intraday series as well.

One can see that the concept of volatility is very important for financial analysis. In general, *volatility* can be looked upon as the spread of all possible outcomes of an uncertain variable. Volatility is related to, but not exactly the same as, risk. Risk is associated with undesirable outcome, whereas volatility (as the measure strictly for uncertainty) can occur due to a positive outcome. In any case, the volatility is an input to core applications in finance such as investing, portfolio construction, option pricing, hedging, and risk management. Even volatility indices are constructed nowadays so that futures and options on the observed volatility can be traded on exchanges (e.g., the index VIX on the derivative exchange CBOE).

Typically, in financial markets, one is often concerned with the spread of asset returns and measures their volatility simply as the sample variance, e.g.:

$$\widehat{\sigma}^2 = \frac{1}{T-1}\sum_{t=1}^{T}(r_t - \widehat{\mu})^2, \tag{8.4}$$

where r_t is the log return on day t, and $\widehat{\mu}$ is the average return over a longer period of T trading days (the standard deviation $\widehat{\sigma}$ may be sometimes preferred to the variance in (8.4), since it has the same unit of measure as the observations in sample). As the volatility does not remain constant through time, the *conditional* volatility $\widehat{\sigma}_t^2$ may be a better instrument for asset pricing and risk management in time t, e.g.:

$$\widehat{\sigma}_t^2 = \frac{1}{k-1}\sum\nolimits_{\tau=t-k}^{t-1}(r_\tau - \widehat{\mu}_\tau)^2, \quad \text{where} \quad \widehat{\mu}_t = \frac{1}{k}\sum\nolimits_{\tau=t-k}^{t-1} r_\tau \tag{8.5}$$

(it is constructed conditionally using information relevant for time t, e.g., data over several days for risk management, over several months for option pricing, or over several years for investment analysis).

The volatility is obviously a latent (i.e., non-observable) matter. In Sect. 8.3, several models are given that enable to estimate it. However, besides model approaches to volatility one can also use so-called proxy approaches which are based on replacing the non-observable volatility by an observable *proxy* of it (see, e.g., Poon (2005)):

- The most usual proxy for the volatility in time t is the square of log return in this time, i.e., r_t^2 (surprisingly, taking deviations around zero instead of centering them by means of the sample mean typically increases the accuracy of volatility prediction; see Poon (2005)).
- Another approach consists in applying H-L measure (so-called *high-low measure* by Parkinson (1980))

$$\widehat{\sigma}_t^2 = \frac{(\ln H_t - \ln L_t)^2}{4 \cdot \ln 2}, \tag{8.6}$$

  where H_t and L_t denote, respectively, the highest and the lowest prices on day t (the H-L proxy assumes that the price process follows a geometric Brownian motion).
- If one disposes of intraday data at short intervals such as 5 or 15 min (so-called *tick data*), then the *realized volatility* can be constructed by integrating squared log returns, i.e.:

$$RV_{t+1} = \sum_{j=1}^{m} r_{m,t+j/m}^2 \tag{8.7}$$

(there are m log returns in one unit of time). If the log returns are serially uncorrelated, then one can show (see, e.g., Karatzas and Shreve (1988)) that

$$\operatorname*{plim}_{m\to\infty}\left(\int_{t}^{t+1}\sigma_s^2 \mathrm{d}s - \sum_{j=1}^{m} r_{m,t+j/m}^2\right), \tag{8.8}$$

i.e., the realized volatility converges in probability to so-called *integrated volatility* that is based on the concept of volatility σ_s^2 continuous in time.

8.2 Classification of Nonlinear Models of Financial Time Series

If we confine ourselves to *purely stochastic* models of time series (i.e., without deterministic trends, periodicities, and other deterministic components), then the general nonlinear model of time series can be written as

$$y_t = f(e_t, e_{t-1}, e_{t-2}, \ \ldots), \tag{8.9}$$

where f is a (nonlinear) function of uncorrelated random variables e_t with zero mean value and constant variance (a stronger assumption requiring that variables e_t are even *iid* is often applied guaranteeing the existence of models of this type; see, e.g., Wu (2005)). The random variables e_t have the meaning of *prediction errors* (or equivalently *deviations from the conditional mean value*) in (8.9) (some authors call them also *shocks* or *innovations*). The model (8.9) is a natural nonlinear extension of the linear process (6.17) from Sect. 6.2

$$y_t = \varepsilon_t + \sum_{i=1}^{\infty} \psi_i \varepsilon_{t-i} \tag{8.10}$$

(the linear process is a general scheme of linear models, which in comparison to (8.9) uses explicitly white noise values ε_t as the corresponding generators).

The nonlinear process in the form (8.9) is too general to be applied practically. Therefore, one prefers a more specific form of it written by means of the first and second conditional moments. It should not be surprising, since, e.g., the simple stationary process AR(1) introduced in (6.36) can be written by means of the *conditional* mean value as

$$\mathrm{E}(y_t|y_{t-1}) = \varphi_1 y_{t-1}. \tag{8.11}$$

Generally, one can condition in time t by the entire information Ω_{t-1} known till time $t-1$. More specifically, we can imagine that the past information is generated by all past values $\{y_{t-1}, y_{t-2}, \ldots\}$ and $\{e_{t-1}, e_{t-2}, \ldots\}$ using a suitable function of these values (one usually uses the term σ-algebra in such a situation). Due to the restriction to the first and second moments only, one usually models the conditional mean value μ_t and the conditional variance σ_t^2 by means of simple (nonlinear) functions of information in Ω_{t-1}

$$\mu_t = \mathrm{E}(y_t|\Omega_{t-1}) = g(\Omega_{t-1}), \quad \sigma_t^2 = h_t = \mathrm{var}(y_t|\Omega_{t-1}) = h(\Omega_{t-1}), \tag{8.12}$$

where g and h are suitable functions ($h(\cdot) > 0$). Although the time index should distinguish the conditional moments from the unconditional ones, it would be more correct to write, e.g., $\mu_{t|t-1}$ and $\sigma_{t|t-1}^2$ instead of simplified symbols μ_t and σ_t^2 in (8.12) (in fact, these are one-step-ahead predictions of mean value and variance of the given process). Then one can write

$$y_t = \mu_t + e_t, \tag{8.13}$$

since e_t are the prediction errors or equivalently the deviations from conditional mean value (if $y_t = r_t$, then one also uses for e_t the term *mean-corrected returns*). Moreover, as it holds

$$\sigma_t^2 = \mathrm{var}(y_t|\Omega_{t-1}) = \mathrm{var}(e_t|\Omega_{t-1}), \tag{8.14}$$

one addresses σ_t^2 as *volatility* of given time series in time t (see also Sect. 8.1). The final form of nonlinear process, which is applied in this context most often, is then

$$y_t = \mu_t + \sigma_t \cdot \varepsilon_t = \mu_t + \sqrt{h_t} \cdot \varepsilon_t = g(\Omega_{t-1}) + \sqrt{h(\Omega_{t-1})} \cdot \varepsilon_t, \tag{8.15}$$

where ε_t are *iid* random variables with zero mean value and unit variance. Obviously, it holds

$$e_t = \sigma_t \varepsilon_t. \tag{8.16}$$

It is worth noting that the random variables e_t are uncorrelated, but in contrast to ε_t they do not need to be generally independent.

Overall, the considered model is given by two equations in (8.12): the first one is the *mean equation* and the second one is the *volatility equation*. According to the type of these equations, the nonlinear processes can be classified to

- *Nonlinear in mean*: they have the nonlinear function g.
- *Nonlinear in variance*: they have the nonlinear function h; this function is often invariant in time so that such processes are frequently addressed as the processes with *conditional heteroscedasticity*.

Both categories can be combined and sorted to plenty of more specific processes (see further). The linear Box–Jenkins models from Chap. 6 present a special case of (8.15) for a linear function g and a constant function h.

8.3 Volatility Modeling

Modeling and forecasting of volatility is nowadays a very important topic of financial analysis (both theoretical and practical ones). It is no way surprising because the volatility considered as the standard deviation of various indicators of profit and loss rates represents the basic measure of risk, see, e.g., the methodology of capital adequacy for banks denoted as Basel III and the solvency in insurance companies denoted as Solvency II which are based on the concept of value-at-risk (VaR) and on similar measures of risk (including commercial software products of the type Risk Metrics (1996)).

In this section, we shall present various methods for estimating and predicting volatility in financial time series (even if approaches based on proxies are also possible; see Sect. 8.1).

8.3.1 Historical Volatility and EWMA Models

Historical volatility is the original approach to volatility estimating it simply as the sample variance or the sample standard deviation over a chosen historical period (hence *historical volatility*), i.e., in the simplest case as

$$\widehat{\sigma}_t^2 = \frac{\sum_{\tau=t-k}^{t-1} (y_\tau - \widehat{\mu}_t)^2}{k-1}, \qquad \text{where} \quad \widehat{\mu}_t = \frac{\sum_{\tau=t-k}^{t-1} y_\tau}{k} \tag{8.17}$$

for a suitable length of sample period k (see also (8.5)). Moreover, the value (8.17) is often used in practice as the prediction constructed in time t for short prediction horizons. Even though previously the historical volatility has been applied broadly in practice (e.g., in order to estimate the volatility of underlying assets when calculating option premiums according to Black–Scholes formula), nowadays its meaning is reduced to determination of benchmarks when assessing the effectiveness of more complex models of volatility.

A pragmatic extension of the historical volatility approach are EWMA models. The most frequent *model EWMA* (*exponentially weighted moving average*) is an analogy of simple exponential smoothing (see Sect. 3.3.1) for volatility. In contrast to the historical volatility calculation, in EWMA models the averaged squares in (8.17) are weighted with weights which decrease exponentially to the past. Such a modification is advantageous practically:

- In practice, volatility is usually influenced more by current values which are distinguished by higher weights from the values farther in the past.
- Moreover, in EWMA models the influence of high deviations persists during longer time periods than in (8.17) with smaller k, where high deviations leaving the sample range can cause even jumps in the estimated volatility.

Due to analogy of EWMA models to the simple exponential smoothing (see (3.75) and (3.77)), the volatility can be estimated using EWMA as

$$\widehat{\sigma}_t^2 = (1-\lambda)\sum_{j=0}^{\infty}\lambda^j(y_{t-1-j} - \overline{y})^2 = (1-\lambda)(y_{t-1} - \overline{y})^2 + \lambda\,\widehat{\sigma}_{t-1}^2, \qquad (8.18)$$

where the estimated volatility $\widehat{\sigma}_t^2$ presents the volatility prediction from time $t-1$, $\overline{y}$ is an average level of given time series, and λ $(0 < \lambda < 1)$ is a discount constant chosen in advance. If one calculates the volatility of time series of *log returns* r_t (see (8.1)), then the average return is often nearly zero (particularly for higher frequencies of observations, e.g., for daily returns), then (8.18) transfers to the form

$$\widehat{\sigma}_t^2 = (1-\lambda)\sum_{j=0}^{\infty}\lambda^j\, r_{t-1-j}^2 = (1-\lambda)r_{t-1}^2 + \lambda\,\widehat{\sigma}_{t-1}^2. \qquad (8.19)$$

In financial practice (see, e.g., RiskMetrics (1996)) due to broad experience with volatility estimation, one recommends for constant λ routinely the value 0.94.

Example 8.1 Table 8.1 and Fig. 8.1 show daily log returns r_t of index PX in 2016 (250 values for 251 trading days calculated as differences of logarithmic index values $r_t = \ln PX_t - \ln PX_{t-1}$).

This time series $\{r_t\}$ shows typical features of financial time series (see Sect. 8.1):

- Volatility clustering (see the volatility bunches in the beginning and in the middle of time series in Fig. 8.1)
- Leptokurtic distribution (see the histogram, the kurtosis coefficient $4.997 - 3 = 1.997 > 0$, and the test of normality Jarque–Bera in Fig. 8.4).

By means of the recursive formula (8.19) of EWMA model with zero initial value, one has estimated the corresponding volatility (see the graphical plot in Fig. 8.5). The EWMA estimation justifies the previous subjective conclusion (namely the occurrence of increased volatility in the beginning and in the middle of $\{r_t\}$).

◇

Table 8.1 Daily log returns of index PX in 2016 (250 values for 251 trading days written in columns) from Example 8.1 (see also Fig. 8.1 and Table 6.10)

	1	2	3	4	5	6	7	8	9	10
1	0.0030	−0.0227	−0.0043	0.0018	0.0011	0.0107	0.0147	−0.0111	0.0078	0.0021
2	−0.0052	0.0125	0.0002	0.0011	−0.0009	0.0092	−0.0065	−0.0021	−0.0012	−0.0034
3	−0.0217	−0.0193	−0.0131	−0.0079	0.0025	0.0023	0.0028	−0.0044	0.0075	−0.0066
4	0.0086	−0.0015	0.0182	0.0028	0.0012	−0.0143	0.0023	−0.0043	0.0076	0.0021
5	−0.0059	0.0347	−0.0070	−0.0013	−0.0062	−0.0041	0.0031	0.0012	−0.0014	0.0043
6	0.0011	−0.0166	−0.0081	−0.0052	0.0002	0.0111	−0.0030	0.0043	−0.0160	0.0024
7	−0.0053	0.0186	−0.0119	0.0077	−0.0101	0.0085	−0.0058	−0.0009	0.0068	−0.0006
8	−0.0178	0.0011	−0.0017	−0.0073	0.0144	−0.0014	−0.0064	0.0128	−0.0040	0.0092
9	−0.0197	−0.0083	0.0048	−0.0140	0.0033	0.0224	0.0002	−0.0021	−0.0142	−0.0007
10	−0.0152	0.0175	0.0085	−0.0109	−0.0239	0.0218	−0.0040	−0.0057	−0.0065	0.0059
11	0.0070	−0.0085	−0.0010	−0.0001	−0.0072	0.0077	−0.0015	−0.0032	0.0075	0.0013
12	−0.0209	−0.0274	−0.0078	−0.0217	−0.0325	0.0068	0.0052	0.0095	−0.0176	0.0052
13	0.0046	0.0119	0.0044	−0.0041	−0.0261	0.0067	0.0086	−0.0126	0.0058	0.0062
14	0.0308	−0.0003	−0.0082	0.0061	−0.0125	0.0061	−0.0026	0.0058	−0.0012	−0.0089
15	−0.0033	−0.0090	−0.0058	−0.0029	0.0115	0.0044	−0.0030	0.0075	−0.0002	0.0099
16	0.0101	0.0165	−0.0001	0.0052	−0.0021	0.0012	−0.0003	0.0068	0.0044	0.0061
17	0.0109	0.0085	0.0169	0.0028	0.0186	−0.0034	0.0069	0.0094	−0.0008	−0.0056
18	0.0076	0.0042	−0.0064	0.0100	0.0093	0.0045	0.0004	−0.0041	0.0046	0.0055
19	0.0127	0.0075	−0.0074	−0.0152	0.0040	−0.0069	0.0084	0.0059	−0.0005	0.0001
20	−0.0069	−0.0019	0.0097	0.0047	0.0115	−0.0062	0.0107	0.0040	−0.0168	0.0000
21	−0.0138	0.0051	0.0000	−0.0002	0.0038	−0.0019	0.0058	−0.0043	−0.0059	−0.0009
22	−0.0173	−0.0066	0.0122	0.0023	−0.0426	−0.0043	0.0018	−0.0046	−0.0005	0.0039
23	0.0121	0.0106	−0.0022	−0.0004	−0.0366	−0.0212	−0.0011	0.0140	0.0093	0.0035
24	0.0081	0.0127	0.0116	0.0016	0.0205	−0.0021	0.0038	0.0086	−0.0040	−0.0043
25	−0.0363	0.0070	0.0019	0.0159	0.0022	−0.0074	−0.0056	0.0092	−0.0027	0.0022

Source: Calculated by EViews

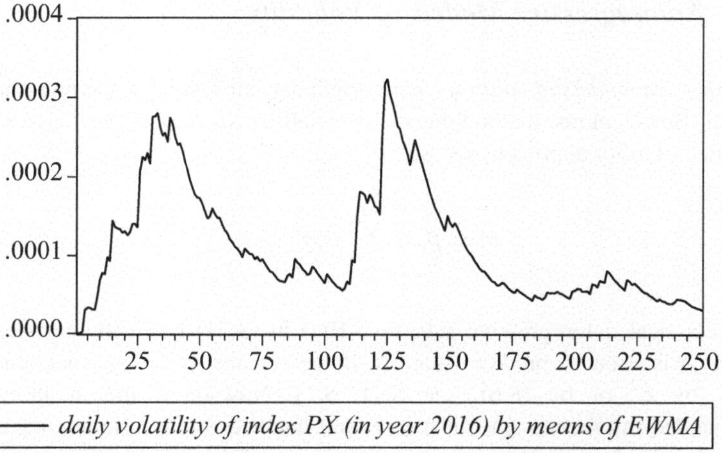

Fig. 8.5 Volatility of daily log returns of index PX in 2016 (250 values for 251 trading days) estimated by means of EWMA model in Example 8.1

8.3.2 *Implied Volatility*

In finance one exploits some relations using the volatility as one of explaining factors. The best known in this context is Black–Scholes formula mentioned in Sect. 8.3.1, which expresses analytically the call or put option premium as a function of five factors: S_t (spot price of underlying asset, e.g., a stock), X (exercise price of option), $T - t$ (time to maturity of option), σ (volatility of underlying asset), and i (risk-free interest rate in the given capital environment). For example, the premium of European call option C_t in time t is

$$C_t = S_t \cdot \Phi(d_1) - X \cdot e^{-i(T-t)} \cdot \Phi(d_2), \tag{8.20}$$

where $d_1 = \dfrac{\ln(S_t/X) + (i + \sigma^2/2)(T - t)}{\sigma\sqrt{T - t}}$, $d_2 = d_1 - \sigma\sqrt{T - t}$,

$\Phi(\cdot) \sim$ distribution function $N(0, 1)$ (see, e.g., Hull (1993)). If we observe the prices of traded (i.e., quoted) options for known values of explained factors (excluding the volatility), then we are capable of applying a suitable numerical procedure to calculate just this volatility factor, which is then usually called *implied volatility* (more exactly, it is the prediction of volatility of underlying asset price in time t with prediction horizon equal to time $T - t$ to the maturity of option).

However, the implied volatility is derived under assumptions that do not need to be fulfilled in practice (e.g., the lognormal distribution of price of underlying asset), and therefore, it can be significantly different from the real volatility. The practical experience shows that the implied volatility is usually higher than the volatility derived, e.g., by means of the GARCH models (see Sect. 8.3.5).

8.3.3 Autoregressive Models of Volatility

Autoregressive models of volatility were originally introduced as a direct implementation of Box–Jenkins methodology for volatility (they can be classified as a stochastic volatility approach; see Sect. 8.3.6):

$$\sigma_t^2 = \beta_0 + \sum_{j=1}^{s} \beta_j \sigma_{t-j}^2 + \varepsilon_t. \tag{8.21}$$

The classical autoregressive scheme AR(s) in (8.21) (with the classical white noise $\{\varepsilon_t\}$) is used to predict volatility if we replace $\{\sigma_t^2\}$ by a suitable proxy (usually by r_t^2 or by (8.6); see Sect. 8.1). Nowadays, this method is not recommended in practice since it has several handicaps (e.g., the nonnegativity of the right-hand side of (8.21) is not guaranteed, even if the logarithmic transformation can solve this problem; see also (8.79)).

8.3.4 ARCH Models

A significant breakthrough to model the volatility systematically has been just the model ARCH (*autoregressive conditional heteroscedasticity*) applied by Engle (1982) to model the inflation in the UK. The models of this type (and particularly their generalization to the GARCH models; see Sect. 8.3.5) are apparently one of the most successful instruments of modeling financial time series (so far without significant competitors). Their principle is based on two predicates, namely

- The models of financial time series are heteroscedastic, i.e., their volatility changes in time.
- The volatility is a simple quadratic function of past prediction errors e_t (deviations from the conditional mean value).

Only the second predicate needs an explanation (the first one is sufficiently supported by financial empirical experience): Due to the phenomenon of volatility clustering according to which high (low) deviations of returns can be expected rather after higher (lower) previous deviations, respectively, one can assume that the particular volatilities are positively correlated and make use of the autoregressive model as the simplest scheme to model them. Moreover, according to (8.14) it holds

$$\sigma_t^2 = \operatorname{var}(e_t|\Omega_{t-1}) = \mathrm{E}\left(e_t^2|\Omega_{t-1}\right) \approx e_t^2 \tag{8.22}$$

(obviously $\mathrm{E}(e_t) = 0$) so that the squared errors e_t^2 can be used as natural approximations of volatilities σ_t^2. Therefore, if we express the volatility as the following quadratic function of delayed values e_t^2:

$$\sigma_t^2 = \alpha_0 + \alpha_1 e_{t-1}^2 + \ldots + \alpha_r e_{t-r}^2, \tag{8.23}$$

then we obtain a realistic model (when the order r is chosen in a suitable size). It is worth noting that (8.23) is a "nonstochastic" relation, i.e., without a random residual component.

Due to the previous discussion and respecting the general form of nonlinear model (8.15), one formulates the model ARCH(r) of order r as

$$y_t = \mu_t + e_t, \qquad e_t = \sigma_t \varepsilon_t, \qquad \sigma_t^2 = \alpha_0 + \alpha_1 e_{t-1}^2 + \ldots + \alpha_r e_{t-r}^2, \tag{8.24}$$

where ε_t are *iid* random variables with zero mean value and unit variance (moreover, they are frequently assumed to have the normal distribution, i.e., $\varepsilon_t \sim N(0, 1)$, or the t-distribution which is standardized to have also zero mean value and unit variance). In any case, increased past values of volatility imply the increased present volatility in the model (8.24) which can be also rewritten as

$$y_t = e_t, \qquad e_t = \sqrt{h_t}\varepsilon_t, \qquad h_t = \alpha_0 + \alpha_1 e_{t-1}^2 + \ldots + \alpha_r e_{t-r}^2. \tag{8.25}$$

The conditional mean value μ_t is modeled by means of a suitable mean equation that is often linear: one can apply the conditional mean value corresponding to a linear regression model (sometimes it can be even reduced to the intercept only) or a process ARMA. More specifically, the following forms may serve as examples (some models nonlinear in mean value will be described later in Sect. 9.1):

$$y_t = e_t, \qquad e_t = \sigma_t \varepsilon_t, \qquad \sigma_t^2 = \alpha_0 + \alpha_1 e_{t-1}^2 + \ldots + \alpha_r e_{t-r}^2 \tag{8.26}$$

or equivalently

$$y_t = \sigma_t \varepsilon_t, \qquad \sigma_t^2 = \alpha_0 + \alpha_1 y_{t-1}^2 + \ldots + \alpha_r y_{t-r}^2 \tag{8.27}$$

(i.e., with *zero mean value* μ_t) or

$$y_t = \gamma_0 + \gamma_1 x_{t1} + \ldots + \gamma_k x_{tk} + e_t, \;\; e_t = \sigma_t \varepsilon_t, \;\; \sigma_t^2 = \alpha_0 + \alpha_1 e_{t-1}^2 + \ldots + \alpha_r e_{t-r}^2 \tag{8.28}$$

(i.e., with *exogenous variables* x_1, ..., x_k) or

$$y_t = \varphi_1 y_{t-1} + \ldots \varphi_p y_{t-p} + e_t, \;\; e_t = \sigma_t \varepsilon_t, \;\; \sigma_t^2 = \alpha_0 + \alpha_1 e_{t-1}^2 + \ldots + \alpha_r e_{t-r}^2 \tag{8.29}$$

(i.e., with conditional mean value μ_t corresponding to the *process* AR(p)).

Moreover, the parameters of the model ARCH(r) must fulfill the following constraints:

$$\alpha_0 > 0, \ \ \alpha_1 \geq 0, \ \ \ldots, \ \ \alpha_r \geq 0 \tag{8.30}$$

and

$$\alpha_1 + \ \ldots \ + \alpha_r < 1. \tag{8.31}$$

The first constraint (8.30) guarantees that the sign of volatility σ_t^2 in (8.24) is positive: it is a sufficient (but not necessary) condition for this natural property of the model. The second constraint (8.31) is not so clear but is also important: it guarantees that the model ARCH(r) has constant (finite) *unconditional* variance (see its derivation for ARCH(1) in (8.36)).

Remark 8.1 One should stress once more that in general the random variables e_t are only uncorrelated (see (8.35)), while ε_t are independent. The graphical plots of correlogram and partial correlogram of model ARCH correspond to this fact: e.g., the estimated correlogram of time series y_t in the model (8.27) should have all values insignificant as a white noise while the estimated partial correlogram of squared time series y_t^2 should have the truncation point equal to r.

◇

Remark 8.2 The model (8.24) can be generalized by means of matrix calculus:

$$y_t = \mu_t + e_t, \ \ e_t = \sigma_t \varepsilon_t, \ \ \sigma_t^2 = \alpha_0 + (e_{t-1}, \ldots, e_{t-r})\mathbf{A}(e_{t-1}, \ldots, e_{t-r})', \tag{8.32}$$

where the matrix **A** of unknown parameters must be positive semidefinite and $\alpha_0 > 0$. The original model (8.24) is a special parsimonious version of (8.32), where the matrix **A** is diagonal with nonnegative diagonal elements.

◇

The main properties of the ARCH models will be derived only for the process ARCH(1), i.e., for the model

$$y_t = \mu_t + e_t, \qquad e_t = \sigma_t \varepsilon_t, \qquad \sigma_t^2 = \alpha_0 + \alpha_1 e_{t-1}^2 \qquad (\alpha_0 > 0, \ \ \alpha_1 \geq 0) \tag{8.33}$$

(the derivation for higher orders is analogous):

1. *Zero unconditional mean value of* e_t :

$$\mathrm{E}(e_t) = \ \mathrm{E}(\mathrm{E}(e_t|\Omega_{t-1})) = \mathrm{E}(\sigma_t \mathrm{E}(\varepsilon_t|\Omega_{t-1})) = 0. \tag{8.34}$$

2. *Uncorrelated* e_t *and* e_{t-k} $(k > 0)$:

$$\mathrm{cov}(e_t, e_{t-k}) = \mathrm{E}(e_t e_{t-k}) = \mathrm{E}(\mathrm{E}(\sigma_t \varepsilon_t e_{t-k} | \Omega_{t-1})) = \mathrm{E}(\sigma_t e_{t-k} \mathrm{E}(\varepsilon_t | \Omega_{t-1})) = 0. \tag{8.35}$$

3. *Constant unconditional variance of* e_t :

$$\begin{aligned}\mathrm{var}(e_t) = \mathrm{E}(e_t^2) &= \mathrm{E}\big(\mathrm{E}(e_t^2 | \Omega_{t-1})\big) = \mathrm{E}(\sigma_t^2) = \mathrm{E}(\alpha_0 + \alpha_1 e_{t-1}^2) \\ &= \alpha_0 + \alpha_1 \mathrm{var}(e_{t-1}).\end{aligned} \tag{8.36}$$

Since the variance of prediction errors e_t should be constant, it must hold

$$\mathrm{var}(e_t) = \frac{\alpha_0}{1 - \alpha_1} \tag{8.37}$$

under the constraint

$$0 \le \alpha_1 < 1 \tag{8.38}$$

(the properties 1–3 mean that the time series $\{e_t\}$ is a white noise (in particular, weakly stationary) under the sufficient condition (8.38)). Interestingly, despite the changing *conditional* variance, i.e., the changing volatility, the *unconditional* variance of $\{e_t\}$ remains constant over time.

4. *Constant nonnegative kurtosis of* e_t : If $\varepsilon_t \sim N(0, 1)$, then it holds

$$\begin{aligned}\mathrm{E}(e_t^4) &= \mathrm{E}\big(\mathrm{E}(e_t^4 | \Omega_{t-1})\big) = 3\mathrm{E}\Big(\big(\alpha_0 + \alpha_1 e_{t-1}^2\big)^2\Big) \\ &= 3\big(\alpha_0^2 + 2\alpha_0\alpha_1 \mathrm{var}(e_{t-1}) + \alpha_1^2 \mathrm{E}(e_{t-1}^4)\big).\end{aligned} \tag{8.39}$$

Hence analogously as in 3 one obtains

$$\mathrm{E}(e_t^4) = \frac{3\alpha_0^2(1 + \alpha_1)}{(1 - \alpha_1)(1 - 3\alpha_1^2)} \tag{8.40}$$

under the sufficient condition

$$0 \le \alpha_1 < \sqrt{1/3}. \tag{8.41}$$

Finally using (8.37), the (unconditional) kurtosis coefficient of e_t is

$$\gamma_2 = \frac{\mathrm{E}(e_t^4)}{(\mathrm{var}(e_t))^2} - 3 = 3\frac{1-\alpha_1^2}{1-3\alpha_1^2} - 3 = \frac{6\alpha_1^2}{1-3\alpha_1^2} \geq 0, \tag{8.42}$$

which is in accord with modeling leptokurtic distributions (in particular, the deviations e_t in the conditionally normal model ARCH(1) produce outliers with higher probabilities than a normal white noise).

Remark 8.3 The previous results can be extended easily to the model ARCH(r). The sufficient condition of weak stationarity of its deviations e_t demands that all roots of the autoregressive polynomial $1 - \alpha_1 z - \ldots - \alpha_r z^r$ lie outside the unit circle in complex plane. Since the parameters $\alpha_1, \ldots, \alpha_r$ must be nonnegative, this sufficient condition can be rewritten in a more comfortable form (8.31). Under this condition, the variance of deviations e_t can be expressed as

$$\mathrm{var}(e_t) = \frac{\alpha_0}{1-\alpha_1 - \ldots - \alpha_r}. \tag{8.43}$$

◇

After the theoretical description of ARCH models, we can deal briefly with their practical construction (since the construction of GARCH models is analogical, we will skip this technical topic in Sect. 8.3.5 devoted to these models). Here for simplicity, we confine ourselves to the model ARCH(r) in the form (8.26), i.e., with zero conditional mean value μ_t, where the deviations e_t are directly observable (this assumption is fulfilled in practice for the financial time series of log returns r_t). In the opposite case, one eliminates the deviations e_t at first, e.g., in the model (8.29) as

$$e_t = y_t - \varphi_1 y_{t-1} - \ldots - \varphi_p y_{t-p}. \tag{8.44}$$

8.3.4.1 Identification of Order of Model ARCH

The order r can be identified as the truncation point of estimated partial correlogram in model

$$e_t^2 = \alpha_0 + \alpha_1 e_{t-1}^2 + \ldots + \alpha_r e_{t-r}^2 + u_t, \tag{8.45}$$

where u_t is the classical white noise (i.e., in the same way as for the classical AR model in the framework of Box–Jenkins methodology; see Sect. 6.3.1). If the order r is too high, then the nonnegativity of large number of parameters can be a problem (see the constraint (8.30)). In his seminal work, Engle (1982) suggested to apply a parsimonious model only with two parameters (but with $r = 4$) instead of (8.23), namely

$$\sigma_t^2 = \delta_0 + \delta_1\left(0.4e_{t-1}^2 + 0.3e_{t-2}^2 + 0.2e_{t-3}^2 + 0.1e_{t-4}^2\right). \tag{8.46}$$

8.3.4.2 Estimation of Model ARCH

Due to various reasons, the estimation methods based on the principle of least squares are not suitable for models with conditional heteroscedasticity. Therefore, one recommends for these models the method of maximum likelihood. The corresponding probability density fulfills obviously the relation

$$f(e_1, \ \ldots, \ e_n) = f(e_n|\Omega_{n-1}) \cdot \ldots \cdot f(e_{r+1}|\Omega_r) f(e_1, \ \ldots, \ e_r). \tag{8.47}$$

Therefore assuming $\varepsilon_t \sim N(0, 1)$, one can write the (conditional) log likelihood function as

$$l(\alpha_0, \ \ldots, \ \alpha_r) = \sum_{t=r+1}^{n}\left(-\frac{1}{2}\ln(2\pi) - \frac{1}{2}\ln\left(\sigma_t^2\right) - \frac{1}{2}\frac{e_t^2}{\sigma_t^2}\right) \tag{8.48}$$

(we have omitted the last factor in (8.47) since we condition by initial values $e_1, \ldots, e_r$ in (8.48)). The values e_t necessary for the construction of (8.48) are calculated recursively for each choice of arguments $\alpha_0, \ldots, \alpha_r$ including the volatilities

$$\sigma_t^2 = \alpha_0 + \alpha_1 e_{t-1}^2 + \ldots + \alpha_r e_{t-r}^2, \qquad t = r+1, \ldots, n. \tag{8.49}$$

If the normal distribution of ε_t does not fit heavy tails of modeled financial data properly, then one can apply other distributions. For instance, if the standardized t-distribution is a better choice for ε_t (with the unit variance and the degrees of freedom ν ($\nu > 2$)), then (8.48) must be replaced by

$$l(\alpha_0, \ \ldots, \ \alpha_r) = -\sum_{t=r+1}^{n}\left(\frac{\nu+1}{2}\ln\left(1 + \frac{e_t^2}{(\nu-2)\sigma_t^2}\right) + \frac{1}{2}\ln\left(\sigma_t^2\right)\right) \tag{8.50}$$

(the probability behavior of tails can be controlled by means of ν : if $\nu \to \infty$, then the applied t-distribution transfers back to the normal one). If even the t-distribution is not enough for the analyzed heavy-tailed data, then some software systems (e.g., EViews) offer the generalized error distribution GED or other possibilities (see, e.g., McNeil et al. (2005)).

The maximization of log likelihood function to obtain the final ML estimates is the concern of software optimization algorithms. As a matter of fact, one usually estimates the entire model including its mean equation (e.g., including the parameters φ, ..., φ_p in the expression (8.44)) and including the variance matrix of estimated parameters.

Remark 8.4 If the assumption of conditional normal distribution (tj. $\varepsilon_t \sim N(0, 1)$) is used improperly in a (correctly identified) model ARCH then the corresponding ML estimates of its parameters remain consistent, but their estimated variance matrix should be repaired. For such a case, Bollerslev and Wooldridge (1992) suggested a possible approach denoted as *heteroscedasticity consistent covariances*, which is robust against non-normal distributions (it is based on the QML estimation (*quasi-maximum likelihood*; see Example 8.2). Nowadays, nonparametric estimation of conditional heteroscedasticity is also recommended (see, e.g., Fan and Yao (2005)).

◇

8.3.4.3 Verification of Model ARCH

Most estimation procedures for ARCH models enable to obtain "by-products" which can be used consequently to the verification of constructed model, namely:

- The estimated deviation $\widehat{e}_t$ (the one-step-ahead prediction error in given time series for time t): e.g., in the model (8.29) it can be estimated as

$$\widehat{e}_t = y_t - \widehat{\varphi}_1 y_{t-1} - \ldots - \widehat{\varphi}_p y_{t-p}. \tag{8.51}$$

- The estimated volatility $\widehat{\sigma}_t^2$ (the variance of deviation e_t estimated in time $t-1$ for time t or equivalently the one-step-ahead prediction of variance in given time series denoted as $\widehat{\sigma}_t^2(t-1)$ using such an interpretation; see (8.53)).
- The estimated standardized deviation $\widetilde{e}_t$:

$$\widetilde{e}_t = \frac{\widehat{e}_t}{\widehat{\sigma}_t}. \tag{8.52}$$

To verify a constructed model ARCH(r) the following procedures are mostly applied which explore the properties of estimated standardized deviation (8.52):

- Verification of estimated mean equation: Q-tests of the type (6.67) for time series $\{\widetilde{e}_t\}$.
- Verification of estimated volatility equation: Q-tests of the type (6.67) for time series $\{\widetilde{e}_t^2\}$ or special tests for time series $\{\widetilde{e}_t\}$ (e.g., LM-tests based on Lagrange multipliers) testing a potential remaining ARCH structure in $\{\widetilde{e}_t\}$.
- Verification of normality of conditional ARCH model: Jarque–Bera test (or the numerical value of kurtosis coefficient only) for time series $\{\widetilde{e}_t\}$.

8.3.4.4 Prediction of Volatility in Model ARCH

Volatility can be predicted by means of the relation (8.23) in the same way as we construct predictions in the linear models of Box–Jenkins methodology (see Sect. 6.6), i.e.:

$$\widehat{\sigma}_t^2(t-1) = \widehat{\sigma}_t^2 = \widehat{\alpha}_0 + \widehat{\alpha}_1 \widehat{e}_{t-1}^2 + \widehat{\alpha}_2 \widehat{e}_{t-2}^2 + \ldots + \widehat{\alpha}_r \widehat{e}_{t-r}^2, \tag{8.53}$$

$$\widehat{\sigma}_{t+1}^2(t-1) = \widehat{\alpha}_0 + \widehat{\alpha}_1 \widehat{\sigma}_t^2(t-1) + \widehat{\alpha}_2 \widehat{e}_{t-1}^2 + \ldots + \widehat{\alpha}_r \widehat{e}_{t+1-r}^2 \tag{8.54}$$

etc.

8.3.5 GARCH Models

The model ARCH(r) from the previous section has some drawbacks, e.g.:

- One must often use a high order r to describe the volatility of given time series in an adequate way.
- If r is high, then it is necessary to estimate a large number of parameters under the condition of nonnegativeness (8.30) and stationarity (8.31).
- ARCH models cover the volatility clustering but not the leverage effect (i.e., asymmetry of the impact of past positive and negative deviations e_t on the current volatility).

These drawbacks can be reduced by applying the model GARCH (*generalized* ARCH) suggested by Bollerslev (1986). In this model and in its various modifications (see Sect. 8.3.6), the volatility (i.e., the conditional variance) may also depend on its previous (lagged) values. Specially the model GARCH(1,1), which is the simplest representative of this class of models, is very popular model instrument for financial time series nowadays: it is capable of managing very general volatility structures by applying three parameters only (the GARCH models of higher orders are used in routine practice rarely).

The model GARCH(r, s) has the form

$$y_t = \mu_t + e_t, \qquad e_t = \sigma_t \varepsilon_t, \qquad \sigma_t^2 = \alpha_0 + \sum_{i=1}^{r} \alpha_i e_{t-i}^2 + \sum_{j=1}^{s} \beta_j \sigma_{t-j}^2, \tag{8.55}$$

where ε_t are *iid* random variables with zero mean value and unit variance (again they are mostly assumed to have the normal or t-distribution) and the parameters of model fulfill

$$\alpha_0 > 0, \quad \alpha_i \geq 0, \quad \beta_j \geq 0, \quad \sum_{i=1}^{\max\{r,s\}} (\alpha_i + \beta_i) < 1 \tag{8.56}$$

(one puts $\alpha_i = 0$ for $i > r$ and $\beta_j = 0$ for $j > s$; if $s = 0$, then we go back to the model ARCH(r)). The last inequality in (8.56) is the sufficient condition for the existence of variance

$$\mathrm{var}(e_t) = \frac{\alpha_0}{1 - \sum_{i=1}^{\max\{r,s\}} (\alpha_i + \beta_i)}. \tag{8.57}$$

Remark 8.5 If we put $u_t = e_t^2 - \sigma_t^2$ in (8.55), then u_t has the property of white noise, and it holds

$$e_t^2 = \alpha_0 + \sum_{i=1}^{\max\{r,s\}} (\alpha_i + \beta_i) e_{t-i}^2 + u_t - \sum_{j=1}^{s} \beta_j u_{t-j}. \tag{8.58}$$

Hence the volatility equation of the model GARCH can be looked upon as the model ARMA for the time series of squared deviations $\{e_t^2\}$. As the (non-squared) process $\{e_t\}$ is concerned, under the assumptions as (8.31) or (8.56) it is weakly (second order moments) stationary. The strict (distribution) stationarity demands other type of assumptions than (8.56), e.g., $\mathrm{E}\{\ln(\alpha_1 \varepsilon_t^2 + \beta_1)\} < 0$ for GARCH(1,1) in (8.59); see Francq and Zakoian (2010).

◇

In particular, the model GARCH(1,1) has a simpler form

$$y_t = \mu_t + e_t, \qquad e_t = \sigma_t \varepsilon_t,$$
$$\sigma_t^2 = \alpha_0 + \alpha_1 e_{t-1}^2 + \beta_1 \sigma_{t-1}^2 \qquad (\alpha_0 > 0,\ \alpha_1, \beta_1 \geq 0,\ \alpha_1 + \beta_1 < 1). \tag{8.59}$$

Its kurtosis coefficient fulfills

$$\gamma_2 = \frac{3\left(1 - (\alpha_1 + \beta_1)^2\right)}{1 - 2\alpha_1^2 - (\alpha_1 + \beta_1)^2} - 3 = \frac{6\alpha_1^2}{1 - 2\alpha_1^2 - (\alpha_1 + \beta_1)^2} \geq 0 \tag{8.60}$$

under the validity of sufficient condition

$$1 - 2\alpha_1^2 - (\alpha_1 + \beta_1)^2 > 0 \tag{8.61}$$

(compare with (8.41) and (8.42) for the model ARCH(1)).

As the construction of the models of type GARCH is concerned, we premised in Sect. 8.3.4 that it is quite analogical to ARCH models. For instance, the volatility in the model GARCH(1,1) (see (8.59)) can be predicted as

$$\widehat{\sigma}_t^2(t-1) = \widehat{\sigma}_t^2 = \alpha_0 + \alpha_1 e_{t-1}^2 + \beta_1 \sigma_{t-1}^2 \tag{8.62}$$

(in practice one uses estimated parameters). Since it holds

$$\sigma_{t+1}^2 = \alpha_0 + (\alpha_1 + \beta_1)\sigma_t^2 + \alpha_1 \sigma_t^2 (\varepsilon_t^2 - 1) \text{ and } \mathrm{E}(\varepsilon_t^2 - 1|\Omega_{t-1}) = 0, \tag{8.63}$$

one can also write

$$\widehat{\sigma}_{t+1}^2(t-1) = \alpha_0 + (\alpha_1 + \beta_1)\widehat{\sigma}_t^2(t-1) \tag{8.64}$$

and generally

$$\widehat{\sigma}_{t+\tau}^2(t) = \alpha_0 + (\alpha_1 + \beta_1)\widehat{\sigma}_{t+\tau-1}^2(t), \qquad \tau > 1. \tag{8.65}$$

Repeating this procedure, one obtains finally

$$\widehat{\sigma}_{t+\tau}^2(t) = \frac{\alpha_0\left(1 - (\alpha_1 + \beta_1)^{\tau-1}\right)}{1 - (\alpha_1 + \beta_1)} + (\alpha_1 + \beta_1)^{\tau-1}\widehat{\sigma}_{t+1}^2(t) \to \frac{\alpha_0}{1 - (\alpha_1 + \beta_1)} \tag{8.66}$$

for $\tau \to \infty$. Obviously, the volatility prediction converges with increasing prediction horizon to the unconditional variance of prediction errors e_t (see (8.57)).

Example 8.2 In Example 8.1, we have estimated the volatility of daily log returns r_t of index PX in 2016 (see 250 values in Table 8.1) by means of the model EWMA. For comparison, the model GARCH(1,1) is estimated by means of EViews (see Table 8.2) for the same data as

$$r_t = e_t, \qquad e_t = \sigma_t \varepsilon_t, \qquad \sigma_t^2 = 0.1259 e_{t-1}^2 + 0.8521 \sigma_{t-1}^2$$

or (since the conditional mean value is insignificant according to Table 8.2)

$$r_t = \sigma_t \varepsilon_t, \qquad \sigma_t^2 = 0.1259 r_{t-1}^2 + 0.8521 \sigma_{t-1}^2$$

(the method by Bollerslev and Wooldridge from Remark 8.4 has been applied to estimate the variance matrix of estimated parameters to be robust against non-normal distributions). The results of the verification procedures are not presented here, but Q tests for the estimated standardized deviation (8.52) and for its square (see Sect.

Table 8.2 Estimation of the process GARCH(1, 1) from Example 8.2 (*index PX in year 2016*)

Dependent Variable: log returns of PX (in 2016)				
Included observations: 250 after adjustments				
Convergence achieved after 12 iterations Bollerslev–Wooldrige robust standard errors and covariance				
GARCH = C(2) + C(3)∗RESID(-1)^2 + C(4)∗GARCH(-1)				
	Coefficient	Std. Error	*z*-Statistic	Prob.
C	0.000286	0.000538	0.531451	0.5951
	Variance Equation			
C	2.48E-06	2.00E-06	1.244526	0.2133
RESID(-1)^2	0.125878	0.036311	3.466680	0.0005
GARCH(-1)	0.852148	0.037461	22.74740	0.0000

Source: Calculated by EViews

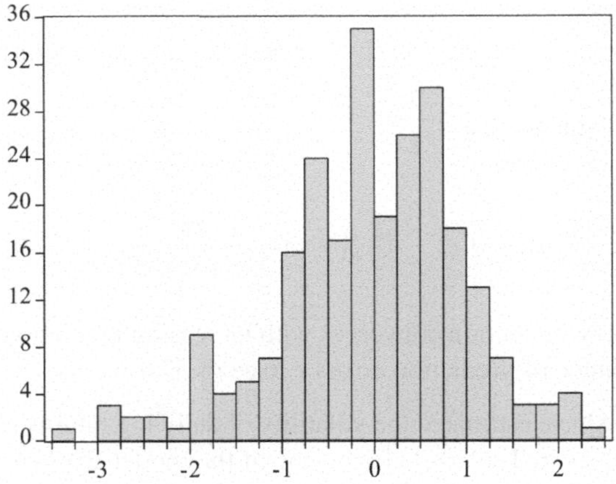

Series: Standardized Residuals	
Observations 250	
Mean	-0.038325
Median	-0.007027
Maximum	2.375025
Minimum	-3.470062
Std. Dev.	1.006280
Skewness	-0.532023
Kurtosis	3.667134
Jarque-Bera	16.42982
Probability	0.000271

Fig. 8.6 Histogram of estimated standardized deviation $\widetilde{e}_t$ (see (8.52)) for daily log returns of index PX in 2016 from Example 8.2. Source: calculated by EViews

8.3.4.3) verify statistically the constructed model (also the LM test mentioned in Sect. 8.3.4.3 does not find in these estimated deviations any remaining ARCH structure). On the other hand, the histogram of estimated $\{\widetilde{e}_t\}$ and Jarque–Bera test shown in Fig. 8.6 indicate the non-normality with higher kurtosis so that one should apply t or GED distributions when constructing this GARCH model (see Sect. 8.3.4.2).

Finally, Fig. 8.7 plots the volatility which is constructed by means of the estimated model GARCH(1, 1) (one can compare it with its EWMA estimate from Example 8.1 in Fig. 8.5).

◇

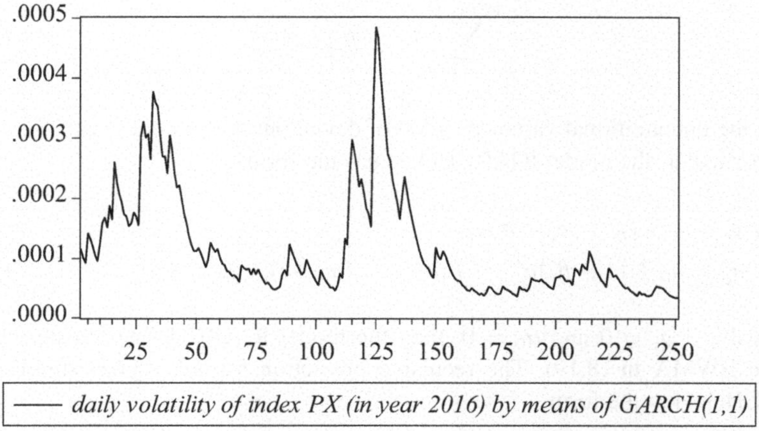

Fig. 8.7 Volatility of daily log returns of index PX in 2016 (250 values for 251 trading days) estimated by means of model GARCH(1, 1) in Example 8.2 (compare with its EWMA estimate from Example 8.1 in Fig. 8.5)

8.3.6 Various Modifications of GARCH Models

Analysis of financial (or nonlinear) time series is a very progressive sector. The offer of various models is really enormous including a flood of nonsystematic acronyms of the type FIEGARCH with tens of references in various sources each year (therefore, it has no sense to survey the bibliography in this section). Typical examples are just various modifications of GARCH models motivated mostly by an effort to repair various drawbacks of the classical GARCH models from Sect. 8.3.5: some of them are briefly described just in this section (respecting the fact that practical calculations mostly suppose the application of specialized software instruments).

8.3.6.1 IGARCH

Integrated GARCH model denoted as IGARCH(r, s) is the model GARCH(r, s) with unit roots of the autoregressive polynomial in volatility equation (it is an analogy of ARIMA models but for volatility modeling). Its typical feature is the so-called *volatility persistence*: while the classical process GARCH is stationary in volatility (so that the volatility prediction converges with increasing horizon to the unconditional variance of this process; see (8.66)), in the model IGARCH the current information persists even for very long prediction horizons.

The model IGARCH(r, s) is defined as the model GARCH(r, s) (see (8.55)), where in addition

$$\sum_{i=1}^{\max\{r,s\}} (\alpha_i + \beta_i) = 1, \tag{8.67}$$

so that the unconditional variance (8.57) of deviations e_t does not exist.

In particular, the model IGARCH(1,1) has the form

$$\begin{aligned} y_t &= \mu_t + e_t, \qquad e_t = \sigma_t \varepsilon_t, \\ \sigma_t^2 &= \alpha_0 + (1-\beta_1)e_{t-1}^2 + \beta_1 \sigma_{t-1}^2 \qquad (\alpha_0 > 0, \quad 0 \le \beta_1 \le 1). \end{aligned} \tag{8.68}$$

Obviously, if $\mu_t = 0$ and $\alpha_0 = 0$, then the model IGARCH(1,1) transfers to the scheme EWMA in (8.19). The recursive prediction formula (8.65) simplifies in IGARCH(1,1) to the form

$$\widehat{\sigma}_{t+\tau}^2(t) = \alpha_0 + \widehat{\sigma}_{t+\tau-1}^2(t), \qquad \tau > 1, \tag{8.69}$$

so that

$$\widehat{\sigma}_{t+\tau}^2(t) = (\tau - 1)\alpha_0 + \widehat{\sigma}_{t+1}^2(t) = (\tau - 1)\alpha_0 + \widehat{\sigma}_{t+1}^2, \qquad \tau > 1. \tag{8.70}$$

One can see that the influence of current volatilities on predictions of future volatilities really persists and that these predictions follow a line with the slope α_0.

8.3.6.2 GJR GARCH

The classical GARCH model is not capable of modeling the leverage effect, i.e., the asymmetry in the impact of past positive and negative deviations e_t on the current volatility (the volatility is prone to increase more after price drops than after price growths of the same size; see also Sect. 8.1). Glosten et al. (1993) suggested a successful modification of GARCH model correcting this drawback (see also Zakoian (1994)), which is usually denoted as GJR GARCH according to its authors (sometimes the denotation *threshold* GARCH or acronym TARCH also appears):

$$\begin{aligned} y_t &= \mu_t + e_t, \qquad e_t = \sigma_t \varepsilon_t, \\ \sigma_t^2 &= \alpha_0 + \sum_{i=1}^{r} \alpha_i e_{t-i}^2 + \sum_{j=1}^{s} \beta_j \sigma_{t-j}^2 + \sum_{k=1}^{n} \gamma_k e_{t-k}^2 I_{t-k}^-, \quad I_t^- = \begin{cases} 1 & \text{for } e_t < 0, \\ 0 & \text{for } e_t \ge 0. \end{cases} \end{aligned} \tag{8.71}$$

This model can be interpreted in such a way that the impact of "good news" ($e_{t-i} \geq 0$) modeled by means of α_i differs from the impact of "bad news" ($e_{t-i} < 0$) modeled by means of $\alpha_i + \gamma_i$. If $\gamma_i > 0$, then the bad news induce the growth of volatility so that the leverage effect works with delay i. In any case, the model behaves asymmetrically for $\gamma_i \neq 0$.

The most frequent form of the model GJR GARCH in practice (see also Example 8.3) is simply

$$y_t = \mu_t + e_t, \qquad e_t = \sigma_t \varepsilon_t, \qquad \sigma_t^2 = \alpha_0 + \alpha_1 e_{t-1}^2 + \beta_1 \sigma_{t-1}^2 + \gamma_1 e_{t-1}^2 I_{t-1}^-,$$
$$I_t^- = \begin{cases} 1 & \text{for } e_t < 0, \\ 0 & \text{for } e_t \geq 0. \end{cases} \tag{8.72}$$

8.3.6.3 EGARCH

EGARCH is another approach to asymmetry suggested by Nelson (1991). After various simplifications (e.g., originally Nelson recommended only the probability distribution GED for variables ε_t), the *exponential model* GARCH (denoted by acronym EGARCH) has the form

$$y_t = \mu_t + e_t, \;\; e_t = \sigma_t \varepsilon_t, \;\; \ln \sigma_t^2 = \alpha_0 + \sum_{i=1}^{r} \alpha_i \left| \frac{e_{t-i}}{\sigma_{t-i}} \right| + \sum_{j=1}^{s} \beta_j \ln \sigma_{t-j}^2 + \sum_{k=1}^{n} \gamma_k \frac{e_{t-k}}{\sigma_{t-k}}. \tag{8.73}$$

The application of logarithmic volatilities enables us to remove the constraints for parameter signs (e.g., $\alpha_0 > 0$). Further the leverage effect is exponential (and not quadratic) in (8.73). Obviously, the asymmetry occurs, if $\gamma_i \neq 0$ for a delay i (particularly, the leverage effect occurs for $\gamma_i < 0$).

The most frequent form of the model EGARCH in practice (see also Example 8.3) is

$$y_t = \mu_t + e_t, \;\; e_t = \sigma_t \varepsilon_t, \;\; \ln \sigma_t^2 = \alpha_0 + \alpha_1 \left| \frac{e_{t-1}}{\sigma_{t-1}} \right| + \beta_1 \ln \sigma_{t-1}^2 + \gamma_1 \frac{e_{t-1}}{\sigma_{t-1}}. \tag{8.74}$$

Moreover, before constructing asymmetric models of the type GJR GARCH and EGARCH, the asymmetry should be tested statistically (see, e.g., Engle and Ng (1993)). One usually uses the residuals $\widehat{e}_t$ obtained from the estimated (symmetric) GARCH model (see, e.g., (8.51)) and tests by means of classical t, F, or LM tests the significance of parameters in the linear model of the type

$$\widehat{e}_t^2 = \delta_0 + \delta_1 S_{t-1}^- + u_t, \quad S_{t-1}^- = \begin{cases} 1 & \text{for } \widehat{e}_{t-1} < 0 \ , \\ 0 & \text{for } \widehat{e}_{t-1} \geq 0 \ ; \end{cases} \tag{8.75}$$

$$\widehat{e}_t^2 = \delta_0 + \delta_1 S_{t-1}^- \widehat{e}_{t-1} + u_t; \tag{8.76}$$

$$\widehat{e}_t^2 = \delta_0 + \delta_1 S_{t-1}^- + \delta_2 S_{t-1}^- \widehat{e}_{t-1} + \delta_3 S_{t-1}^+ \widehat{e}_{t-1} + u_t, \quad S_{t-1}^+ = 1 - S_{t-1}^-, \tag{8.77}$$

where u_t is the classical white noise. The significant parameters

- δ_1 in the model (8.75) justifies the asymmetry of volatility in the given time series.
- δ_0 and δ_1 in the model (8.76) justify the asymmetry of volatility and the impact of size of negative deviations e_t on volatility in the given time series.
- δ_0 and δ_3 in the model (8.77) justify the asymmetry of volatility and the impact of size of positive and negative deviations e_t on volatility in the given time series.

Example 8.3 Table 8.3 and Fig. 8.8 show log returns $r_t = \ln KB_t - \ln KB_{t-1}$ of daily closing prices of stocks KB (the bank in *Société Générale Group)* in 2005 (see also prices KB_t in CZK in Fig. 8.9 for 253 trading days).

The model GJR GARCH(1,1) was estimated for r_t by means of EViews (see Table 8.4) as

$$r_t = e_t, \quad e_t = \sigma_t \varepsilon_t, \quad \sigma_t^2 = -0.0291 e_{t-1}^2 + 0.7622 \sigma_{t-1}^2 + 0.2727 e_{t-1}^2 I_{t-1}^-,$$
$$I_t^- = \begin{cases} 1 & \text{for } e_t < 0, \\ 0 & \text{for } e_t \geq 0, \end{cases}$$

where the significantly positive estimate 0.2727 of parameter γ_1 confirms the occurrence of leverage effect in the given time series.

Similarly, the model EGARCH(1,1) was estimated for r_t by means of EViews (see Table 8.5) as

$$r_t = e_t, \ e_t = \sigma_t \varepsilon_t, \ \ln \sigma_t^2 = -0.9683 + 0.1531 \left| \frac{e_{t-1}}{\sigma_{t-1}} \right| + 0.8958 \ \ln \sigma_{t-1}^2 - 0.1661 \frac{e_{t-1}}{\sigma_{t-1}},$$

where this once the significantly negative estimate –0.166 1 of parameter γ_1 confirms again the occurrence of leverage effect.

Finally, Figs. 8.10 and 8.11 plot the volatilities which are constructed by means of the estimated model GJR GARCH(1, 1) and EGARCH(1, 1) (they can be compared mutually).

◇

Table 8.3 Daily log returns of stocks KB in 2005 (252 values for 253 trading days written in columns) from Example 8.3 (see also Fig. 8.9)

	1	2	3	4	5	6	7	8	9	10	11
1	–	0.0215	–0.0281	0.0041	–0.0268	0.0068	–0.0057	0.0081	–0.0417	–0.0039	–0.0055
2	0.0255	–0.0014	0.0116	–0.0154	–0.0181	–0.0023	–0.0048	0.0526	0.0121	0.0039	0.0128
3	–0.0062	–0.0043	–0.0794	–0.0083	–0.0052	0.0100	–0.0015	–0.0029	0.0000	0.0104	–0.0087
4	–0.0194	0.0043	–0.0073	0.0146	0.0172	0.0096	0.0150	0.0060	–0.0124	–0.0021	
5	0.0238	0.0239	0.0285	–0.0098	0.0196	0.0182	–0.0060	0.0082	–0.0129	0.0208	
6	–0.0074	0.0179	–0.0303	–0.0013	0.0149	–0.0182	–0.0030	0.0145	0.0029	–0.0138	
7	0.0118	–0.0014	0.0422	–0.0341	0.0131	0.0032	–0.0060	0.0179	0.0303	0.0088	
8	0.0003	0.0216	–0.0413	–0.0353	–0.0078	–0.0095	–0.0091	0.0173	0.0230	–0.0070	
9	–0.0003	–0.0244	0.0294	0.0220	–0.0020	0.0127	0.0119	–0.0110	0.0125	0.0218	
10	0.0058	–0.0110	0.0051	0.0277	0.0114	–0.0111	0.0093	–0.0027	0.0210	0.0029	
11	0.0176	0.0308	–0.0265	–0.0250	0.0132	0.0079	–0.0060	0.0027	0.0290	–0.0218	
12	–0.0060	–0.0016	–0.0225	–0.0384	0.0156	–0.0032	–0.0106	–0.0219	0.0060	0.0088	
13	–0.0087	–0.0097	0.0289	0.0290	0.0072	–0.0160	0.0076	0.0227	–0.0086	0.0072	
14	–0.0205	–0.0071	–0.0015	0.0101	–0.0041	0.0032	–0.0122	–0.0103	0.0158	0.0066	
15	0.0059	0.0257	–0.0107	0.0040	0.0016	–0.0129	0.0003	–0.0055	0.0014	–0.0034	
16	–0.0068	–0.0048	0.0227	–0.0178	–0.0238	0.0058	–0.0058	–0.0125	0.0045	0.0012	
17	0.0286	0.0075	0.0015	–0.0150	0.0159	0.0290	–0.0006	0.0193	–0.0217	0.0086	
18	0.0242	–0.0180	0.0006	–0.0314	0.0281	0.0094	–0.0127	0.0082	–0.0043	–0.0071	
19	–0.0343	–0.0145	0.0318	–0.0316	0.0000	–0.0091	0.0019	0.0014	–0.0090	0.0043	
20	0.0093	–0.0041	–0.0087	–0.0248	0.0015	0.0141	0.0276	–0.0571	–0.0023	–0.0086	
21	–0.0136	0.0296	0.0035	0.0259	–0.0287	0.0068	0.0105	–0.0276	–0.0106	0.0029	
22	0.0058	–0.0011	0.0020	0.0568	0.0144	0.0147	0.0228	0.0132	–0.0149	0.0000	
23	–0.0087	–0.0081	–0.0218	0.0336	–0.0281	0.0166	–0.0035	0.0116	0.0149	–0.0116	
24	0.0186	–0.0081	–0.0256	–0.0064	–0.0178	0.0089	–0.0032	–0.0253	0.0089	0.0029	
25	–0.0098	–0.0165	–0.0291	0.0166	0.0033	–0.0119	–0.0272	–0.0415	–0.0172	–0.0017	

Source: kurzy.cz (https://akcie-cz.kurzy.cz/akcie/komercni-banka-590/graf_2005)

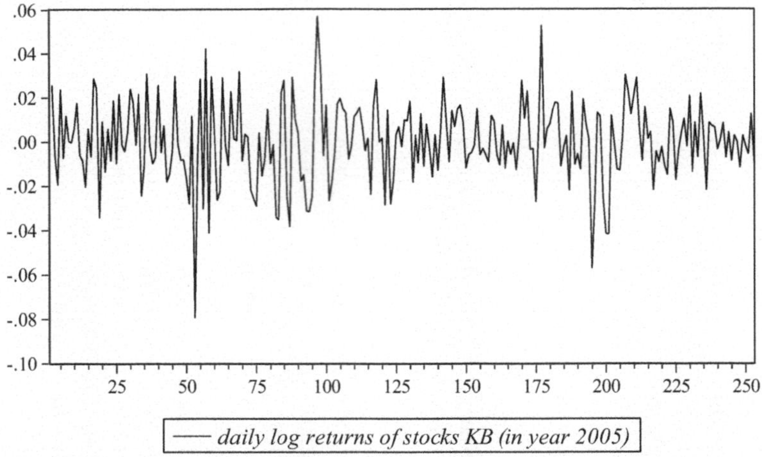

Fig. 8.8 Daily log returns of stocks KB in 2005 (252 values for 253 trading days) from Example 8.3 (see also Table 8.3). Source: calculated by EViews

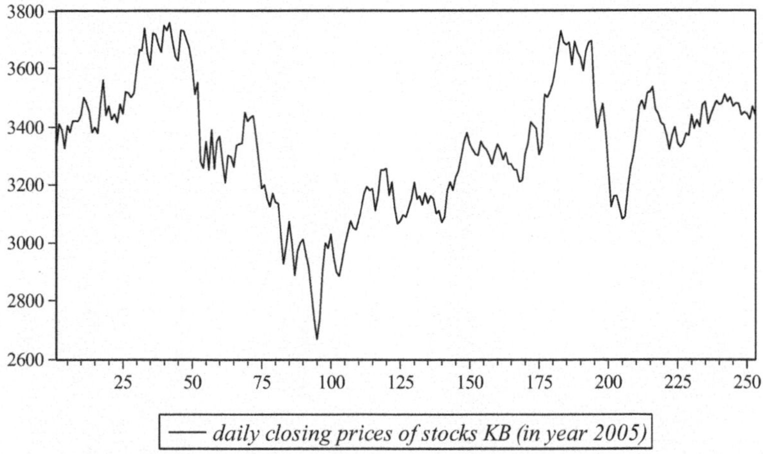

Fig. 8.9 Daily closing prices of stocks KB in year 2005 (values in CZK for 253 trading days) from Example 8.3. Source: kurzy.cz (https://akcie-cz.kurzy.cz/akcie/komercni-banka-590/graf_2005)

Remark 8.6 In financial applications, one combines frequently the asymmetric models for volatility with the classical AR models for mean value. For example, for the daily log returns r_t of stocks IBM from July 1962 to 1999 (9442 values), Tsay (2002) constructed the model AR(2)-GJR GARCH(1,1) of the form

Table 8.4 Estimation of the process GJR GARCH(1, 1) from Example 8.3 (*daily log returns of stocks KB in year 2005*)

Included observations: 252				
Convergence achieved after 26 iterations				
Bollerslev–Wooldrige robust standard errors and covariance				
GARCH = C(2) + C(3)∗RESID(-1)^2 + C(4)∗RESID(-1)^2∗(RESID(-1)<0) + C(5)∗GARCH(-1)				
	Coefficient	Std. Error	*z*-Statistic	Prob.
C	0.000117	0.001024	0.114515	0.9088
	Variance Equation			
C	4.26E-05	1.93E-05	2.210756	0.0271
RESID(-1)^2	−0.029062	0.039569	−0.734468	0.4627
RESID(-1)^2∗(RESID(-1)<0)	0.272706	0.125334	2.175834	0.0296
GARCH(-1)	0.762199	0.083962	9.077869	0.0000

Source: Calculated by EViews

Table 8.5 Estimation of the process EGARCH(1, 1) from Example 8.3 (*daily log returns of stocks KB in year 2005*)

Included observations: 252				
Convergence achieved after 20 iterations				
Bollerslev–Wooldrige robust standard errors and covariance				
LOG(GARCH) = C(2) + C(3)∗ABS(RESID(-1)/@SQRT(GARCH(-1))) + C(4)∗RESID(-1)/ @SQRT(GARCH(-1)) + C(5)∗LOG(GARCH(-1))				
	Coefficient	Std. Error	*z*-Statistic	Prob.
C	−1.34E-05	0.001040	−0.012899	0.9897
	Variance Equation			
C(2)	−0.968342	0.415948	−2.328036	0.0199
C(3)	0.153073	0.069119	2.214631	0.0268
C(4)	−0.166100	0.059934	−2.771389	0.0056
C(5)	0.895763	0.048687	18.39837	0.0000

Source: Calculated by EViews

$$r_t = 0.043 - 0.022 r_{t-2} + e_t,$$
$$e_t = \sigma_t \varepsilon_t, \sigma_t^2 = 0.098 e_{t-1}^2 + 0.954 \sigma_{t-1}^2 + \left(0.060 - 0.052 e_{t-1}^2 - 0.069 \sigma_{t-1}^2\right) I_{t-1}^-,$$
$$I_t^- = \begin{cases} 1 & \text{for } e_t < 0, \\ 0 & \text{for } e_t \geq 0. \end{cases}$$

◇

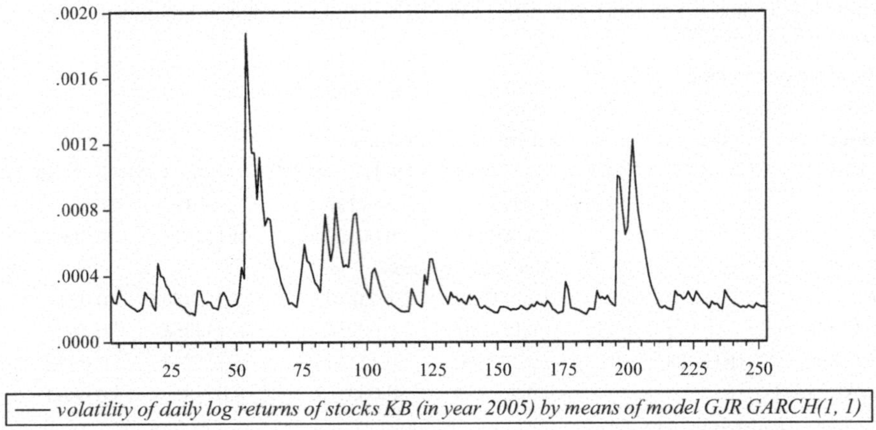

Fig. 8.10 Volatility of daily log returns of stocks KB in 2005 (252 values for 253 trading days) estimated by means of model GJR GARCH(1, 1) in Example 8.3

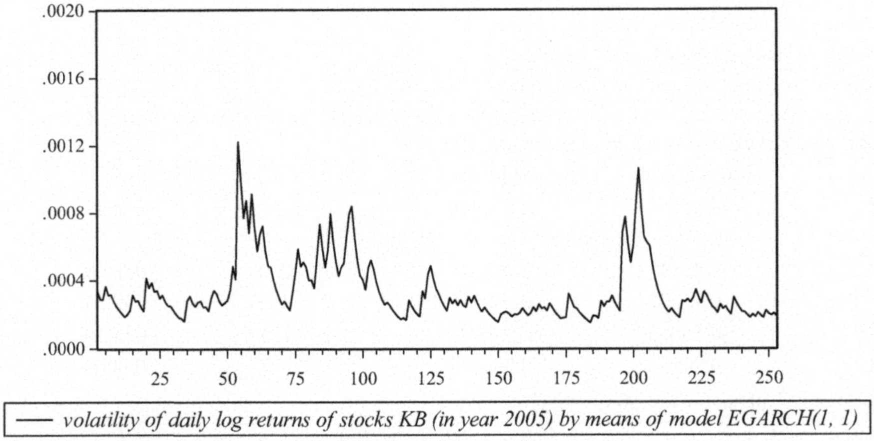

Fig. 8.11 Volatility of daily log returns of stocks KB in 2005 (252 values for 253 trading days) estimated by means of model EGARCH(1, 1) in Example 8.3 (compare with volatility estimated by means of GJR GARCH(1,1) in Fig. 8.10)

8.3.6.4 GARCH-M

The return of financial asset often depends on its volatility (e.g., investors are compensated for higher risk by higher return). Therefore, Engle et al. (1987) suggested a modification of ARCH models (and later GARCH models), where the volatility or its square root enters the mean equation (so-called ARCH-M models).

For instance, the model GARCH(1,1)-M (i.e., GARCH-*in-mean*) has the form

$$\begin{aligned} y_t &= \mu_t + \gamma_1 \sigma_t^2 + e_t \quad (\text{or } y_t = \mu_t + \gamma_1 \sigma_t + e_t), \quad e_t = \sigma_t \varepsilon_t, \\ \sigma_t^2 &= \alpha_0 + \alpha_1 e_{t-1}^2 + \beta_1 \sigma_{t-1}^2. \end{aligned} \tag{8.78}$$

If the parameter γ_1 is significantly positive, then the increased risk manifests itself by increased volatility, which causes the increased level of time series (i.e., the increased mean).

8.3.6.5 Models of Stochastic Volatility SV

The volatility equation of GARCH model is obviously fully deterministic (in the sense of conditioning by past information). The denotation *stochastic volatility* is used in this context only in such a case, when the volatility equation contains additional error term which remains random even if one conditions by the past information. Although simple examples of such models are the autoregressive models of volatility from Sect. 8.3.3, the general SV model (see, e.g., Taylor (1994)) is presented as

$$y_t = \mu_t + e_t, \qquad e_t = \sigma_t \varepsilon_t, \qquad \ln \sigma_t^2 = \alpha_0 + \sum_{j=1}^{s} \beta_j \ln \sigma_{t-j}^2 + u_t, \tag{8.79}$$

where $\{u_t\}$ is another white noise (mostly *iid* with normal distribution) that is independent on $\{\varepsilon_t\}$ (the formulation by means of logarithmic volatility enables to ignore the condition of nonnegativeness similarly as in EGARCH model). The models of the type SV turned out well, e.g., in the context of option pricing, where the volatility of underlying asset enters the famous Black–Scholes formula (see Sect. 8.3.2). On the other hand, the difficult estimation is one of drawbacks of these models.

Remark 8.7 We have mentioned in the beginning of this section that there are many modifications of GARCH (and many new ones probably will appear in future), e.g.:

- FIGARCH (*fractionally* IGARCH) are FI models (i.e., the long-memory processes from Sect. 6.7), but for volatility, an analogical character has the model FIEGARCH and others.
- QGARCH (*quadratic* GARCH) models (see Sentana (1995)) reflect asymmetry in such a way that the delayed deviations e_{t-i} figure directly on the right-hand side of volatility equation (in addition to the squared delayed deviations e_{t-i}^2).
- APARCH (*asymmetric power* ARCH) models (see Ding et al. (1993)) induce the long-memory property by means of a suitable power transformation of volatilities and are capable of expressing well the fat tails, excess kurtosis, and leverage effects.

◇

8.4 Exercises

Exercise 8.1 Repeat the analysis from Examples 8.1 and 8.2 (daily log returns of index PX in 2016), but only for last 100 values of time series $\{r_t\}$ (*hint:* $r_t = \sigma_t \varepsilon_t$, $\sigma_t^2 = 0.0168 r_{t-1}^2 + 0.5419 \sigma_{t-1}^2$).

Exercise 8.2 Repeat the analysis from Example 8.3 (daily log returns of stocks KB in 2005), but only for last 203 values of time series $\{r_t\}$ $\Big($*hint:* $r_t = e_t,\ e_t = \sigma_t \varepsilon_t,$

$$\sigma_t^2 = -0.0641 e_{t-1}^2 + 0.8062 \sigma_{t-1}^2 + 0.2629 e_{t-1}^2 I_{t-1}^-,\ I_t^- = \begin{cases} 1 & \text{for } e_t < 0, \\ 0 & \text{for } e_t \geq 0, \end{cases};$$

$$r_t = e_t,\ \ e_t = \sigma_t \varepsilon_t,\ \ \ln \sigma_t^2 = -0.9576 + 0.0559 \left| \frac{e_{t-1}}{\sigma_{t-1}} \right| + 0.8862 \ln \sigma_{t-1}^2 - 0.1811 \frac{e_{t-1}}{\sigma_{t-1}} \Big).$$

Chapter 9
Other Methods for Financial Time Series

9.1 Models Nonlinear in Mean Value

In Sect. 8.2, we presented a general nonlinear scheme

$$y_t = \mu_t + e_t = \mu_t + \sigma_t \cdot \varepsilon_t = \mu_t + \sqrt{h_t} \cdot \varepsilon_t = g(\Omega_{t-1}) + \sqrt{h(\Omega_{t-1})} \cdot \varepsilon_t, \qquad (9.1)$$

where $\varepsilon_t = e_t/\sigma_t$ are standardized shocks e_t (ε_t are usually *iid* in contrast to the uncorrelated e_t, which may be possibly dependent). In this framework, we dealt so far (see Sect. 8.3) with the volatility equation $\sigma_t^2 = h(\Omega_{t-1})$ only (an exception was the model GARCH-M). Now on the contrary, we focus on various nonlinear models for the (conditional) mean equation $\mu_t = g(\Omega_{t-1})$, even though we shall present only the most important ones from the point of view of applications in finance (see also monographs by Priestley (1988), Tong (1990), and others). These models are mostly specific cases (acceptable from the computational point of view) of the general model

$$y_t = f(e_t, e_{t-1}, e_{t-2}, \ldots) \qquad (9.2)$$

(see also (8.9)), where e_t is an (uncorrelated) white noise with the variance σ_e^2.

More specifically, the models in this section have been motivated by some nonlinear characteristics of data from practice (not necessarily from financial practice only). Examples are the asymmetry between increase and decrease of time series, the limit cycle (i.e., limit form of the process in regular cycles if one excludes all random elements), and the dependence of frequency on amplitude in periodic behavior of some processes (e.g., the frequency increases with decreasing amplitude and decreases with increasing amplitude, or on the contrary).

T. Cipra, *Time Series in Economics and Finance*,
https://doi.org/10.1007/978-3-030-46347-2_9

9.1.1 Bilinear Models

While the linear process (6.17) can be looked upon in such a way that it originates by means of Taylor expansion of the function $f(\cdot)$ in (9.2) to the first order, in the case of bilinear models it should be the expansion to the second order

$$y_t = \alpha + \sum_{i=1}^{p} \varphi_i y_{t-i} + \sum_{j=1}^{q} \theta_j e_{t-j} + \sum_{n=1}^{P} \sum_{m=1}^{Q} \beta_{mn} y_{t-n} e_{t-m} + e_t. \tag{9.3}$$

Some special cases of (9.3) belong to the class of models with conditional heteroscedasticity assuming usually $e_t \sim iid\ (0, \sigma_e^2)$: e.g., if we consider the model

$$y_t = \mu + \sum_{m=1}^{Q} \beta_m e_t e_{t-m} + e_t \tag{9.4}$$

then it holds

$$\mu_t = \mathrm{E}(y_t|\Omega_{t-1}) = \mu, \qquad \sigma_t^2 = \mathrm{var}(y_t|\Omega_{t-1}) = \left(1 + \sum_{m=1}^{Q} \beta_m e_{t-m}\right)^2 \sigma_e^2. \tag{9.5}$$

The most frequent model of this type is so-called *completely bilinear model* of the form

$$y_t = \sum_{n=1}^{P} \sum_{m=1}^{Q} \beta_{mn} y_{t-n} e_{t-m} + e_t \tag{9.6}$$

(here again the white noise values e_t are usually assumed to be independent). If dealing with models of the type (9.6), the form of matrix (β_{mn}) is substantial. Moreover, one distinguishes so-called *superdiagonal* or *diagonal* or *subdiagonal* models depending on whether the matrix (β_{mn}) has zero elements only above the main diagonal or only on the main diagonal or only under the main diagonal, respectively.

The detailed theoretical analysis of some special cases of superdiagonal, diagonal, and subdiagonal models (including conditions of stationarity for these models) has shown some paradoxical results: e.g., the correlation structures of some bilinear models correspond to the correlation structures of simple linear processes ARMA (or even to the one of white noise). It has the practical consequence for the identification of bilinear models when the correlogram does not distinguish them from ARMA models. Moreover, the practical calculation of partial autocorrelation function of bilinear models is so complex that it cannot be recommended for identification of such models. Fortunately, applications show that a sufficient distinguishing criterion in questionable cases is the form of autocorrelation function of the squared time series $\{y_t^2\}$. In order to demonstrate it, we will present some results that have been derived for the simplest types of bilinear models (see, e.g., Granger and Andersen (1978)):

1. Example of *superdiagonal* model:

$$y_t = \beta y_{t-n} e_{t-m} + e_t, \qquad m < n, \tag{9.7}$$

where the sufficient and necessary condition of stationarity has the form $\lambda^2 < 1$ for $\lambda = \beta\sigma_e$. Then the corresponding time series $\{y_t\}$ has the zero mean value, the variance $\sigma_e^2/(1 - \lambda^2)$, and the autocorrelation function

$$\rho_k = 0 \quad \text{for } k \neq 0. \tag{9.8}$$

Moreover, the autocorrelation function $\rho_k^{(2)}$ of the squared time series $\{y_t^2\}$ fulfills

$$\rho_k^{(2)} = \lambda^n \rho_{k-n}^{(2)} \quad \text{for } k > m, \tag{9.9}$$

so that these autocorrelation functions identify the time series $\{y_t\}$ as the white noise, while the time series $\{y_t^2\}$ as the process ARMA(n, m) (see (6.46)).

2. Example of *diagonal* model:

$$y_t = \beta y_{t-1} e_{t-1} + e_t, \tag{9.10}$$

where the sufficient and necessary condition of stationarity has again the form $\lambda^2 < 1$. The corresponding time series $\{y_t\}$ has the mean value $\beta\sigma_e^2$, the variance $\sigma_e^2(1 + \lambda^2 + \lambda^4)/(1 - \lambda^2)$, and the autocorrelation function

$$\rho_k = \frac{\lambda^2(1-\lambda^2)}{1+\lambda^2+\lambda^4} \quad \text{for } k = 1, \qquad \rho_k = 0 \quad \text{for } k > 1 \tag{9.11}$$

(under stronger assumption $e_t \sim iid\ N(0, \sigma_e^2)$). The autocorrelation function $\rho_k^{(2)}$ of $\{y_t^2\}$ fulfills

$$\rho_k^{(2)} = \lambda^2 \rho_{k-1}^{(2)} \quad \text{for } k > 1, \tag{9.12}$$

so that these autocorrelation functions identify the time series $\{y_t\}$ as the process MA(1), while the time series $\{y_t^2\}$ as the process ARMA(1, 1).

3. Example of *subdiagonal* model:

$$y_t = \beta y_{t-2} e_{t-3} + e_t, \tag{9.13}$$

where the sufficient and necessary condition of stationarity has again the form $\lambda^2 < 1$. The corresponding time series $\{y_t\}$ has the zero mean value, the variance $\sigma_e^2/(1 - \lambda^2)$, and the autocorrelation function

$$\rho_k = 0 \quad \text{for } k \neq 0. \tag{9.14}$$

The autocorrelation function $\rho_k^{(2)}$ of $\{y_t^2\}$ fulfills

$$\rho_k^{(2)} = \lambda^2 \rho_{k-2}^{(2)} \quad \text{for} \quad k > 3, \tag{9.15}$$

so that these autocorrelation functions identify the time series $\{y_t\}$ as white noise, while the time series $\{y_t^2\}$ as the process ARMA(2, 3). Anyway, the analysis of theoretical properties of subdiagonal models is usually much more complex than for the superdiagonal and diagonal models.

The bilinear models can be estimated similarly as the linear models applying a recursive calculation of values e_t in dependence on the model parameters (see Sect. 6.3.2). Also the predictions can be constructed analogously as in the linear case. For example in the diagonal model (9.10), it is possible to derive a necessary condition of invertibility in the form (see Granger and Newbold (1986))

$$\frac{\lambda^2 (2\lambda^2 + 1)}{|1 - \lambda^2|} < 1 \tag{9.16}$$

(i.e., $|\lambda| < 0.605$). The prediction in this model can be constructed as

$$\widehat{y}_{t+1}(t) = \widehat{\beta}\, y_t \widehat{e}_t, \quad \widehat{y}_{t+\tau}(t) = \mathrm{E}(y_{t+\tau}) = \widehat{\beta}\, \widehat{\sigma}_e^2 \quad \text{for } \tau > 1. \tag{9.17}$$

Remark 9.1 Granger and Andersen (1978) present the following financial application of bilinear models. A time series $\{y_t\}$ of stock prices of a big corporation with length of 169 observations was originally estimated by means of a linear model $y_t = e_t + 0.26 e_{t-1}$ with the variance of white noise estimated as 24.8. However, the time series $\{e_t\}$ originally looked on as a (linear) white noise was identified and estimated as the bilinear model of the form $e_t = 0.02 e_{t-1} u_{t-1} + u_t$, where u_t denotes a white noise with estimated variance 23.5. Even though the reduction of white noise variance seems insignificant (from 24.8 to 23.5), the mean squared error *MSE* of the one-step-ahead prediction (see (2.11)) calculated for the last fifteen out-of-sample observations (i.e., h=15) decreased by 11% when applying the bilinear scheme.

◇

9.1.2 Threshold Models SETAR

Threshold models SETAR replaces linear relations by a piecewise linear function $f(\cdot)$ in (9.2), the changes of this function being controlled not from the time space but from the state space of function values. More specifically, these models are constructed applying some critical limits (thresholds) and change when the observed time series exceeds these thresholds (a similar principle has been applied for GJR GARCH processes from Sect. 8.3.6, but for their conditional variance (volatility), and not for the conditional mean which is just the case of models SETAR). A more general framework are *switching regimes models*, where the particular regimes can be controlled by fixed (deterministic) thresholds (see just the models SETAR) or by a stochastic way (see, e.g., MSW models later in this section).

Let us consider a very simple model SETAR of the form

$$y_t = \begin{cases} -1.8y_{t-1} + e_t & \text{for } y_{t-1} < 0, \\ 0.5y_{t-1} + e_t & \text{for } y_{t-1} \geq 0, \end{cases} \tag{9.18}$$

where e_t are *iid* N(0, 1) (obviously, this model has a single threshold in zero, where the past value y_{t-1} with time delay $d = 1$ controls the current value y_t). Figure 9.1 plots one of simulations of this process with length 200 and zero starting value (trajectories of other simulations are very similar). At first glance, one can see some interesting properties of this process:

- The process is stationary (even though the first autoregressive polynomial has a root lying significantly inside the unit circle in complex plane).
- The process is (geometrically) ergodic, i.e., its sample mean converges (in a specific way) to the theoretical mean.

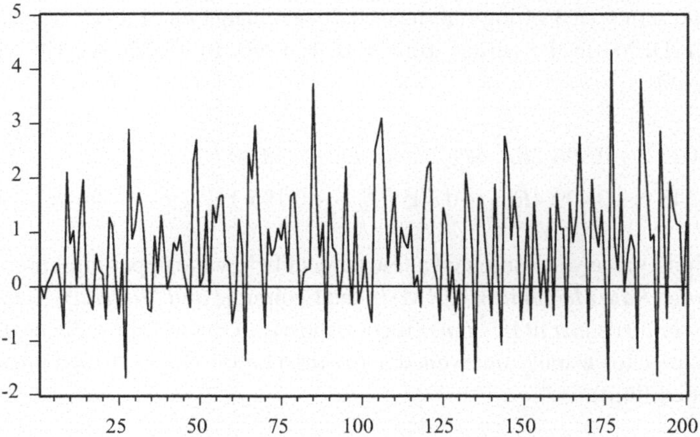

Fig. 9.1 Simulation of threshold model (9.18) with one zero threshold

- The given realization shows an asymmetry between its upward and downward jumps: if $y_{t-1} < 0$, then the process tends to turn over immediately to a positive (i.e., opposite) value due to the significantly negative value of the autoregressive parameter –1.8, while if $y_{t-1} > 0$, then the turnover to negative (i.e., opposite) values usually takes more time units. It implies directly that the process attains more values above the zero threshold than below it and that it shows immediate significant jumps upward to the positive values, as soon as it becomes negative.
- The sample mean of the given realization is 0.75 with standard deviation 0.08 of this sample estimate so that it lies significantly above the zero threshold (the theoretical mean of the given process is the weighted average of its conditional mean values for both threshold regions with weights corresponding to the probabilities of both regions from the point of view of stationary distribution of the process).

In general, the process $\{y_t\}$ denoted usually as the *thresholds autoregressive model* with r autoregressive regimes of orders p_j and controlling delay d (the acronym SETAR comes from *self-exciting threshold* AR to stress the self-regulation of process) has the form

$$y_t = \alpha^{(j)} + \varphi_1^{(j)} y_{t-1} + \ldots + \varphi_{p_j}^{(j)} y_{t-p_j} + e_t^{(j)} \quad \text{for} \quad P_{j-1} \leq y_{t-d} < P_j, \ j = 1, \ldots, r, \tag{9.19}$$

where d and r are given natural numbers, thresholds P_j are real numbers fulfilling inequalities $-\infty = P_0 < P_1 < \ldots < P_r = \infty$, and $\{e_t^{(j)}\}$ are mutually independent white noises usually of the type *iid* with variances σ_j^2.

The identification and (simultaneous) estimation of the models SETAR is mostly realized by applying information criteria of the type AIC from Sect. 6.3.1 (see Tong (1983, 1990)).

Remark 9.2 Chappell et al. (1996) estimated the model SETAR with one threshold for the time series $\{E_t\}$ of log returns of daily exchange rate French franc / German mark (FRF/DEM) in the period from May 1, 1990, to March 30, 1992 (i.e., 450 observations)

$$E_t = \begin{cases} 0.022\,2 + 0.996\,2E_{t-1} + e_t^{(1)} & \text{for } E_{t-1} < 5.830\,6\,, \\ 0.348\,6 + 0.439\,4E_{t-1} + 0.305\,7E_{t-2} + 0.195\,1E_{t-3} + e_t^{(2)} & \text{for } E_{t-1} \geq 5.830\,6. \end{cases}$$

The threshold value was estimated a few percent below the upper limit prescribed by the *Exchange Rate Mechanism* (ERM) in the Economic *and Monetary Union (EMU) which was in force just at this time* (it corresponds to reality, since the central banks of particular states usually intervened some time before the exchange rate achieved the permitted limit).

◇

Remark 9.3 As the conditional mean value of models SETAR is not continuous (indeed, the thresholds are points of discontinuity of μ_t), one has suggested the models STAR (*smooth transition* AR *model*; see Chan and Tong (1986) and others). For instance, in the case of model with two regimes it can be

$$y_t = \alpha^{(1)} + \sum_{i=1}^{p} \varphi_i^{(1)} y_{t-i} + F\left(\frac{y_{t-d} - \Delta}{s}\right)\left(\alpha^{(2)} + \sum_{i=1}^{p} \varphi_i^{(2)} y_{t-i}\right) + e_t, \qquad (9.20)$$

where parameters Δ and s and a *transition function* $F(\cdot)$ determine the way of transition between both regimes (a usual choice of F in practice is the distribution function of logistic or exponential distribution). Even though the corresponding conditional mean value μ_t is assumed to be differentiable in continuous time, the consistency of parameter estimation is usually problematic (particularly for the location Δ and scale s).

◇

9.1.3 Asymmetric Moving Average Models

Another approach to asymmetry is represented by *asymmetric moving average processes* (see Wecker (1981))

$$y_t = e_t^+ + \theta_1^+ e_{t-1}^+ + \ldots + \theta_q^+ e_{t-q}^+ + e_t^- + \theta_1^- e_{t-1}^- + \ldots + \theta_q^- e_{t-q}^-, \qquad (9.21)$$

where e_t is a normal white noise with variance σ_e^2, $e_t^+ = \max(0, e_t)$, $e_t^- = \min(0, e_t)$, and $\theta_1^+, \ldots, \theta_q^-$ are parameters. If $\theta_j^+ = \theta_j^-$ ($j = 1, \ldots, q$), then (9.21) is the classical "symmetric" moving average process MA(q) (see (6.24)).

In contrast to the symmetric moving average models, the asymmetric ones may not have zero mean value, even though their correlation structure is similar to the symmetric case with truncation point in q. For example, the asymmetric process MA(1) fulfills

$$\mu = \mathrm{E}(y_t) = \frac{(\theta_1^+ - \theta_1^-)\sigma_e}{\sqrt{2\pi}}, \qquad (9.22)$$

$$\gamma_0 = \mathrm{var}(y_t) = \frac{\left(1 + (\theta_1^+)^2\right)\sigma_e^2}{2} + \frac{\left(1 + (\theta_1^-)^2\right)\sigma_e^2}{2} - \mu^2,$$

$$\rho_1 = \frac{(\theta_1^+ + \theta_1^-)\sigma_e^2}{2}, \quad \rho_k = 0 \quad \text{for } k > 1. \tag{9.23}$$

If $\theta_1{}^+ = -\theta_1{}^-$, then such a process has obviously the same correlation structure as a white noise.

9.1.4 Autoregressive Models with Random Coefficients RCA

Autoregressive processes with random coefficients RCA were originally suggested to model more appropriately the conditional mean value of the autoregressive process $\{y_t\}$ regarding its parameters as random variables (see Nicholls and Quinn (1982)), but one could classify them also as models with conditional heteroscedasticity. The basic form is

$$y_t = \alpha + \sum_{i=1}^{p} (\varphi_i + \delta_{it}) y_{t-i} + e_t, \tag{9.24}$$

where $\{\boldsymbol{\delta}_t\} = \{(\delta_{1t}, \ldots, \delta_{pt})'\}$ is a sequence of independent random vectors with zero (vector) mean and variance matrix $\boldsymbol{\Sigma}_{\boldsymbol{\delta\delta}}$ independent of the white noise $\{e_t\}$. The conditional mean and variance of (9.24) fulfill

$$\mu_t = \mathrm{E}(y_t | \Omega_{t-1}) = \alpha + \sum_{i=1}^{p} \varphi_i y_{t-i},$$

$$\sigma_t^2 = \mathrm{var}(y_t | \Omega_{t-1}) = \sigma_e^2 + (y_{t-1}, \ldots, y_{t-p}) \boldsymbol{\Sigma}_{\boldsymbol{\delta\delta}} (y_{t-1}, \ldots, y_{t-p})'. \tag{9.25}$$

9.1.5 Double Stochastic Models

Double stochastic models extend the principle of RCA by modeling the parameters of an ARMA process (or another classical linear process) by means of other random processes (see, e.g., Chen and Tsay (1993); Tjøstheim (1986)). Special cases are *functional-coefficient autoregressive processes* FAR of the form

$$y_t = f_1(y_{t-1}, \ \ldots, \ y_{t-k}) \cdot y_{t-1} + \ldots + f_p(y_{t-1}, \ \ldots, \ y_{t-k}) \cdot y_{t-p} + e_t, \tag{9.26}$$

where functions f_1, ..., f_p should be differentiable to the second order, e.g., the *exponential autoregressive model* (see Haggan and Ozaki (1981))

$$\begin{aligned} y_t = & \left(\varphi_1 + \pi_1 \exp\left(-\gamma \cdot y_{t-1}^2\right)\right) \cdot y_{t-1} + \ldots \\ & + \left(\varphi_p + \pi_p \exp\left(-\gamma \cdot y_{t-1}^2\right)\right) \cdot y_{t-p} + e_t, \end{aligned} \tag{9.27}$$

where the autoregressive parameters depend exponentially on the amplitude $|y_{t-1}|$: obviously, these parameters are close to φ_i for high amplitudes, while they are close to $\varphi_i + \pi_i$ for low amplitudes (moreover, the amplitude effect is modified by means of the parameter $\gamma > 0$). Such *amplitude-dependent* models possess some typical features of physical process, e.g., the dependence of frequency on amplitude, the limit cycle, and others (see introduction to this Sect. 9.1).

9.1.6 Switching Regimes Models MSW

In contrast to the models, the regimes of which are controlled by observable variables (e.g., by the location of a past value of the given time series between thresholds of SETAR), the models denoted as MSW change particular regimes in an unobservable (latent) way, namely by means of a Markov mechanism (MSW is the acronym for *Markov switching*).

Particularly, the simplest case of so-called process MSA (*Markov-switching autoregressive*) with two regimes has the form

$$y_t = \begin{cases} \alpha^{(1)} + \sum_{i=1}^{p_1} \phi_i^{(1)} y_{t-i} + e_t^{(1)} & \text{for } s_t = 1, \\ \alpha^{(2)} + \sum_{i=1}^{p_2} \phi_i^{(2)} y_{t-i} + e_t^{(2)} & \text{for } s_t = 2, \end{cases} \tag{9.28}$$

where s_t is a Markov chain with values 1 and 2 and transition probabilities

$$\mathrm{P}(s_t = 2 \mid s_{t-1} = 1) = w_1, \qquad \mathrm{P}(s_t = 1 \mid s_{t-1} = 2) = w_2 \tag{9.29}$$

and $\{e_t^{(1)}\}$ and $\{e_t^{(2)}\}$ are (mutually independent) *iid* white noises. Obviously, a small value of the transit probability w_i means that the process remains a longer time in the state i (the reciprocal value $1/w_i$ is equal to the mean period of stay (the mean holding time) in this state. One can see that the process MSA makes use of the Markov probability mechanism to control the transits among particular conditional mean values.

Due to the stochastic (i.e., latent) control of regime switching, the construction of models MSW is not simple (see, e.g., Hamilton (1989, 1994)). One can make use of some estimation techniques based on simulations, e.g., MCMC method (*Markov Chain Monte Carlo*). The construction of prediction is more complex as well, combining linearly the predictions constructed for particular regimes (in contrast to predicting in a threshold model, where the observed past value y_{t-d} unambiguously determines in which regime the prediction will be constructed).

Remark 9.4 In financial practice, the models MSW are popular for modeling time series of gross domestic products GDP_t. For example, Tsay (2002) constructed the following model for the quarterly (seasonally adjusted) time series of GDP growth (in %) in the USA in years 1947–1990

$$y_t = \begin{cases} 0.909 + 0.265y_{t-1} + 0.029y_{t-2} - 0.126y_{t-3} - 0.110y_{t-4} + e_t^{(1)} & \text{for } s_t = 1, \\ -0.420 + 0.216y_{t-1} + 0.628y_{t-2} - 0.073y_{t-3} - 0.097y_{t-4} + e_t^{(2)} & \text{for } s_t = 2, \end{cases}$$

where the standard deviations $\sigma_e^{(1)}$ and $\sigma_e^{(2)}$ of white noises were estimated as 0.816 and 1.017. Hence the mean values of the process $\{y_t\}$ for the first and second state can be evaluated as 0.965 and –1.288 (evidently, the first state corresponds to the quarters of economic growth, while the second state to the quarters of economic decline). Finally, the transit probabilities w_1 and w_2 were estimated as 0.118 and 0.286 which can be interpreted as follows: to get out of recession is more probable than to enter it. More specifically, the mean length of recession period is $1/0.286 = 3.50$ quarters, i.e., less than 1 year, while the mean length of boom period is $1/0.118 = 8.47$ quarters, i.e., more than 2 years).

◇

9.2 Further Models for Financial Time Series

There are further approaches to nonlinear models of financial time series that cannot be classified analogously as in Sect. 9.1 (due to their philosophy, due to the type of analysis, etc.). Two examples of such approaches will be given here as illustrations.

9.2.1 Nonparametric Models

Let in the simplest case two financial variables y_t and x_t be linked by the relation

$$y_t = m(x_t) + e_t, \tag{9.30}$$

where $m(\cdot)$ is an unknown (nonlinear, but smooth) function and $\{e_t\}$ is a white noise. A natural task is to estimate the function $m(\cdot)$ as truly as possible by means of observed data $y_1, ..., y_T$ and $x_1, ..., x_T$.

As the arithmetic average of values $e_1, ..., e_T$ converges to zero with increasing T (according to the law of large numbers), it seems that the natural estimate of $m(\cdot)$ is the arithmetic average of values $y_1, ..., y_T$

$$\frac{1}{T}\sum_{t=1}^{T} y_t. \tag{9.31}$$

However, there is a problem consisting in the fact that one should estimate the function $m(x)$ for a given value of argument x, but the observed values $x_1, ..., x_T$ may differ from x significantly. Therefore, it is reasonable to replace (9.31) by the weighted average

$$\widehat{m}(x) = \frac{1}{T}\sum_{t=1}^{T} w_t(x) \cdot y_t, \tag{9.32}$$

where the weights $w_t(x)$ are large (or small) for such indices t, for which the observed values x_t lie close to x (or far from x), respectively.

In other words, one weighs the values y_t by means of locally weighted averages using weights of described properties. A unifying principle in this context consists in application of so-called *kernel* (see, e.g., Härdle (1990)), which is a suitable function $K(\cdot)$ with properties of a probability density

$$K(x) \geq 0, \qquad \int_{-\infty}^{\infty} K(z)\ \mathrm{d}z = 1. \tag{9.33}$$

Specialized statistical software systems offer various choices of kernels $K(x)$ (in the module usually titled as *nonparametric regression*). Instead of (9.32), one then puts

$$\widehat{m}(x) = \frac{\sum_{t=1}^{T} K_h(x - x_t) \cdot y_t}{\sum_{t=1}^{T} K_h(x - x_t)}, \tag{9.34}$$

where $h > 0$ is so-called *bandwidth* and $K_h(x) = (1/h) \cdot K(x/h)$ is the kernel calibrated by the scale h. The estimate (9.34) known in nonparametric regression as *Nadaraya–Watson estimate* reflects the distances of particular regressors x_t from x in a systematic way.

9.2.2 Neural Networks

Neural networks enable to parametrize any continuous (nonlinear) function. Therefore, this approach may be useful also for more complex financial models (see, e.g., Ripley (1993); Tsay (2002)).

A simple neural network can be looked upon as a system that connects an *input layer x* over possible *hidden layers h* to an *output layer o*. Each layer is represented by a given number of *nodes* (*neurons*). A neural network processes information from one layer to the next one by means of *activation functions* (it holds for the neural networks of the type *feed-forward*, while backward connections are also allowable in the networks of the type *feed-back*). For example, the network of the type 3-2-1 has three nodes in the input layer, two nodes in the hidden layer, and one node in the output layer. A typical activation function has the form

$$o_j = f_j\left(\alpha_{0j} + \sum_{i \to j} w_{ij} x_i\right), \tag{9.35}$$

where f_j is the logistic function of the type

$$f_j(z) = \frac{\exp(z)}{1 + \exp(z)}, \tag{9.36}$$

x_i is the value of the ith input node, o_j is the value of the jth output node, α_{0j} is called *bias*, w_{ij} are *weights*, and the summation $i \to j$ means summing over all input nodes feeding to j. If a node has an activation function of the form

$$f_j(z) = \begin{cases} 1 & \text{for } z > 0, \\ 0 & \text{for } z \leq 0, \end{cases} \tag{9.37}$$

then one calls it a *threshold node*, with “1” denoting that the node fires (revitalizes) its message. The final connection from inputs to outputs can be more complex, if there are hidden layers (due to compounding gradually the activation functions), e.g.:

$$o = f\left(\alpha_0 + \sum_{i \to o} w_i x_i + \sum_{k \to o} w_{ko} f_k\left(\alpha_{0k} + \sum_{i \to k} w_{ik} x_i\right)\right), \tag{9.38}$$

where not only a direct connection from the input to the output layer is possible, but also an indirect one by means of a hidden layer (with summing index k in (9.38)).

A typical application of neural networks for financial time series is the following one. One observes data $\mathbf{x}_t$ and y_t ($t = 1, ..., T$), where $\mathbf{x}_t$ is a vector of input values at time t and y_t is an observation of given time series at time t. In addition to it, we have also model output values o_t expressed analytically for particular times t by means of relations of the type (9.38). Then by minimizing a simple criterion, e.g., the sum of squares

$$\sum_{t=1}^{T} (y_t - o_t)^2, \tag{9.39}$$

one estimates the biases α and weights w in (9.38). The neural network calibrated in this way can be used, e.g., for construction of predictions in the given time series. Moreover, the *hold-out sample* approach (see Sect. 2.2.3.4) enables us to evaluate the prediction qualities of this model.

Remark 9.5 Tsay (2002) applied the model (9.38) with three nodes in the input layer, two nodes in the hidden layer, and one node in the output layer for 864 daily log returns r_t of IBM stocks. Choosing the vector of input values as $\mathbf{x}_t = (r_{t-1}, r_{t-2}, r_{t-3})$, one estimated the neural network as

$$\widehat{r}_t = 3.22 - 1.81 f_1(\mathbf{x}_t) - 2.28 f_2(\mathbf{x}_t) - 0.09 r_{t-1} - 0.05 r_{t-2} - 0.12 r_{t-3},$$

where

$$f_1(\mathbf{x}_t) = \frac{\exp(-8.34 - 18.97 r_{t-1} + 2.17 r_{t-2} - 19.17 r_{t-3})}{1 + \exp(-8.34 - 18.97 r_{t-1} + 2.17 r_{t-2} - 19.17 r_{t-3})},$$

$$f_2(\mathbf{x}_t) = \frac{\exp(39.25 - 22.17 r_{t-1} - 17.34 r_{t-2} - 5.98 r_{t-3})}{1 + \exp(39.25 - 22.17 r_{t-1} - 17.34 r_{t-2} - 5.98 r_{t-3})}.$$

◇

9.3 Tests of Linearity

An important part of analysis of financial time series concerns verifying nonlinearity of analyzed data. It can be done using financial theory that often confirms that nonlinear relations are the most acceptable ones for given variables. An alternative approach is based on statistical tests denoted in this context as *tests of linearity*. These tests are mostly applied not for the original time series but for residuals calculated by means of the model that should be verified since it is frequent in practice that after applying a linear model of the type ARMA or a nonlinear model of the type GARCH, an unexplained nonlinear structure may remain in the analyzed data outside the systematic part of the model. Therefore, we shall often use the symbol $\{e_t\}$ for the tested time series (and not $\{y_t\}$) in the following text.

The null hypothesis of such tests of linearity is mostly the acceptability of a linear model for the analyzed time series. In particular, one often verifies the null hypothesis that the values $\{e_t\}$ are independent or even *iid*: any violation of independence of the calculated residuals usually indicates an inadequacy of constructed model including the assumption of linearity. In practical analysis of (financial) time series, one recommends particularly the following tests of linearity (see, e.g., Tsay (2002)):

- *Q tests* using, e.g., Ljung–Box statistics with the critical region of the form (applying the significance level α)

$$Q^* = n(n+2)\sum_{k=1}^{K} \frac{1}{n-k}(r_k(e_t))^2 \geq \chi^2_{1-\alpha}(K-p-q), \tag{9.40}$$

  where $\{e_t\}$ are residuals constructed by estimating a model ARMA(p, q) for the given time series (see (6.67) and the verification of models ARCH in Sect. 8.3.4.3).

- RESET tests were suggested for various regression problems in statistics (*Regression Equation Specification Error Tests*, see Ramsey (1969)). When applying them to test the linearity in time series, one tests, e.g., the null hypothesis of the form

$$H_0: \ \beta_1 = 0, \quad \ldots, \quad \beta_s = 0 \tag{9.41}$$

  in the following model:

$$y_t = \alpha + \varphi_1 y_{t-1} + \ldots + \varphi_p y_{t-p} + \beta_1 \cdot \widehat{y}_t^2 + \ldots + \beta_s \cdot \widehat{y}_t^{s+1} + \varepsilon_t, \tag{9.42}$$

  where the values $\widehat{y}_t$ are calculated in the original model AR(p) for the original time series $\{y_t\}$ (i.e., under constraints (9.41) in (9.42)).

- *BDS test* is a widely used (nonparametric) test of independence H_0: $e_t \sim iid$ (it is called according to its authors Brock, Dechert, and Scheinkman). This test is also recommended as an effective test of linearity of $\{e_t\}$ in the framework of modeling financial time series. Numerous applications show that the test has a high power when detecting various violations of independence (these violations may have different forms of linear or nonlinear dependence of all types, deterministic chaos, etc.). Let us describe this test in more details:

The BDS test starts by choosing a distance $\varepsilon > 0$. If really $e_t \sim iid$, then the probability that the distance $|e_s - e_t|$ does not exceed ε for a pair e_s and e_t is the same for an arbitrary choice of such pairs. Let us denote this probability as $c_1(\varepsilon)$. The index 1 in this symbol is used due to the fact that we shall consider more generally also m such pairs ordered in time as (e_s, e_t), (e_{s+1}, e_{t+1}), ..., (e_{s+m-1}, e_{t+m-1}), and in such a case, the symbol $c_m(\varepsilon)$ will denote the probability that for all pairs in this group the corresponding distances do not exceed ε. If the null hypothesis on independence holds, then it must be

$$c_m(\varepsilon) = (c_1(\varepsilon))^m. \tag{9.43}$$

To perform the test practically, one must dispose of sample versions (estimates) of these values (they are called *correlation integrals* in the theory of chaos)

$$c_{m,T}(\varepsilon) = \frac{2}{(T-m+1)(T-m)} \sum_{s=1}^{T-m+1} \sum_{t=s+1}^{T-m+1} \prod_{j=0}^{m-1} I_\varepsilon(e_{s+j}, e_{t+j}), \tag{9.44}$$

where

$$I_\varepsilon(x,y) = \begin{cases} 1 & \text{for } |x-y| \le \varepsilon, \\ 0 & \text{otherwise} . \end{cases} \tag{9.45}$$

Then the test of hypothesis H$_0$: $e_t \sim iid$ consists in testing whether the deviation

$$b_{m,T}(\varepsilon) = c_{m,T}(\varepsilon) - (c_{1,T-m+1}(\varepsilon))^m \tag{9.46}$$

is not significantly different from zero (one removes deliberately $m - 1$ observations in the second term on the right-hand side of (9.46) to base its calculation on the same number of observations as for the first term). Brock et al. (1996) show that (under the null hypothesis H$_0$: $e_t \sim iid$), the following asymptotic result holds for increasing T:

$$\sqrt{T-m+1}\,\frac{b_{m,T}(\varepsilon)}{\sigma_{m,T}(\varepsilon)} \to N(0,1), \tag{9.47}$$

where

$$\sigma^2_{T,m}(\varepsilon) = 4\Bigg((k_T(\varepsilon))^m + 2\sum_{j=1}^{m-1} (k_T(\varepsilon))^{m-j}(c_{1,T}(\varepsilon))^{2j} + (m-1)^2(c_{1,T}(\varepsilon))^{2m}$$
$$-m^2 k_T(\varepsilon)(c_{1,T}(\varepsilon))^{2m-2}\Bigg),$$
$$k_T(\varepsilon) = \frac{2}{T(T-1)(T-2)}$$
$$\times \sum_{t=1}^{T} \sum_{s=t+1}^{T} \sum_{r=s+1}^{T} (I_\varepsilon(e_t,e_s)I_\varepsilon(e_s,e_r) + I_\varepsilon(e_t,e_r)I_\varepsilon(e_r,e_s) + I_\varepsilon(e_s,e_t)I_\varepsilon(e_t,e_r)). \tag{9.48}$$

This result is then used in the (asymptotic) BDS test with a given significance level. Table 9.1 presents the application of BDS test for 100 simulated values of the type *iid* $N(0, 1)$. The test was performed for particular values $m = 2, ..., 6$ and for the distance limit $\varepsilon = 1.378$ which is set up optimally by software (moreover, if the sample size

Table 9.1 BDS test for 100 simulated values of type *iid* $N(0, 1)$

BDS Test for Y					
Included observations: 100					
Dimension	BDS Statistic	Std. Error	z-Statistic	Normal Prob.	Bootstrap Prob.
2	0.001753	0.006313	0.277743	0.7812	0.7352
3	0.002676	0.010054	0.266189	0.7901	0.7128
4	0.003480	0.011994	0.290146	0.7717	0.6772
5	0.009605	0.012523	0.767018	0.4431	0.4236
6	0.013696	0.012097	1.132115	0.2576	0.2848
Raw epsilon		1.378340			

Source: Calculated by EViews

T is smaller, software systems often enable us to calculate the corresponding critical values of the test with regard to their asymptotic character by means of bootstrap simulations, see, e.g., Table 9.1). Obviously, due to high p-values in Table 9.1 the null hypothesis on independence of simulated data cannot be rejected (see also Example 9.1 with real data).

Example 9.1 Table 9.2 presents the application of BDS test when verifying a model construction for daily log returns of index PX50 of Prague Exchange in year 2004 (249 values for 250 trading days; see Table 9.2 and Fig. 9.2). The estimated model GARCH(1,1) has the form

$$r_t = 0.0023 + e_t, \qquad e_t = \sigma_t \varepsilon_t, \qquad \sigma_t^2 = 0.0707 e_{t-1}^2 + 0.8496 \sigma_{t-1}^2.$$

The test was performed again for particular values $m = 2, ..., 6$ and for the distance limit $\varepsilon = 0.013$ set up optimally by software. Due to high p-values in Table 9.3, the null hypothesis on independence of residuals estimated by means of this model cannot be rejected (more specifically, the test confirms that no unexplained nonlinear structure is remaining in these residuals).

◇

9.4 Duration Modeling

Typical data in finance are *transactions data* (usually prices or volumes of traded financial assets) with values observed at times of particular transactions (i.e., at time t_i for the ith transaction). Financial time series originating in this way have some special features:

- As the non-aggregated information has often the form of *high-frequency data* in this context, the timescale must reflect this fact (minutes or even seconds for big stock and derivative exchanges or multinational foreign exchange markets). Moreover, the technical support of trading becomes very important including

Table 9.2 Daily log returns of index PX50 in 2004 (249 values for 250 trading days written in columns) from Example 9.1 (see also Fig. 9.2 and Table 10.1)

	1	2	3	4	5	6	7	8	9	10
1	–	0.0090	–0.0049	0.0114	–0.0045	0.0047	0.0074	–0.0001	–0.0064	0.0016
2	0.0123	0.0084	–0.0047	0.0017	0.0137	–0.0034	–0.0131	0.0054	–0.0107	0.0208
3	0.0057	0.0126	0.0129	–0.0139	0.0076	0.0088	–0.0052	0.0057	0.0250	0.0010
4	–0.0098	0.0171	0.0076	0.0022	0.0111	–0.0051	0.0089	0.0025	0.0088	–0.0108
5	–0.0036	–0.0081	0.0036	–0.0031	0.0022	–0.0082	–0.0101	0.0087	0.0067	0.0067
6	–0.0029	0.0007	–0.0092	0.0111	–0.0100	–0.0044	–0.0037	0.0013	–0.0055	0.0117
7	0.0175	–0.0043	0.0080	0.0073	0.0040	–0.0029	0.0027	0.0037	0.0022	0.0004
8	0.0030	0.0040	0.0111	–0.0093	–0.0003	0.0037	0.0075	–0.0119	–0.0014	–0.0089
9	0.0028	0.0016	0.0053	–0.0295	0.0040	0.0081	–0.0085	0.0057	–0.0148	0.0031
10	0.0082	0.0019	–0.0049	–0.0016	0.0134	–0.0075	–0.0020	0.0056	0.0211	–0.0114
11	0.0050	0.0008	–0.0010	–0.0212	–0.0028	–0.0100	0.0037	–0.0122	0.0054	–0.0317
12	0.0010	–0.0065	0.0132	–0.0065	–0.0113	0.0065	–0.0001	–0.0026	0.0214	0.0113
13	–0.0004	0.0150	0.0004	–0.0067	–0.0027	0.0022	0.0084	–0.0002	–0.0010	0.0019
14	0.0075	0.0029	0.0077	–0.0213	–0.0040	0.0008	0.0070	0.0325	–0.0029	0.0211
15	–0.0014	0.0184	0.0082	–0.0410	0.0000	–0.0082	–0.0013	0.0000	0.0084	0.0042
16	–0.0067	0.0019	0.0103	0.0295	0.0056	–0.0128	0.0063	–0.0038	–0.0080	0.0006
17	–0.0007	–0.0004	–0.0044	–0.0004	0.0052	0.0108	0.0037	–0.0092	0.0058	–0.0070
18	0.0038	0.0012	0.0079	–0.0188	0.0212	0.0065	0.0038	0.0091	0.0112	–0.0034
19	0.0026	0.0103	0.0058	0.0021	0.0036	0.0017	0.0107	0.0074	0.0081	–0.0076
20	0.0003	0.0096	–0.0040	–0.0158	–0.0117	0.0006	0.0020	0.0148	–0.0026	0.0051
21	0.0091	0.0238	0.0048	0.0080	–0.0040	0.0101	–0.0007	–0.0080	0.0037	0.0039
22	0.0071	–0.0019	–0.0227	0.0231	0.0130	–0.0049	0.0026	0.0199	0.0026	0.0046
23	0.0091	–0.0098	0.0036	–0.0021	–0.0065	–0.0037	0.0149	–0.0036	0.0348	0.0047
24	0.0066	–0.0138	–0.0079	–0.0022	–0.0014	0.0089	–0.0030	0.0016	0.0067	0.0101
25	0.0073	0.0037	0.0059	0.0054	–0.0079	0.0009	0.0106	–0.0123	–0.0030	0.0011

Source: Calculated by EViews

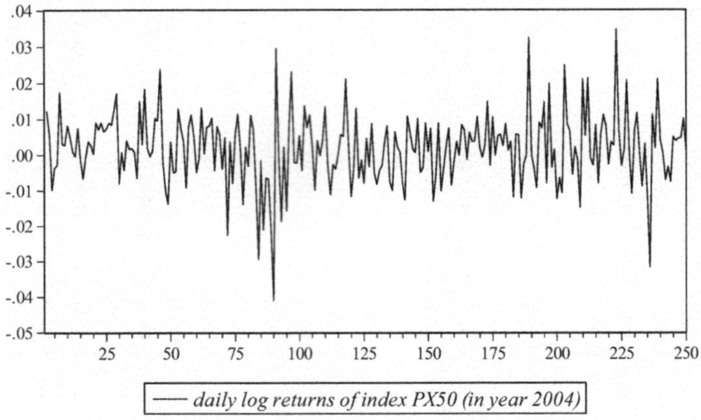

Fig. 9.2 Daily log returns of index PX50 in 2004 (249 values for 250 trading days)

Table 9.3 BDS test applied to residuals estimated by means of model GARCH(1,1) from Example 9.1 (*daily log returns of index PX50 in year 2004*)

BDS Test for RESID				
Included observations: 249				
Dimension	BDS Statistic	Std. Error	z-Statistic	Prob.
2	–0.002106	0.005460	–0.385828	0.6996
3	0.003350	0.008682	0.385813	0.6996
4	0.012462	0.010345	1.204570	0.2284
5	0.017166	0.010790	1.590984	0.1116
6	0.017386	0.010413	1.669745	0.0950
Raw epsilon		0.013311		

Source: Calculated by EViews

the corresponding trading rules, e.g., the standardized or minimum admissible traded volumes, the minimum upward or downward movement in prices (so-called *tick*), regulation rules in the case of extreme or critical price changes, the difference between purchase and sale price (*bid–ask spread*), and the system of charges (*fee*) for trade agents. Moreover, one should eliminate so-called *microstructure noise* in such data; see, e.g., Ait-Sahalia and Jacod (2014), Hautsch (2012), and others.

- Transactions such as stock trading do not occur at equally spaced time intervals, but typically these are *unequally* or *irregularly spaced data*. In such a case, the time duration between trades becomes important and might contain useful information about market microstructure.
- The high-frequency data often exhibit characteristic daily periodicity (*diurnal pattern*), where the transactions are heavier at the beginning and closing of the traded hours and thinner during the lunch hours (typically it results in a U-shape transaction intensity).

- The price is mostly a discrete-valued variable in transactions data, since the price change from one transaction to the next occurs only in multiples of tick size (see above). For example, the NYSE traded gradually in eights, sixteenths, and decimals of dollar.
- In periods of heavy trading, more transactions may occur (even with different prices) within a single second or another very small time unit of transaction recording (so-called *multiple transactions*).

In this section, we focus on modeling durations between particular transactions. To be more specific, let t_i denote the (calendar) time measured in seconds since midnight till the moment of the ith transaction (obviously, the index i describes the order of transactions in time, not the calendar time). The corresponding duration between the $(i-1)$th and ith transaction will be then denoted as $\Delta t_i = t_i - t_{i-1}$ (and sometimes for simplicity even in the abbreviated form as $z_i = \Delta t_i$).

One of the most successful approaches to the duration modeling copies the philosophy of GARCH models, but for time durations z_i (and not for values of the given time series). The corresponding models based on this analogy are called *autoregressive conditional duration processes* ACD(r, s) (see Engle and Russell (1998)):

$$z_i = \tau_i \varepsilon_i, \qquad \tau_i = \alpha_0 + \sum_{j=1}^{r} \alpha_j \tau_{i-j} + \sum_{k=1}^{s} \beta_k z_{i-k}, \tag{9.49}$$

where ε_i are *iid* nonnegative random variables generally with unit mean value and specifically with exponential distribution in the model EACD(r, s), or Weibull distribution in the model WACD(r, s), or gamma distribution in the model GACD (r, s).

Analogously as in the model GARCH, if the following sufficient condition holds

$$\alpha_0 > 0, \quad \alpha_j \geq 0, \quad \beta_k \geq 0, \qquad \sum_{j=1}^{\max\{r,s\}} \left(\alpha_j + \beta_j\right) < 1, \tag{9.50}$$

then the model ACD is stationary with mean value of the form

$$\mathrm{E}(z_i) = \frac{\alpha_0}{1 - \sum_{j=1}^{\max\{r,s\}} \left(\alpha_j + \beta_j\right)}. \tag{9.51}$$

In particular, let us consider the model EACD(1,1), where ε_t have standardized exponential distribution with unit mean value and unit variance. Moreover, let the following sufficient condition of stationarity be fulfilled:

$$\alpha_1^2 + 2\beta_1^2 + 2\alpha_1\beta_1 < 1. \tag{9.52}$$

Then the process $\{z_i\}$ is stationary with variance

$$\operatorname{var}(z_i) = \left(\frac{\alpha_0}{1-\alpha_1-\beta_1}\right)^2 \frac{1-\alpha_1^2-2\alpha_1\beta_1}{1-\alpha_1^2-2\beta_1^2-2\alpha_1\beta_1}. \tag{9.53}$$

The models of the type EACD, WACD, and GACD can be estimated by the method of maximum likelihood due to specified distributions of ε_i (see, e.g., Tsay (2002)).

Remark 9.6 Tsay (2002) constructed the following model WACD(1,1) for durations in trading the stocks IBM during five trading days (in total, 3534 durations after eliminating the diurnal pattern of daily periodicity were used in this construction)

$$z_i = \tau_i \varepsilon_i, \qquad \tau_i = 0.169 + 0.885\tau_{i-1} + 0.064 z_{i-1},$$

where Weibull distribution of ε_i (standardized by unit mean value) has the probability density

$$f(x \mid \lambda) = \begin{cases} \lambda\left[\Gamma\left(1+\frac{1}{\lambda}\right)\right]^{\lambda} x^{\lambda-1} \exp\left\{-\left[\Gamma\left(1+\frac{1}{\lambda}\right)\cdot x\right]^{\lambda}\right\} & \text{for } x \geq 0, \\ 0 & \text{otherwise} \end{cases} \tag{9.54}$$

with the estimated parameter λ of size 0.879. Then the estimated mean duration (9.51) after elimination of daily periodicity is 3.31 seconds, which coincides with the duration estimated directly as the sample mean of the time series of periodically adjusted durations.

◇

9.5 Exercises

Exercise 9.1 Apply the BDS test from Sect. 9.3 for (a) 100 simulated values of the type *iid* $N(0, 1)$; (b) the daily log returns of index PX in 2016 estimated as GARCH (1, 1) in Example 8.2.

Chapter 10
Models of Development of Financial Assets

Important relations in modern finance and the corresponding econometric calculations (e.g., calibrations of models in given financial environment, simulations for various financial scenarios, and the like) necessitate to model developments of financial assets (prices, volumes, returns) in continuous time (see also random processes in continuous time in Sects. 2.4 and 2.5). Such models mostly require relatively complex theoretical background denoted generally as *stochastic calculus*, which is the classical calculus of derivations and integrals modified for random variables (including special instruments such as stochastic integrals, martingales, Ito's lemma, Wiener process, risk neutral probabilities, and others). In this chapter, we outline basic principles of this methodology, but only within the scope that enables to cope with simple practical applications for financial times series (in any case, there are references with technical and theoretical details, e.g., Campbell et al. (1997), Cipra (2010), Duffie (1988), Elliot and Kopp (2004), Enders (1995), Franke et al. (2004), Franses and van Dijk (2000), Gourieroux and Jasiak (2001), Hendry (1995), Hull (1993), Karatzas and Shreve (1988), Kwok (1998), Lim (2011), Malliaris and Brock (1982), McNeil et al. (2005), Mills (1993), Musiela and Rutkowski (2004), Neftci (2000), Poon (2005), Rachev et al. (2007), Ruppert (2004), Steele (2001), Taylor (1986), Tsay (2002), Wang (2003), and Wilmott (2000)).

10.1 Financial Modeling in Continuous Time

In Sect. 2.1, we have defined the random (or stochastic) process $\{Y_t,\ t \in T\}$ in continuous time as a set of random variables in the same probability space $(\Omega, \mathfrak{I}, \mathrm{P})$ indexed by means of values t from the set $T = \langle 0, \infty)$ interpreted as time. The continuous time is the necessary assumption for various financial schemes that model (in a practically acceptable way) the price changes of financial assets, even though in reality these prices are observed in discrete time moments only (however,

T. Cipra, *Time Series in Economics and Finance*,
https://doi.org/10.1007/978-3-030-46347-2_10

an awareness of the fact that the analyzed prices exist continually and change in time unceasingly may serve as a motivation for their analysis).

The models of Box–Jenkins methodology in discrete time from Chap. 6 (e.g., the linear process (6.17)) are based on unpredictable discrete increments (innovations, shocks) in the form of white noise. The analogy for models in continuous time can be based on increments of *Wiener process* $\{W_t, t \geq 0\}$. The properties (2.56) of Wiener process can be rewritten by means of its increments $\Delta W_t = W_{t+\Delta t} - W_t$ as

$$\begin{cases} (i) & W_0 = 0; \\ (ii) & \text{the particular trajectories are continuous in time;} \\ (iii) & \Delta W_t = \varepsilon\sqrt{\Delta t} \ \text{ with a random variable } \varepsilon \sim N(0,1); \\ (iv) & \Delta W_t \text{ is independent on } W_s \text{ for arbitrary } 0 \leq s < t. \end{cases} \tag{10.1}$$

The property (*i*) can be formulated more generally as $P(W_0 = 0) = 1$. If we delete the property (*ii*), then such a process is called *standard Brownian motion* (however, it is possible to show that every Brownian motion has a modification with continuous trajectories). The property (*iii*) can be also written as

$$\Delta W_t \sim N(0, \ \Delta t). \tag{10.2}$$

In particular, the standard deviation of the process increment is equal to the square root of the corresponding time increment. Finally, the assumption (*iv*) is so-called Markov property (see also (2.48)), i.e., at time t, any future value W_{t+h} $(h > 0)$ depends only on the present value W_t, and not on previous values W_s $(s < t)$. It means consequently that the increments of the process are mutually independent for *non-overlapping time intervals* (see Sect. 2.5.2). From the financial interpretation point of view, this Markov property corresponds to so-called weakly efficient markets. There are other specific properties of Wiener process suitable for financial modeling, e.g., it holds (if we consider a "long" time increment from zero to t)

$$W_t \sim N(0,t) \quad \text{for } t \geq 0 \tag{10.3}$$

so that the variance of Wiener process accrues linearly with increasing time.

As the trajectories of Wiener process are not differentiable in any point of time (i.e., they have nowhere derivations, even though they are continuous), one cannot integrate them in the classical way and has to make use of so-called stochastic (Ito's) calculus.

10.1.1 Diffusion Process

The general scheme of financial models in continuous time is the *diffusion process* (*Ito's process*). If we denote small changes of a variable x as $\mathrm{d}x$, then the usual form of this process $\{Y_t, t \geq 0\}$ is

$$\mathrm{d}Y_t = \mu(Y_t, t)\,\mathrm{d}t + \sigma(Y_t, t)\,\mathrm{d}W_t \quad \text{for } t \geq 0, \tag{10.4}$$

where W_t is Wiener process. This model has the drift component $\mu(Y_t, t)\mathrm{d}t$ for modeling the trend and the diffusion component $\sigma(Y_t, t)\mathrm{d}W_t$ for modeling the volatility. Since the *drift coefficient* $\mu(Y_t, t)$ and the *diffusion coefficient* $\sigma(Y_t, t)$ may change in time (they depend on the time and even on the value of the process), one has to use integration when solving the differential equation (10.4), i.e.,

$$Y_t = Y_0 + \int_0^t \mu(Y_s, s)\mathrm{d}s + \int_0^t \sigma(Y_s, s)\mathrm{d}W_s \quad \text{for } t \geq 0, \tag{10.5}$$

where the second integral is stochastic (i.e., one integrates with respect to random processes; see also Sect. 10.1.2) assuming so-called *previsibility* of process $\sigma(Y_t, t)$ (i.e., independence on the future in terms of measurability of this process with respect to the current and past information).

An important special case of Ito's process is *Wiener process with drift* μ *and volatility* σ (*generalized Wiener process, arithmetic Wiener process*) of the form

$$\mathrm{d}Y_t = \mu\,\mathrm{d}t + \sigma\,\mathrm{d}W_t \quad \text{for } t \geq 0 \tag{10.6}$$

($\sigma \geq 0$), which has the constant drift and diffusion coefficient. It means that it holds (assuming $Y_0 = 0$)

$$Y_t = \mu \cdot t + \sigma \cdot W_t \quad \text{for } t \geq 0 \tag{10.7}$$

and hence according to (10.3)

$$\mathrm{E}(Y_t) = \mu \cdot t, \qquad \mathrm{var}(Y_t) = \sigma^2 \cdot t. \tag{10.8}$$

Remark 10.1 When testing various financial scenarios of development of prices and returns of financial assets, one often uses simulations based on diffusion processes. If, e.g., the time is measured in years and one chooses the daily step $\Delta t = 1/252$ (for 252 trading day in 1 year), then one can generate an *iid* sequence ε_1, ε_2, ... ~ $N(0, 1)$ and construct (recursively) a simulated generalized Wiener process (10.7) as

$$Y_{k\cdot\Delta t} = \mu \cdot k \cdot \Delta t + \sigma \cdot \sqrt{\Delta t} \cdot \sum_{j=1}^{k} \varepsilon_j \quad \text{for } k = 1,\ 2,\ \ldots. \tag{10.9}$$

◊

Another special case of Ito's process is the exponential Wiener process (sometimes called geometric Brownian motion; see Sects. 2.5.2 and 10.1.3).

10.1.2 Ito's Lemma and Stochastic Integral

The stochastic calculus demands to modify the classic derivatives and integrals to their stochastic variants, which will be briefly described below (the theoretical backgrounds of this complex discipline including a technical construction of stochastic integral can be found in various sources, see, e.g., Baxter and Rennie (1996), Dupačová et al. (2002), Karatzas and Shreve (1988), Kwok (1998), Malliaris and Brock (1982), Musiela and Rutkowski (2004), Neftci (2000), Wilmott (2000), and others):

The basic principle of *random differential* (or equivalently *stochastic differentiation*) is the well-known Ito's lemma. Let us consider a diffusion process $\{Y_t, t \geq 0\}$ according to (10.4). Further let $f(y, t)$ be a continuous (nonrandom) function of variable y and time t with continuous partial derivatives $f_y = \partial f/\partial y$, $f_{yy} = \partial^2 f/\partial y^2$, and $f_t = \partial f/\partial t$. Then the transformed process $f(Y_t, t)$ fulfills (so-called *Ito's lemma*):

$$\mathrm{d}f(Y_t,t) = \left(f_y \cdot \mu(Y_t,t) + f_t + \frac{1}{2} f_{yy} \cdot \sigma^2(Y_t,t) \right) \mathrm{d}t + f_y \cdot \sigma(Y_t,t)\, \mathrm{d}W_t \quad \text{for } t \geq 0. \tag{10.10}$$

If, e.g., $f(W_t, t) = W_t^2$, then $f_y = 2W_t$, $f_{yy} = 2$ and $f_t = 0$, so that the differential of squared Wiener process fulfills

$$\mathrm{d}W_t^2 = \mathrm{d}t + 2W_t\, \mathrm{d}W_t \quad \text{for } t \geq 0, \tag{10.11}$$

which is different significantly from the classical (nonrandom) differential $\mathrm{d}(x^2) = 2x\, \mathrm{d}x$.

The *random integral* represents an inverse operation to the random differential (stochastic differentiation) described above, i.e.,

$$\int_0^t \mathrm{d}Y_s = Y_t - Y_0. \tag{10.12}$$

In particular, due to the zero initial value of Wiener process, it holds

$$\int_0^t \mathrm{d}W_s = W_t. \tag{10.13}$$

When integrating both sides of relation (10.11), one receives

$$W_t^2 = t + 2\int_0^t W_s \mathrm{d}W_s, \tag{10.14}$$

which implies another important formula of stochastic integration of Wiener process

$$\int_0^t W_s \mathrm{d}W_s = \frac{1}{2}\left(W_t^2 - t\right) \tag{10.15}$$

(again it is different significantly from the classical (nonrandom) integral $\int_0^t x \ \mathrm{d}x = t^2/2$).

10.1.3 Exponential Wiener Process

One of the most utilized processes for modeling prices $\{P_t, t \geq 0\}$ of financial assets (e.g., stocks) in continuous time is *exponential Wiener process* (*geometric Brownian motion*) defined as Ito's process of the form

$$\mathrm{d}P_t = \mu \cdot P_t \ \mathrm{d}t + \sigma \cdot P_t \ \mathrm{d}W_t \quad \text{for } t \geq 0 \tag{10.16}$$

($\sigma \geq 0$). The discretized version of (10.16)

$$\frac{\Delta P_t}{P_t} = \mu \cdot \Delta t + \sigma \cdot \Delta W_t \tag{10.17}$$

indicates that one models, as a matter of fact, returns of given asset by means of (deterministic) drift component $\mu \cdot \Delta t$ and (random) diffusion component $\sigma \cdot \Delta W_t \sim N(0, \sigma^2 \Delta t)$.

However, in practice we often model the logarithmic price

$$p_t = \ln P_t, \tag{10.18}$$

since then we obtain easily by discrete differencing the log return $r_t = p_t - p_{t-1}$ (see (8.1)). Obviously, it holds

$$\frac{\partial \ln P_t}{\partial P_t} = \frac{1}{P_t}, \qquad \frac{\partial^2 \ln P_t}{\partial P_t^2} = -\frac{1}{P_t^2}, \qquad \frac{\partial \ln P_t}{\partial t} = 0,$$

so that the logarithmic transformation of the process P_t from (10.16) fulfills according to Ito's lemma

$$\mathrm{d}p_t = \mathrm{d}\ \ln P_t = \left(\mu - \frac{\sigma^2}{2}\right) \mathrm{d}t + \sigma\, \mathrm{d}W_t \quad \text{for } t \geq 0. \tag{10.19}$$

Therefore, the logarithmic price has the drift coefficient $\mu - \sigma^2/2$ and the diffusion coefficient (or volatility) σ. The differential equation (10.19) may be solved by integrating its both sides (see Sect. 10.1.2)

$$\int_0^t \mathrm{d}p_s = \left(\mu - \frac{\sigma^2}{2}\right) \int_0^t \mathrm{d}s + \sigma \int_0^t \mathrm{d}W_s, \tag{10.20}$$

i.e.,

$$p_t = p_0 + \left(\mu - \frac{\sigma^2}{2}\right) \cdot t + \sigma \cdot W_t \quad \text{for } t \geq 0 \tag{10.21}$$

or equivalently for the original price P_t (i.e., after removing the logarithms)

$$\begin{aligned} P_t &= \exp\left(p_0 + \left(\mu - \frac{\sigma^2}{2}\right) \cdot t + \sigma \cdot W_t\right) \\ &= P_0 \exp\left(\left(\mu - \frac{\sigma^2}{2}\right) \cdot t + \sigma \cdot W_t\right) \quad \text{for } t \geq 0. \end{aligned} \tag{10.22}$$

The exponential Wiener process is usually presented just in this exponential form, which justifies its name (the form (10.22) can be also compared with the reparameterized version (2.59) of this process).

The formulas (10.21) and (10.22) imply that it holds (conditionally for the price $p_0 = \ln P_0$ at time $t = 0$)

$$p_t \sim N\left(p_0 + \left(\mu - \frac{\sigma^2}{2}\right) \cdot t,\ \sigma^2 \cdot t\right), \quad P_t \sim LN\left(p_0 + \left(\mu - \frac{\sigma^2}{2}\right) \cdot t,\ \sigma^2 \cdot t\right), \tag{10.23}$$

where $LN(\mu, \sigma^2)$ denotes a random variable with *lognormal distribution*, which has the form of exponential function $\exp(X)$ of a random variable $X \sim N(\mu, \sigma^2)$. Then one can show (again conditionally for the price $p_0 = \ln P_0$) that, e.g.:

$$\mathrm{E}(P_t) = P_0 \exp(\mu \cdot t), \quad \mathrm{var}(P_t) = P_0^2 \exp(2\mu \cdot t) \cdot \left(\exp(\sigma^2 \cdot t) - 1\right). \quad (10.24)$$

In particular, the drift coefficient μ represents the average annual log return due to the price changes of given asset (see (8.1) rewritten to the form $P_t = P_{t-1}\exp(r_t)$ for log returns $\{r_t\}$).

Remark 10.2 The relations (10.21) and (10.22) can be used when simulating the development of prices p_t or P_t (see also Remark 10.1). Note also according to (10.21) that if we apply for modeling the price of given financial asset the exponential Wiener process, then the log return $r_t = p_t - p_{t-1}$ is white noise with probability distribution

$$r_t \sim N\left(\mu - \frac{\sigma^2}{2}, \ \sigma^2\right). \quad (10.25)$$

◇

In financial practice, both unknown parameters μ and σ^2 of exponential Wiener process can be statistically estimated by means of observed data (so-called calibration of model in a given financial environment):

Let, e.g., $r_1, \ldots, r_n$ be the corresponding log returns measured using regular time intervals of length Δ, which is a given fraction of year (mostly daily log returns are used, i.e., $\Delta \approx 1/250$ for 250 trading day in 1 year). Then similarly as in (10.25) it holds

$$\mathrm{E}(r_t) = \left(\mu - \frac{\sigma^2}{2}\right) \cdot \Delta, \qquad \mathrm{var}(r_t) = \sigma^2 \cdot \Delta. \quad (10.26)$$

The given data set usually enables to estimate the sample mean and sample variance as

$$\overline{r} = \frac{1}{n}\sum_{t=1}^{n} r_t, \qquad s_r^2 = \frac{1}{n-1}\sum_{t=1}^{n}(r_t - \overline{r})^2. \quad (10.27)$$

Comparing the theoretical values (10.26) and sample values (10.27) for higher n (the sample values can be obviously used as consistent estimate of theoretical values), then one finally obtains the estimates of μ and σ as

$$\widehat{\sigma} = \frac{s_r}{\sqrt{\Delta}}, \qquad \widehat{\mu} = \frac{\overline{r}}{\Delta} + \frac{\widehat{\sigma}^2}{2} = \frac{\overline{r}}{\Delta} + \frac{s_r^2}{2\Delta} \quad (10.28)$$

(moreover, e.g., the standard deviation of the estimate $\widehat{\sigma}$, which serves as an error of this estimate, can be estimated by $\widehat{\sigma}/\sqrt{2n}$).

Example 10.1 In this example, we estimate the exponential Wiener process for index PX50 (250 trading days in year 2004; see Table 10.1 and Fig. 10.1 on the left).

By means of the formulas (10.27), one estimates easily for data sample in Table 10.1

$$\bar{r} = 0.001\,782, \ s_r = 0.009\,895.$$

If $\Delta = 1/250$, then using the formulas (10.28) one obtains further

$$\widehat{\sigma} = \frac{0.009\,895}{\sqrt{(1/250)}} = 0.156\,454 \approx 15.6\ \%, \qquad \widehat{\mu} = \frac{0.001\,782}{1/250} + \frac{0.156\,454^2}{2}$$
$$= 0.457\,739 \approx 45.8\ \%$$

(in particular, the error of volatility estimate is 0.7%), so that the average annual log return of the index PX50 was 45.8%.

Using the estimated parameters μ and σ^2 in the relation (10.22), simulations of the index PX50 are possible similarly to that in Remark 10.1. One of such simulations is plotted in Fig. 10.1 on the right (including the true observations on the left to compare both plots). It is apparent that in such simulations the high drift coefficient 45.8% prevails over the volatility 15.6% so that the simulated trajectories are often very distinctly increasing.

◇

10.2 Black–Scholes Formula

The previous models in continuous time enable to evaluate so-called financial derivatives (options, futures, swaps, and others). *Financial derivatives* (or *derivative securities*) are securities, whose values are dependent on ("derived" from) the values of other more basic underlying variables, which are often traded prices of so-called *underlying assets* (stocks or bonds, currencies, commodities, stock market indices, and the like). For instance, the *options* are securities in the form of contracts agreed at time t, which guarantee to their holder the right (but not the duty) to do a specified asset transaction (e.g., to buy a stock for a fixed price) till a future date T ($t < T$). The well-known *Black–Scholes formula* is broadly used in practice to evaluate *option premiums* (i.e., the prices of options; see, e.g., Cipra (2010), Hull (1993), and others). The key principle of the corresponding methodology (including the derivation of Black–Scholes formula) consists in the assumption of *no arbitrage* (the arbitrage consists in opportunities of riskless yield due to different simultaneous prices of possible buy or sell positions for underlying assets).

One can derive Black–Scholes formula solving the following general problem. We look for an explicit form of unknown function F evaluating a given financial derivative (e.g., an option) under the following assumptions:

Table 10.1 Index PX50 in the year 2004 (values for 250 trading days written in columns) from Example 10.1 (see also Fig. 10.1 on the left)

	1	2	3	4	5	6	7	8	9	10
1	662.10	726.00	789.70	846.00	759.70	796.20	800.90	837.60	883.80	999.90
2	670.30	732.10	786.00	847.40	770.20	793.50	790.50	842.10	874.40	1020.90
3	674.10	741.40	796.20	835.70	776.10	800.50	786.40	846.90	896.50	1021.90
4	667.50	754.20	802.30	837.50	784.80	796.40	793.40	849.00	904.40	1010.90
5	665.10	748.10	805.20	834.90	786.50	789.90	785.40	856.40	910.50	1017.70
6	663.20	748.60	797.80	844.20	778.70	786.40	782.50	857.50	905.50	1029.70
7	674.90	745.40	804.20	850.40	781.80	784.10	784.60	860.70	907.50	1030.10
8	676.90	748.40	813.20	842.50	781.60	787.00	790.50	850.50	906.20	1021.00
9	678.80	749.60	817.50	818.00	784.70	793.40	783.80	855.40	892.90	1024.20
10	684.40	751.00	813.50	816.70	795.30	787.50	782.20	860.20	911.90	1012.60
11	687.80	751.60	812.70	799.60	793.10	779.70	785.10	849.80	916.80	981.00
12	688.50	746.70	823.50	794.40	784.20	784.80	785.00	847.60	936.60	992.10
13	688.20	758.00	823.80	789.10	782.10	786.50	791.60	847.40	935.70	994.00
14	693.40	760.20	830.20	772.50	779.00	787.10	797.20	875.40	933.00	1015.20
15	692.40	774.30	837.00	741.50	779.00	780.70	796.20	875.40	940.90	1019.50
16	687.80	775.80	845.70	763.70	783.40	770.80	801.20	872.10	933.40	1020.10
17	687.30	775.50	842.00	763.40	787.50	779.20	804.20	864.10	938.80	1013.00
18	689.90	776.40	848.70	749.20	804.40	784.30	807.30	872.00	949.40	1009.60
19	691.70	784.40	853.60	750.80	807.30	785.60	816.00	878.50	957.10	1002.00
20	691.90	792.00	850.20	739.00	797.90	786.10	817.60	891.60	954.60	1007.10
21	698.20	811.10	854.30	744.90	794.70	794.10	817.00	884.50	958.10	1011.00
22	703.20	809.60	835.10	762.30	805.10	790.20	819.10	902.30	960.60	1015.70
23	709.60	801.70	838.10	760.70	799.90	787.30	831.40	899.10	994.60	1020.50
24	714.30	790.70	831.50	759.00	798.80	794.30	828.90	900.50	1001.30	1030.90
25	719.50	793.60	836.40	763.10	792.50	795.00	837.70	889.50	998.30	1032.00

Source: Prague Stock Exchange (https://www.pse.cz/en/indices/index-values/detail/XC0009698371?tab=detail-history)

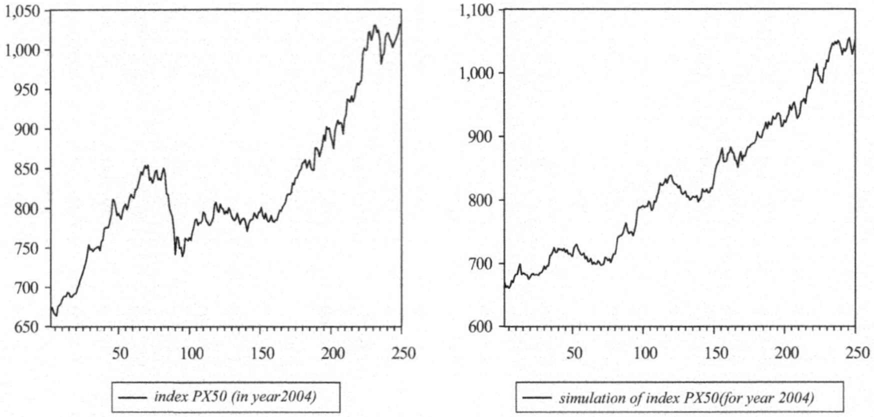

Fig. 10.1 Index PX50 in the year 2004 (on the left) and its simulation by means of estimated exponential Wiener process from Example 10.1 (on the right)

- F is a function of time t and the price P_t of underlying asset at time t (e.g., stock, currency, crude oil, stock index). One will write for simplicity $F_t = F(P_t, t)$.
- The value F_T is determined by a boundary condition at maturity time T ($t < T$) according to the type of financial derivative.

For example, let us consider a (European) call option (simply *call*) which gives its buyer (holder in *long position*) the right to buy at maturity time T an underlying asset for a preset price X (*exercise price* or *strike price*) even though the market price of the given asset at time T is P_T. The call option can be purchased at time t for a price $Call_t$ (so-called *call premium*), where one knows only the asset price P_t at time t, but not the future price P_T at time T. The seller of this call (underwriter in *short position*) must sell the underlying asset at maturity time according to the holder's decision. In this case, the boundary condition for the function F_t ($= Call_t$) has obviously the form

$$F_T = \max(P_T - X, 0) \tag{10.29}$$

(European options can be exercised only at the maturity date, while American options at any time up to the maturity date). Similarly, a (European) put option (simply *put*) gives its buyer (holder in *long position*) the right to sell at maturity time T an underlying asset for a preset price X. The put option can be purchased at time t for a price Put_t (*put premium*). The seller of this put (underwriter in *short position*) must buy the underlying asset at maturity time according to the holder's decision. In this case, the boundary condition for the function F_t ($= Put_t$) has the form

$$F_T = \max(X - P_T, 0). \tag{10.30}$$

Generally, if the price P_t of underlying asset behaves as an exponential Wiener process (10.16), then according to Ito's lemma (10.10) we get

$$dF_t = \left(\frac{\partial F_t}{\partial P_t}\mu \cdot P_t + \frac{\partial F_t}{\partial t} + \frac{1}{2}\frac{\partial^2 F_t}{\partial P_t^2}\sigma^2 \cdot P_t^2\right) dt + \frac{\partial F_t}{\partial P_t}\sigma \cdot P_t \, dW_t. \tag{10.31}$$

One can rewrite the previous relations to the following discrete form:

$$\Delta P_t = \mu \cdot P_t \cdot \Delta t + \sigma \cdot P_t \cdot \Delta W_t, \tag{10.32}$$

$$\Delta F_t = \left(\frac{\partial F_t}{\partial P_t}\mu \cdot P_t + \frac{\partial F_t}{\partial t} + \frac{1}{2}\frac{\partial^2 F_t}{\partial P_t^2}\sigma^2 \cdot P_t^2\right) \Delta t + \frac{\partial F_t}{\partial P_t}\sigma \cdot P_t \, \Delta W_t. \tag{10.33}$$

Let us now construct so-called replicating ("self-financing") portfolio

$$V_t = -F_t + \frac{\partial F_t}{\partial P_t} \cdot P_t. \tag{10.34}$$

In the framework of this portfolio, one owns the underlying asset of size $\partial F_t / \partial P_t$ (with price P_t per unit) and simultaneously owes the considered financial derivative of unit size (with price F_t per unit). It holds

$$\Delta V_t = -\Delta F_t + \frac{\partial F_t}{\partial P_t} \cdot \Delta P_t = \left(-\frac{\partial F_t}{\partial t} - \frac{1}{2}\frac{\partial^2 F_t}{\partial P_t^2}\sigma^2 \cdot P_t^2\right) \cdot \Delta t, \tag{10.35}$$

i.e., the changes of this portfolio do not include the random component ΔW_t. Therefore, the portfolio V_t is riskless in the framework of small time changes Δt and must earn during such time changes equally as other riskless investments with riskless interest rate r_f (*free of risk*), i.e.:

$$\Delta V_t = r_f \cdot V_t \cdot \Delta t \tag{10.36}$$

(otherwise one could break the no arbitrage rule; see above). Substituting (10.34) and (10.35) to (10.36) one obtains

$$\left(\frac{\partial F_t}{\partial t} + \frac{1}{2}\frac{\partial^2 F_t}{\partial P_t^2}\sigma^2 \cdot P_t^2\right) \cdot \Delta t = r_f \cdot \left(F_t - \frac{\partial F_t}{\partial P_t}P_t\right) \cdot \Delta t. \tag{10.37}$$

We derived so-called *Black–Scholes differential equation* for evaluating (pricing) financial derivatives

$$\frac{\partial F_t}{\partial t} + r_f \cdot P_t \cdot \frac{\partial F_t}{\partial P_t} + \frac{1}{2}\sigma^2 \cdot P_t^2 \cdot \frac{\partial^2 F_t}{\partial P_t^2} = r_f \cdot F_t. \tag{10.38}$$

The solution of this equation under the boundary condition (10.29) is the well-known *Black–Scholes formula* for the (European) call premium (see also (8.20)):

$$Call_t = P_t \cdot \Phi(d_+) - X \cdot \mathrm{e}^{-r_f(T-t)} \cdot \Phi(d_-), \tag{10.39}$$

where

$$d_+ = \frac{\ln(P_t/X) + (r_f + \sigma^2/2)(T-t)}{\sigma\sqrt{T-t}}, \quad d_- = \frac{\ln(P_t/X) + (r_f - \sigma^2/2)(T-t)}{\sigma\sqrt{T-t}} = d_+ - \sigma\sqrt{T-t},$$

$\Phi(\cdot)$ is the distribution function of $N(0, 1)$ and the remaining symbols are described in the previous text. Similarly, Black–Scholes formula for the (European) put premium is

$$Put_t = X \cdot \mathrm{e}^{-r_f(T-t)} \cdot \Phi(-d_-) - P_t \cdot \Phi(-d_+). \tag{10.40}$$

Moreover, it holds so-called *put–call parity*

$$Put_t = Call_t + X \cdot \mathrm{e}^{-r_f(T-t)} - P_t. \tag{10.41}$$

Example 10.2 Let us consider a call option to buy a stock with exercise price 50 EUR. Three months before maturity of this European option, the price of stock is 53 EUR. Which is the call premium according to Black–Scholes formula (the price volatility of stock has been estimated as 0.50 EUR and the corresponding riskless interest rate is 5% p.a.)?

According to (10.39) for $P_t = 53$, $X = 50$, $T - t = 3/12 = 0.25$, $\sigma = 0.50$ and $i = 0.05$, one obtains

$$d_+ = \frac{\ln(53/50) + (0.05 + 0.50^2/2) \cdot 0.25}{0.50 \cdot \sqrt{0.25}} = 0.4081, \quad \Phi(0.4081) = 0.6584,$$

$$d_- = 0.4081 - 0.50 \cdot \sqrt{0.25} = 0.1581, \quad \Phi(0.1581) = 0.5628,$$

so that

$$Call_t = 53 \cdot 0.6584 - 50 \cdot \mathrm{e}^{-0.05 \cdot 0.25} \cdot 0.5628 = 7.10 \text{ EUR}.$$

Analogously, the European put option premium according to (10.40) or (10.41) would be $Put_t = 3.48$ EUR.

◇

10.3 Modeling of Term Structure of Interest Rates

The structure of interest rates is usually presented as behavior of *yields to maturity* (YTM) in time for investments that have the generic form of *zero-coupon bonds*. Such bonds may be looked upon as investment loans which enables to trade loans as discount securities.

Even though the structure of interest rates depends on a number of various factors, here we shall deal with its dependence on the time of *maturity* only (so-called *term structure of interest rates*). In this context, one usually denotes the relation between the yield to maturity and the time of maturity as a *yield curve*. The yield curves in practice are mostly constructed for the whole classes of bonds of similar characteristics which are mainly the type of bond issuer (e.g., government bonds) and rating (e.g., AA). Since the yields of government bonds may be usually regarded as riskless (so-called *risk-free yield*), the yield curves of "risk" bonds are often constructed in such a way that one shifts a suitable yield curve of government bonds by *credit spread* which is considered as a risk premium corresponding to the risk rating of the evaluated bond.

Moreover, one distinguishes so-called spot and forward yield curves: *spot yield curve* describes the dependence of the yield to maturity on the time of maturity, where time is measured from the actual (spot) moment. *Forward yield curve* measures this dependence starting from a future moment (which must be exactly specified in advance). In this context, one uses the following denotation:

1. The symbol $P(t, T)$ denotes the price of zero-coupon bond with time of maturity T and unit nominal value (i.e., $P(T, T) = 1$) at time t ($t < T$). The continuous yield to maturity of this bond at time t denoted as $R(t, T)$ is

$$R(t,T) = -\frac{\ln P(t,T)}{T-t}, \tag{10.42}$$

which follows from the formula of continuous discounting (see, e.g., Cipra (2010)):

$$P(t,T) = \mathrm{e}^{-R(t,T)\cdot(T-t)}. \tag{10.43}$$

In financial theory, the yield to maturity $R(t, \cdot)$ considered as a function of argument T ($T \geq t$) is usually called *yield curve at time t* (while the price of bond $P(t, \cdot)$ considered as a function of argument T is called *discount curve at time t*).

2. Particularly the value

$$r_t = R(t,t) = -\frac{\partial \ln P(t,T)}{\partial T}\bigg|_{T=t} = -\frac{\partial \ln P(t,t)}{\partial T} \tag{10.44}$$

is called *instantaneous interest rate* at time t (since r_t is the limit value $R(t, T)$ for $T \to t$). The adjective "instantaneous" expresses the fact that if applying this interest rate to a capital K at time t, then the capital accrual during a short time interval Δt is approximately $\Delta K = K \cdot r_t \cdot \Delta t$, even if more correctly one should write the differential relation $\mathrm{d}K = K \cdot r_t\, \mathrm{d}t$. Obviously, the instantaneous rate r_t presents the interest intensity at time t independently of time of maturity of the corresponding loan (investment).

Most models of term structure of interest rates assume that r_t is the diffusion process of the form (see (10.4))

$$\mathrm{d}r_t = a(r_t, t)\, \mathrm{d}t + b(r_t, t)\, \mathrm{d}W_t. \tag{10.45}$$

The models used in practice have specified forms of drift coefficient $a(r_t, t)$ and diffusion coefficient $b(r_t, t)$. Their main outputs are explicit formulas for instantaneous interest rate r_t and bond price $P(r_t, t, T)$ by means of these coefficients and consequently an explicit formula for yield curve $R(r_t, t, T)$ according to (10.42). If we succeed in estimating the chosen model (10.45) for observed financial data (at time t one observes the bond prices $P(t, T_k)$ with various times of maturity T_k; see also Remark 10.4), then we can:

- Estimate continuous yield curves using only limited volume of data.
- Perform various simulations (for yield curves, instantaneous interest rates, and the like).

In practice, one constructs (see, e.g., Baxter and Rennie (1996), Cipra (2010), Hull (1993)):

- *Single-factor interest rate models* that include only one interest rate factor r_t (e.g., models by *Vasicek, Cox–Ingersoll–Ross, Hull–White, Ho-Lee, Black–Derman–Toy, Black–Karasinski*, and others).
- *Binomial tree models* (e.g., models by *Rendleman–Bartter, Jarrow-Rudd*, and others).
- *Multi-factor interest rate models* that include several interest rate factors (e.g., models by *Brennan–Schwartz, Fong–Vasicek, Longstaff–Schwartz*, and others).

1. Vasicek Model

Vasicek model (also *Ornstein–Uhlenbeck process* or *mean-reverting model*) is based on the diffusion equation (10.45) in the form

$$dr_t = \alpha \cdot (\gamma - r_t) \; dt + b \; dW_t \tag{10.46}$$

with parameters $\alpha > 0$, $\gamma \in R$, $b \in R$. The process r_t fluctuates around a constant level γ, but the trend coefficient $\alpha(\gamma - r_t)$ reverts the values r_t that deviated too much from γ back to this level (it is substantial for such a behavior of r_t that the volatility b is constant).

Remark 10.3 Vasicek (1977) showed that the yield curve corresponding to Vasicek model has the form

$$R(r_t, t, T) = \gamma + \frac{b \cdot q}{\alpha} - \frac{b^2}{2\alpha^2} - \frac{1}{\alpha \cdot T} \cdot \left(1 - e^{-\alpha \cdot T)}\right) \cdot \left(\gamma + \frac{b \cdot q}{\alpha} - \frac{b^2}{2\alpha^2} - r_t\right) + \frac{b^2}{4\alpha^3 T} \cdot \left(1 - e^{-\alpha \cdot T}\right)^2, \tag{10.47}$$

where $q = q(r_t, t)$ is so-called *market price of risk* (if no arbitrage opportunities exist, then q does not depend on the time of maturity T). The interpretation of q can be shown symbolically (not writing arguments of variables for simplicity). If one writes the stochastic differential equations for $P = P(r_t, t, T)$ symbolically as $dP = \mu \cdot P \; dt + \sigma \cdot P \; dW$ (applying Ito's lemma to $P(r_t, t, T)$ and (10.46)), then the market price of risk is defined as $q = (\mu - r)/\sigma$. The equality $\mu - r = q \cdot \sigma$ can be interpreted in such a way that the expected yield μ exceeding r compensates the risk $q \cdot \sigma$. Moreover, $R(r_t, t, T)$ increases, or reverses from increase to decrease, or decreases if $r_t \leq \gamma + b \cdot q/\alpha - 3b^2/(4\alpha^2)$, or $\gamma + b \cdot q/\alpha - 3b^2/(4\alpha^2) < r_t < \gamma + b \cdot q/\alpha$, or $r_t \geq \gamma + b \cdot q/\alpha$, respectively.

◊

2. Model Cox–Ingersoll–Ross

This model denoted briefly as *CIR model* is based on the diffusion equation (10.45) in the form

$$dr_t = \alpha \cdot (\gamma - r_t) \; dt + b \cdot \sqrt{r_t} \; dW_t \tag{10.48}$$

with parameters $\alpha > 0$, $\gamma \in R$, $b \in R$. In contrast to Vasicek model, the volatility in this model is not constant but proportional to the value $\sqrt{r_t}$ so that the corresponding solution $P(r_t, t, T)$ cannot attain negative values.

Remark 10.4 In statistic and econometric literature various methods are suggested how to estimate diffusion processes (10.45) directly from discrete data. Several approaches can be distinguished:

- ML estimate: see, e.g., Lo (1988).
- QML estimate (*quasi-maximum likelihood*): see, e.g., Kessler (1997).
- Moment estimate: see, e.g., Conley et al. (1997).
- Nonparametric estimate: see, e.g., Ait-Sahalia (1996).
- Semiparametric estimate: see, e.g., Gallant and Long (1997).

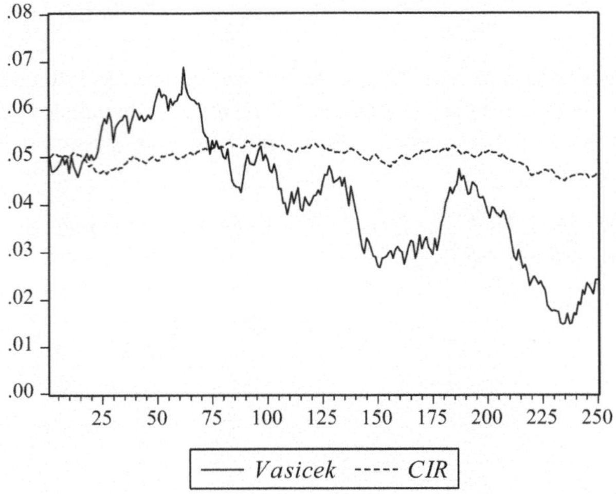

Fig. 10.2 Simulation of instantaneous interest rate r_t by means of Vasicek model (10.46) and CIR model (10.48) ($\alpha = 0.1$, $\gamma = 0.05$, $b = 0.03$)

- MCMC (*Markov chain Monte Carlo*) approach by means of simulations: see, e.g., Elerian et al. (2001), Eraker (2001).

◇

Example 10.3 Figure 10.2 plots one trajectory of simulations of instantaneous interest rate r_t using 250 regular time intervals of length $\Delta = 1/250$ (see Remark 10.2) by means of Vasicek model

$$dr_t = 0.1 \cdot (0.05 - r_t)\ dt + 0.03\ dW_t$$

and CIR model

$$dr_t = 0.1 \cdot (0.05 - r_t)\ dt + 0.03\sqrt{r_t}\ dW_t.$$

Both trajectories fluctuate around the level 5%, but the trajectory of model CIR is much more stable (the trajectory of Vasicek model might sink even to negative rates in longer simulations).

◇

10.4 Exercises

Exercise 10.1 Repeat the simulations analysis from Example 10.3 using different values of coefficients for Vasicek and CIR model and compare their graphs.

Chapter 11
Value at Risk

Methodology VaR (*value at risk*) and its modifications are usual measures of risk in practice (e.g., it is one of the best used approaches to set up capital requirements when regulating *capital adequacy* in so-called *internal models* of banks). More generally, VaR is the key instrument for *financial risk management*, e.g., by means of commercial systems of the type RiskMetrics. This topic is included in the presented text since some methods of VaR construction make use of the analysis of financial time series.

In general, the *financial risk* concerns potential price changes of financial assets, where the corresponding price change (expressed mainly as the rate of return; see Remark 6.20) is looked upon as a random event. If the financial risk is measured as the variance or standard deviation of (log) returns in the form of random process, then it is usually called (conditional) *volatility* (see Chap. 8).

Moreover, the financial risk can be classified into several categories, mainly:

1. *Market risk* is the risk of loss due to changes (variations) of market prices (of securities, commodities, and others) or market rates (interest rates, rates of exchange, and others). Accordingly, it can be sorted to more specific risk sub-categories, e.g.:
   - *Interest risk*
   - *Currency risk*
   - *Stock risk*
   - *Commodity risk*
   - *Credit spread risk* (i.e., the risk of loss due to changes in differences between the yields of various debt instruments)
   - *Correlation risk* (i.e., the risk of loss due to changes in traditional correlations between considered risk categories, e.g., between stocks and bonds) and others.
2. *Credit risk* is the risk that the creditor (lender) may not receive promised repayments on outstanding investments (such as loans, credits, bonds) because

T. Cipra, *Time Series in Economics and Finance*,
https://doi.org/10.1007/978-3-030-46347-2_11

of the default of the debtor (borrower). Defaults may consist in insolvency or reluctance of the debtor, in his refusal to deliver or to buy underlying assets according to contracts, and the like.
3. *Liquidity risk* is the risk of varying levels of convertibility of investments readily into cash. This risk can force investors to sell some assets under very unfavorable conditions.
4. *Operational risk* is the risk of a change in value caused by the fact that actual losses, incurred for inadequate or failed internal processes, people, and systems, or from external events (including legal risk), differ from the expected losses.

11.1 Financial Risk Measures

Risk can be measured and quantified by means of various ways. In some cases (e.g., in various regulatory systems for banks), one prefers deterministic instruments for this purpose, e.g., stress tests constructed in accordance with prescribed instructions without any portion of stochasticity.

In this text, we deal only with stochastic risk measures that respect the random character of potential losses (or profits). Let random variable X represent loss (if X is positive) or profit (if X is negative) accumulated during the given holding period (moreover, X is usually observed in time, i.e., in the form of a time series; see Sect. 11.2). Then a *risk measure* ρ can be defined as a mapping that assigns real values $\rho(X)$ to the random variables X. In particular, a risk measure is called *coherent* if it possesses the following properties (for bounded random variables X and Y denoting losses in the same financial environment):

(*i*) Subadditivity: $\rho(X + Y) \leq \rho(X) + \rho(Y)$
(*ii*) Monotony: if $X \leq Y$, then $\rho(X) \leq \rho(Y)$
(*iii*) Positive homogeneity: $\rho(\lambda \cdot X) = \lambda \cdot \rho(X)$ for arbitrary constant $\lambda > 0$
(*iv*) Translation invariancy: $\rho(X + a) = \rho(X) + a$ for arbitrary constant $a > 0$.

11.1.1 VaR

The methodology *VaR* (*value at risk*) is based on an estimate of the worst loss that can occur with a given probability (confidence) in a given future period (alternatively one can say that with a prescribed confidence α, e.g., 95%, there cannot occur a loss that is higher than *VaR*). For example in the context of capital requirements or capital adequacy of banks, *VaR* represents the smallest capital amount that guarantees the bank solvency with a given confidence. *VaR* is specified by the following factors:

- *Holding period* is the period in which a potential loss can occur. Accordingly, the used terms may be the daily *VaR* (over one business day, e.g., in *RiskMetrics*) or the 10 days *VaR* (over two calendar weeks with 10 business days, e.g., according

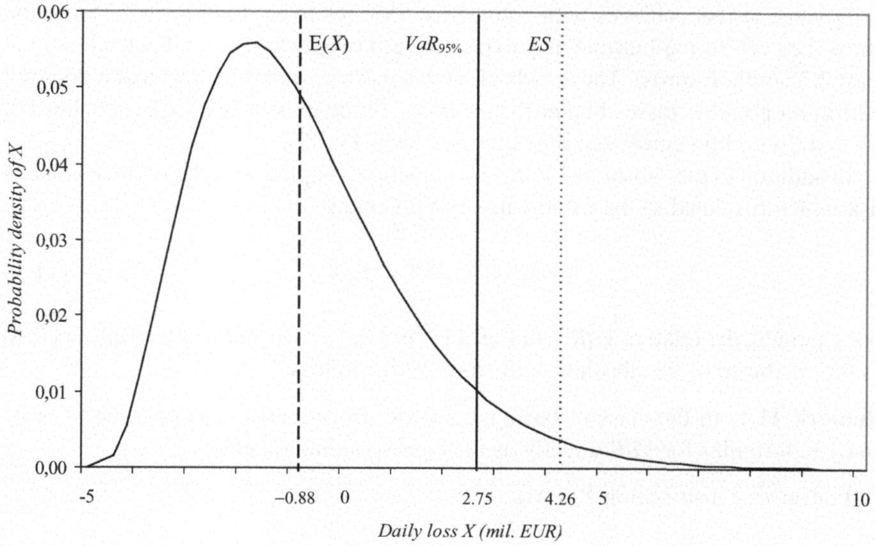

Fig. 11.1 Graphical plot of $VaR_{95\%}$

to the recommendation of the Basel Committee on Banking Supervision) or monthly *VaR*, quarterly *VaR*, or even yearly *VaR* (for various credit portfolios).

- *Confidence level* is the probability that the actual loss does not exceed the value at risk (during the given holding period). In practice, one applies, e.g., the confidence 95% (in RiskMetrics), or 99% (according to Basel Committee on Banking Supervision).

Formally, one can define *VaR* in the following way:

$$\begin{aligned} VaR = VaR_\alpha(X) &= \inf\{x \in (-\infty,\ \infty) :\ \mathrm{P}(X \leq x) \geq \alpha\} \\ &= \inf\{x \in (-\infty,\ \infty) :\ F_X(x) \geq \alpha\}, \end{aligned} \tag{11.1}$$

where random variable X denotes loss (of course the negative loss means profit) accumulated during the given holding period (e.g., during one trading day), α is the corresponding confidence level (e.g., 95% for $\alpha = 0.95$), and $F_X(x) = \mathrm{P}(X \leq x)$ is the probability distribution function of X. When one expresses it in statistical terms, then $VaR_\alpha(X)$ is α-quantile q_α of random variable X. Moreover, if the probability distribution function $F_X(\cdot)$ is increasing and continuous, then it holds simply

$$VaR_\alpha(X) = F_X^{-1}(\alpha) = q_\alpha, \text{ i.e. } \mathrm{P}(X \leq VaR_\alpha(X)) = \alpha. \tag{11.2}$$

Figure 11.1 plots $VaR_{95\%}$ for a daily loss X with given probability density. This random variable X has mean value -0.88 million euros (i.e., one can expect a profit on average) and skewed to the right (i.e., potential losses are not negligible). The daily value at risk achieves with confidence 95% relatively high level 2.75 million euros (i.e., ceteris paribus one may expect in each twentieth trading day the loss of at least 2.75 million euros). The drawback of this risk measure is the fact that it does not inform on possible losses higher than $VaR_{95\%}$ (in contrast to the expected shortfall $ES = 4.26$ million euros; see Fig. 11.1 and Sect. 11.1.2).

In addition to the "absolute" *VaR*, one sometimes applies also the *relative* value at risk which is related to the mean value E(*X*), namely

$$VaR^{rel} = VaR^{abs} - \mathrm{E}(X). \tag{11.3}$$

For example, the relative VaR^{rel} in Fig. 11.1 is $2.75 - (-0.88) = 3.63$ million euros as the "distance of the absolute *VaR* from the mean loss."

Remark 11.1 In the class of basic parametric distributions, it is possible to write analytic formulas for *VaR* directly as the corresponding quantiles q_α, e.g.:

1. For *normal* distribution $X \sim N(\mu, \sigma^2)$:

$$VaR = \mu + \sigma\Phi^{-1}(\alpha), \qquad VaR^{rel} = \sigma\Phi^{-1}(\alpha), \tag{11.4}$$

   where $\Phi(\cdot)$ is the distribution function of standard normal distribution $N(0, 1)$ (e.g., for confidence levels 95% and 99%, it holds $VaR^{rel}_{95\%} = 1.645\sigma$ and $VaR^{rel}_{99\%} = 2.326\sigma$).

2. For *log-normal* distribution $X \sim LN(\mu, \sigma^2)$, i.e., $\ln(X) \sim N(\mu, \sigma^2)$:

$$VaR = \exp\left(\mu + \sigma\Phi^{-1}(\alpha)\right), VaR^{rel} = \exp(\mu)\left\{\exp\left(\sigma\Phi^{-1}(\alpha)\right) - \exp\left(\sigma^2/2\right)\right\} \tag{11.5}$$

   (in this case, the probability density of *X* with a suitable configuration of parameters looks similarly as in Fig. 11.1).

3. For *exponential* distribution $X \sim Exp(\lambda)$, i.e., $F_X(x) = 1 - \exp(-\lambda x)$ for $x \geq 0$:

$$VaR = -\ln(1-\alpha)/\lambda, VaR^{rel} = -(1 + \ln(1-\alpha))/\lambda \tag{11.6}$$

   (this distribution is applicable, e.g., in the case when no profit with negative *X* is possible).

◇

11.1.2 Other Risk Measures

Here we shall give a brief survey of other types of risk measures that are applied in financial practice:

1. Deviation Risk Measures

Deviation risk measures regard the risk as fluctuations around a given value (usually around the mean value E(X) which is interpreted as the average loss). The main representatives (used, e.g., in risk management) are:

- *Standard deviation*:

$$\sigma(X) = \sqrt{\mathrm{E}(X - \mathrm{E}(X))^2} = \sqrt{\mathrm{E}(X^2) - (\mathrm{E}(X))^2} \tag{11.7}$$

($\sigma(X)$ used by Markowitz in his theory of portfolio is a very popular risk measure due to its simplicity; on the other hand, it has some drawbacks, namely (*i*) it is applicable only when the second moment of loss X exists and (*ii*) it does not distinguish positive and negative deviations around E(X) so that it cannot be recommended for asymmetric and skewed loss distributions).

- *Variance*:

$$\mathrm{var}(X) = \sigma^2(X) = \mathrm{E}(X^2) - (\mathrm{E}(X))^2 \tag{11.8}$$

(var(X) is used as the measure of volatility in models of the type ARCH for financial time series in Sect. 8.3).

- *One-sided standard deviations*:

$$\sigma_+(X) = \sqrt{\mathrm{E}(\max\{X - \mathrm{E}(X),\ 0\})^2},\ \sigma_-(L) = \sqrt{\mathrm{E}(\min\{X - \mathrm{E}(X),\ 0\})^2} \tag{11.9}$$

(in contrast to the two-sided standard deviation, $\sigma(X)$ measures only positive or negative deviations from the mean value, respectively).

- *Variance coefficient*:

$$v(X) = \frac{\sigma(X)}{|\mathrm{E}(X)|} \cdot 100\%. \tag{11.10}$$

- *Mean absolute deviation*:

$$MAD(X) = \mathrm{E}|X - \mathrm{E}(X)| \tag{11.11}$$

or *mean absolute semi-deviations*:

$$MAD_+(X) = \mathrm{E}(\max\{X - \mathrm{E}(X),\, 0\}),\ MAD_-(X) = -\mathrm{E}(\min\{X - \mathrm{E}(X),\, 0\}). \tag{11.12}$$

2. Expected Shortfall and Conditional Value at Risk

The drawback of value at risk *VaR* (see Sect. 11.1.1) is the fact that this risk measure does not inform on possible losses higher than *VaR* (moreover, *VaR* is not subadditive so that generally it cannot be coherent; see Sect. 11.1). Therefore, various modifications of *VaR* are now preferred by risk managers in practice, mainly so-called *expected shortfall* defined as

$$ES_\alpha = \frac{1}{1-\alpha} \int_\alpha^1 VaR_u \ \mathrm{d}u, \tag{11.13}$$

Obviously, instead of fixing a particular confidence level α we average VaR_u over all levels $u \geq \alpha$ and thus look further into the tail of the loss distribution (it holds always $ES_\alpha \geq VaR_\alpha$; see Fig. 11.1). In any case, the expected shortfall is the coherent risk measure.

Remark 11.2 For *continuous* loss distribution, an even more intuitive expression for ES_α in (11.13) is possible, namely

$$ES_\alpha = \mathrm{E}(X | X \geq VaR_\alpha), \tag{11.14}$$

which shows that ES_α can be also interpreted as the expected loss that is incurred in the case that VaR_α is exceeded (see, e.g., McNeil et al. (2005)). In general (i.e., including discrete loss distributions), one defines

$$CVaR_\alpha = \mathrm{E}(X | X \geq VaR_\alpha) = \frac{1}{1-\alpha}\ \mathrm{E}\big(X \cdot I_{[X \geq VaR_\alpha]}\big) \tag{11.15}$$

as *conditional value at risk CVaR* (or sometimes also *tail conditional expectation TVaR*).

One can see that the difference between (11.13) and (11.14) consists in the lower bound for averaging the worst losses:

- In (11.13) one averages over the worst scenarios that occur with probability $1-\alpha$.
- In (11.14) one averages over the worst losses which are not lower than VaR_α.

Table 11.1 Probability distribution of losses expected in investment portfolios *A* and B during next year

Portfolio A		Portfolio B	
Loss (million euros)	Probability (%)	Loss (million euros)	Probability (%)
20	4	30	2
10	3	10	98
−100	93	–	–

◇

Remark 11.3 Similarly as in Remark 11.1 one can derive analytic formulas for $ES = ES_\alpha$ under some parametric loss distributions, e.g.:

1. For *normal* distribution $X \sim N(\mu, \sigma^2)$:

$$ES = \mu + \sigma \frac{\varphi\left(\Phi^{-1}(\alpha)\right)}{1-\alpha}, \quad ES^{rel} = \sigma \frac{\varphi\left(\Phi^{-1}(\alpha)\right)}{1-\alpha}, \tag{11.16}$$

   where φ and Φ are the probability density and the distribution function of standard normal distribution $N(0, 1)$, respectively.

2. For *log-normal* distribution $X \sim LN(\mu, \sigma^2)$:

$$\begin{aligned} ES &= \frac{\Phi\left(\sigma - \Phi^{-1}(\alpha)\right)}{1-\alpha} \exp\left(\mu + \sigma^2/2\right), \\ ES^{rel} &= \frac{\alpha - \Phi\left(\Phi^{-1}(\alpha) - \sigma\right)}{1-\alpha} \exp\left(\mu + \sigma^2/2\right). \end{aligned} \tag{11.17}$$

◇

3. Distorted Risk Measures

Distorted risk measures originate by artificially "distorting" the distribution function of loss: the expected value of loss after this adjustment is the risk measure result. The motivation of distortion consists in the fact that in specific situations the risk measures of the type *VaR* and *ES* do not distinguish the risk in an acceptable way. For example, let us have choice between two investment portfolios whose stochastic behavior is described in Table 11.1.

Then it holds

- in the portfolio A:

$$VaR_{0.95} = 10 \text{ million euros}, \qquad ES_{0.95} = \frac{4}{5} \cdot 20 + \frac{1}{5} \cdot 10 = 18 \text{ million euros};$$

- in the portfolio B:

$$VaR_{0.95} = 10 \text{ million euros}, \qquad ES_{0.95} = \frac{2}{5} \cdot 30 + \frac{3}{5} \cdot 10 = 18 \text{ million euros}.$$

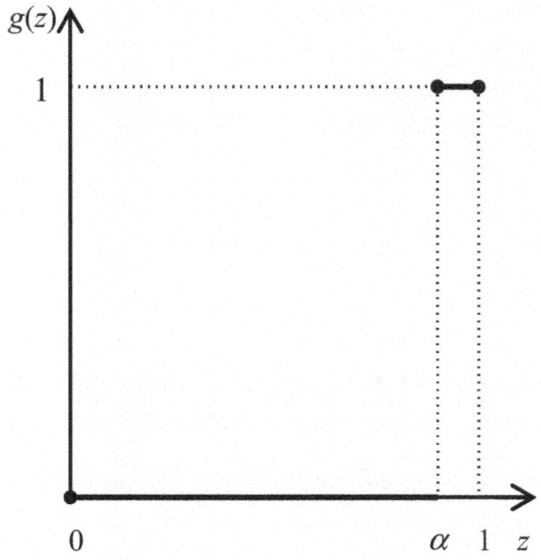

Fig. 11.2 Distorted functions for VaR_α

Even though both the portfolios have the same risk values $VaR_{0.95}$ and $ES_{0.95}$, each investor in practice would prefer to invest to the portfolio A (while the portfolio B is always in loss with maximum possible loss of 30 million euros, the portfolio A is in loss of 10 or 20 million euros only with relatively small probabilities and otherwise it is highly profitable). This reasonable decision can be confirmed by means of risk measures only if we distort them in a suitable way.

In general, let X be a loss with the distribution function $F(x) = \mathrm{P}(X \leq x)$ and the finite mean value $\mathrm{E}(X)$. Then the *distorted risk measure* $\mathrm{E}_g(X)$ of loss X is defined as

$$\mathrm{E}_g(X) = -\int_{-\infty}^{0} F_g(x)\mathrm{d}x + \int_{0}^{\infty} \left(1 - F_g(x)\right)\mathrm{d}x, \tag{11.18}$$

where

$$F_g(x) = g(F(x)) \tag{11.19}$$

for a *distorted function* $g(z)$ with following properties:

- $g(z)$ is nondecreasing for $0 \leq z \leq 1$.
- $g(0) = 0$ and $g(1) = 1$ (in particular, $0 \leq g(z) \leq 1$ for $0 \leq z \leq 1$).

Remark 11.4 Particularly it holds:

- For $g(z) = z$: the relation (11.18) is the well-known general formula for $\mathrm{E}(X)$.
- For $g(z)$ in Fig. 11.2: $\mathrm{E}_g(X) = VaR_\alpha$.
- Similarly $\mathrm{E}_g(X) = ES_\alpha$, etc. for other suitable choices of $g(z)$.

◇

Table 11.2 Values of Wang distorted risk measure for portfolios *A* and B from Table 11.1

λ	Wang distorted risk measure (million euros) for	
	Portfolio A	Portfolio B
−3	11.93	24.84
−2	−17.02	14.36
−1	−62.85	4.38
0	−91.90	0.60
1	−99.24	0.03
2	−99.97	0.00
3	−100.00	0.00

A usual example of $g(z)$ in practice is so-called *Wang distorted function*

$$g_\lambda(z) = \Phi\left(\Phi^{-1}(z) + \lambda\right), \quad 0 \le z \le 1, \tag{11.20}$$

where Φ is the distribution function of standard normal distribution $N(0, 1)$ and $\lambda \in \mathrm{R}$ is a real constant to be chosen. If this *Wang risk measure* (or *Wang transformation*; see Wang (2000)) is used in the numerical example for portfolios A and B from Table 11.1, then one obtains the results in Table 11.2 for chosen values of constant λ. Apparently, the higher values of this risk measure for portfolio B over a broad range of λ indicate the higher riskness of this portfolio in comparison with portfolio A, which corresponds to previous practical conclusions (see above).

4. Spectral Risk Measures

Spectral risk measures (see, e.g., Acerbi (2002)) generalize the concept of expected shortfall ES_α (see (11.13))

$$M_\psi = \int_0^1 \psi(u) \cdot VaR_u \; du, \tag{11.21}$$

where a weight function $\psi(u)$ is nonnegative and nondecreasing for $0 \le u \le 1$ with

$$\int_0^1 \psi(u) \; du = 1. \tag{11.22}$$

The interpretation of (11.21) is obvious: in contrast to ES_α in (11.13), one assigns higher weights to higher values VaR_u in (11.21). The steeper is the function $\psi(u)$ for u going to 1, the higher weights are assigned to extreme (catastrophic) scenarios.

Remark 11.5 Particularly it holds $M_\psi = ES_\alpha$ for the function ψ in Fig. 11.3.

◇

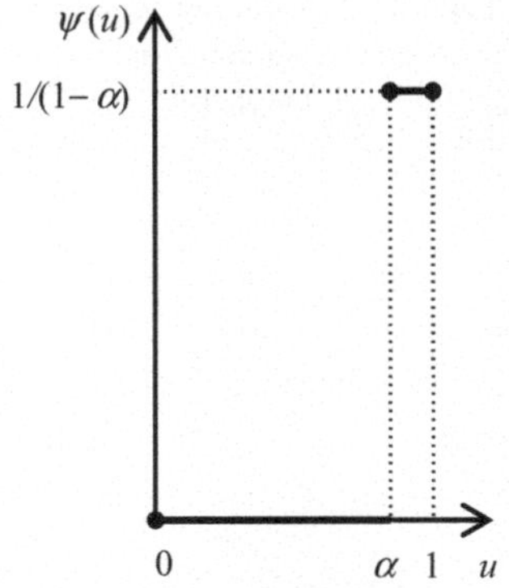

Fig. 11.3 Function ψ for coincidence M_ψ with ES_α

11.2 Calculation of *VaR*

One of the main drawbacks of *VaR* is the fact that different methods of its numerical calculation (or estimation or prediction) may deliver in practice (substantially) different results. In this section, we shall describe several methods for the calculation of *VaR* used in practice which are based on data in the form of time series. Some of them can be classified as parametric methods and others as nonparametric or combined ones.

1. Variance-Covariance Method

This method is frequent in practice among various parametric approaches calculating *VaR*. It offers a direct analytical solution of given problem, but it is based on simplifying assumptions which need not be fulfilled in practice (even in routine situations) and must be taken as drawbacks of this method, namely

(*i*) The loss X_t at time t originates as an aggregate of component losses by m risk sources.

(*ii*) The probability distribution of this aggregate loss X_t can be approximated as

$$X_t \sim N\left(\mu_1 + \cdots + \mu_m, \sigma_1^2 + \cdots + \sigma_d^2 + 2\sigma_{12} + \cdots + 2\sigma_{m-1,m}\right), \qquad (11.23)$$

where $\mu_1, \ldots, \mu_m$ are mean values, ${\sigma_1}^2, \ldots, {\sigma_m}^2$ are variances, and $\sigma_{12}, \sigma_{13}, \ldots, \sigma_{m-1,\,m}$ are covariances of component losses (more generally, these moments can vary in time, but must be estimable from data).

The most usual practical situation for application of this method is the prediction of *VaR* in a portfolio composed of various investment or credit instruments, for which the risk of possible losses must be evaluated or even controlled by management (see Example 11.1). If one has the data information $\mathbf{x}_t, \mathbf{x}_{t-1}, \mathbf{x}_{t-2}, \ldots$ till time

t (i.e., component losses in the form of m-variate time series observed till time t), then, e.g., one can predict *VaR* for next time $t + k$ (e.g., for the next trading day), which may be prescribed by regulators of various financial institution (for banks by Basel III or for insurance companies by Solvency II).

Remark 11.6 In practice, the financial time series $\mathbf{x}_t$ can be often modeled using methods of multivariate volatility modeling and predicting (see Chap. 13, or (8.62) for univariate case). Moreover, one can model (log) returns r_t instead of absolute losses X_t (negative values of r_t can be interpreted as relative losses), namely $r_t = \sum_{i=1}^{m} c_{ti} \cdot r_{ti}$ with moments of the form

$$\mathrm{E}(r_t) = \sum_{i=1}^{m} c_{ti} \cdot \mathrm{E}(r_{ti}), \qquad \mathrm{var}(r_t) = \sum_{i=1}^{m} \sum_{j=1}^{m} c_{ti} c_{tj} \operatorname{cov}(r_{ti}, r_{tj}), \tag{11.24}$$

where r_{ti} and c_{ti} denote (log) returns and portfolio weights of the ith risk component at time t, respectively. Then the formulas of the type $VaR_{95\%}^{rel} = 1.645\widehat{\sigma}_{t+1}(t)$ can be generalized to the multivariate case. The variances and covariances in (11.23) and (11.24) explain the name of this method (sometimes they are not estimated from analyzed data but taken from various published databases).

◇

Example 11.1 (*Calculation of VaR by variance-covariance method*). The calculation of *VaR* by various methods is demonstrated using a real investment portfolio composed of three investment instruments (the Czech Republic in 2013):

- 1000 pieces of the Czech government bonds 3.40/15 (i.e., the face value 10,000 CZK, the annual coupons 340 CZK paid on September 1, 2013, on September 1, 2014, and finally on the maturity date of September 1, 2015; see Fig. 11.4).
- 1 million pieces of the stocks of electricity operator ČEZ (the dividend 40 CZK for each stock paid out in 2013 on June 25; see Fig. 11.5).
- 10 million euros (the deposit priced in CZK using the actual exchange rates EUR/CZK; see Fig. 11.6).

Table 11.3 and Fig. 11.7 present the development of daily portfolio loss in the year 2013 (negative values mean profits). For example, the loss for the first trading day January 2, 2013, is calculated as

$$\begin{aligned}&(1000 \cdot 108.16 + 1000{,}000 \cdot 680.20 + 10{,}000{,}000 \cdot 25.225) - \\ &-(1000 \cdot 108.16 + 1000{,}000 \cdot 682.00 + 10{,}000{,}000 \cdot 25.140) = -1.050 \text{ million CZK}\end{aligned}$$

(the stock dividend and coupon payment were included in such a way that on June 25, 2013, the price of portfolio was increased by the dividend income of $1{,}000{,}000 \cdot 40 = 40$ million CZK and on September 2, 2013, by the coupon income of $1000 \cdot 340 = 0.340$ million CZK, but on the next trading day these incomes are transferred to another account and further are not included in the price of portfolio).

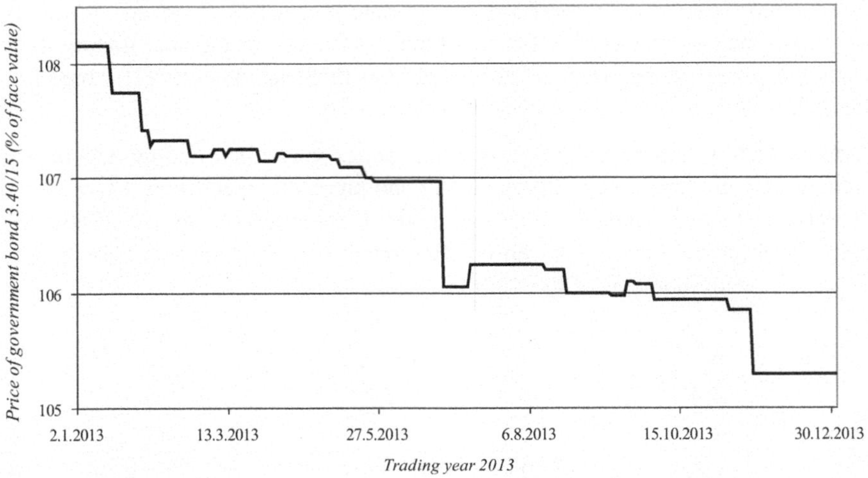

Fig. 11.4 Price of the Czech government bond 3.40/15 in the year 2013 from Example 11.1. Source: kurzy.cz (https://akcie-cz.kurzy.cz/emise/dluhopisy/statni-dluhopisy/2010/)

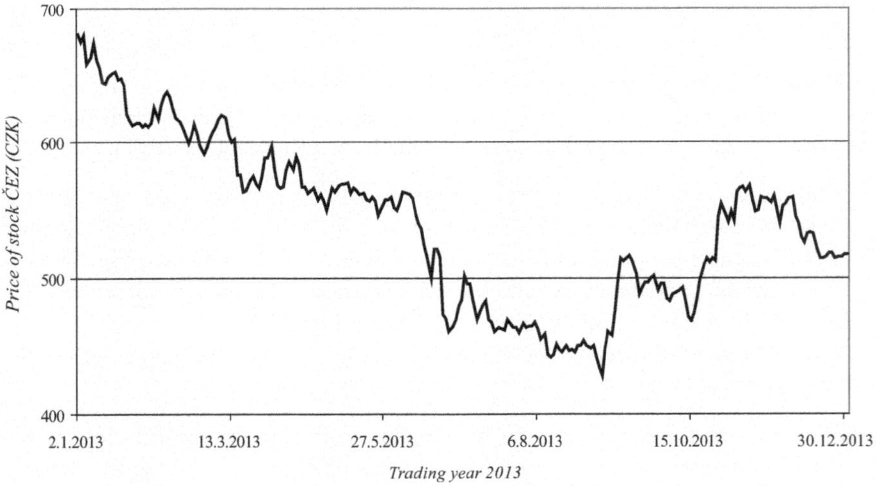

Fig. 11.5 Price of the stock of electricity operator ČEZ in the year 2013 from Example 11.1. Source: kurzy.cz (https://prague-stock.kurzy.cz/akcie/cez-183/graf_2013)

Histogram of portfolio loss in Fig. 11.8 indicates (at least graphically) that the assumption of normality is realistic with negative values denoting profits.

Table 11.4 contains the sample means and the sample covariance matrix of portfolio components (bonds, stocks, euro deposit) which are necessary for the variance-covariance method. Hence one easily calculates by means of (11.23) that

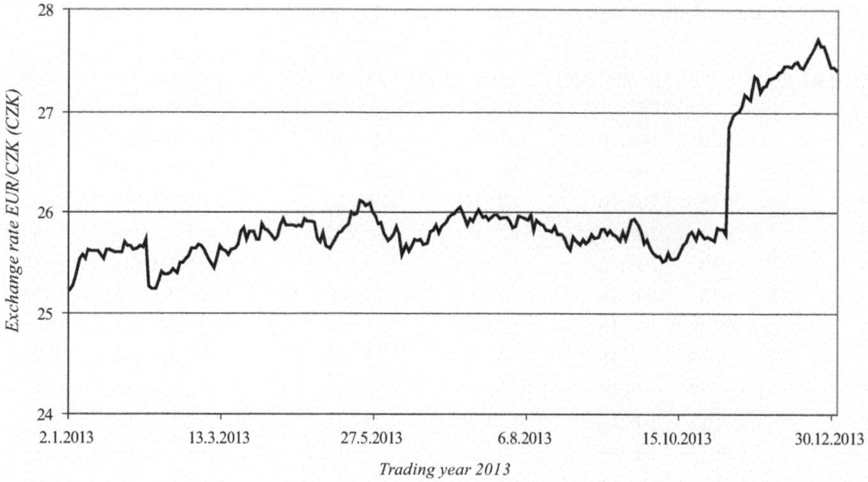

Fig. 11.6 Exchange rates EUR/CZK in the year 2013 from Example 11.1. Source: EUROSTAT (https://ec.europa.eu/eurostat/data/database)

$$X_t \sim N\left(0.670;\ \ 8.595^2\right)$$

so that according to (11.4) one finally obtains

$$VaR_{95\%} = 0.670 + 8.595 \cdot 1.645 = 14.809 \text{ million CZK},$$

$$VaR^{rel}_{95\%} = 8.595 \cdot 1.645 = 14.139 \text{ million CZK}.$$

These values can be interpreted either as risk characteristics of the given portfolio during year 2013 or as VaR predictions for the first trading day of year 2014.

◊

2. Method of Historical Simulation

This method is evidently the most popular in the framework of nonparametric approaches to the calculation of *VaR* in practice, since it is very simple. It ignores entirely the problem of probability distribution or the correlation structure among component losses of portfolio and assumes simply that the character of losses in previous periods (e.g., in the trading days of previous years) will sustain also during a future period. In other words, this method is based on losses, which are simulated by the "history." In this context, one usually applies the following estimate used typically for construction of empirical distribution functions:

Table 11.3 Daily portfolio loss in the year 2013 from Example 11.1 (see also Fig. 11.7)

Trading day	Date	Price of bond 3.40/15 (%)	Price of stock ČEZ (CZK)	Exchange rate EUR/CZK (CZK)	Price of portfolio (CZK)	Loss (million CZK)
0	28.12.2012	108.16	680.00	25.140	2,013,000,000	–
1	2.1.2013	108.16	680.20	25.225	2,014,050,000	−1.050
2	3.1.2013	108.16	675.00	25.260	2,009,200,000	4.850
3	4.1.2013	108.16	680.00	25.355	2,015,150,000	−5.950
4	7.1.2013	108.16	658.50	25.535	1,995,450,000	19.700
5	8.1.2013	108.16	663.50	25.580	2,000,900,000	−5.450
6	9.1.2013	108.16	673.50	25.530	2,010,400,000	−9.500
7	10.1.2013	108.16	661.50	25.630	1,999,400,000	11.000
8	11.1.2013	108.16	655.10	25.615	1,992,850,000	6.550
9	14.1.2013	108.16	644.90	25.615	1,982,650,000	10.200
10	15.1.2013	108.16	644.00	25.610	1,981,700,000	0.950
11	16.1.2013	108.16	648.50	25.580	1,985,900,000	−4.200
12	17.1.2013	107.75	651.70	25.540	1,984,600,000	1.300
13	18.1.2013	107.75	652.90	25.630	1,986,700,000	−2.100
14	21.1.2013	107.75	647.10	25.625	1,980,850,000	5.850
15	22.1.2013	107.75	648.00	25.610	1,981,600,000	−0.750
16	23.1.2013	107.75	643.00	25.600	1,976,500,000	5.100
17	24.1.2013	107.75	622.00	25.595	1,955,450,000	21.050
18	25.1.2013	107.75	617.00	25.605	1,950,550,000	4.900
19	28.1.2013	107.75	613.00	25.700	1,947,500,000	3.050
20	29.1.2013	107.75	615.00	25.660	1,949,100,000	−1.600
21	30.1.2013	107.75	615.00	25.660	1,949,100,000	0.000
22	31.1.2013	107.42	612.10	25.620	1,942,500,000	6.600
⋮	⋮	⋮	⋮	⋮	⋮	⋮
⋮	⋮	⋮	⋮	⋮	⋮	⋮
230	26.11.2013	105.30	540.10	27.330	1,866,400,000	9.600
231	27.11.2013	105.30	553.00	27.340	1,879,400,000	−13.000
232	28.11.2013	105.30	555.00	27.350	1,881,500,000	−2.100
233	29.11.2013	105.30	559.00	27.390	1,885,900,000	−4.400
234	2.12.2013	105.30	560.00	27.405	1,887,050,000	−1.150
235	3.12.2013	105.30	545.00	27.460	1,872,600,000	14.450
236	4.12.2013	105.30	540.80	27.455	1,868,350,000	4.250
237	5.12.2013	105.30	530.00	27.450	1,857,500,000	10.850
238	6.12.2013	105.30	525.50	27.490	1,853,400,000	4.100
239	9.12.2013	105.30	533.00	27.500	1,861,000,000	−7.600
240	10.12.2013	105.30	533.70	27.450	1,861,200,000	−0.200
241	11.12.2013	105.30	532.90	27.435	1,860,250,000	0.950
242	12.12.2013	105.30	520.00	27.480	1,847,800,000	12.450
243	13.12.2013	105.30	514.40	27.535	1,842,750,000	5.050
244	16.12.2013	105.30	513.90	27.595	1,842,850,000	−0.100
245	17.12.2013	105.30	514.80	27.655	1,844,350,000	−1.500

(continued)

Table 11.3 (continued)

Trading day	Date	Price of bond 3.40/15 (%)	Price of stock ČEZ (CZK)	Exchange rate EUR/CZK (CZK)	Price of portfolio (CZK)	Loss (million CZK)
246	18.12.2013	105.30	517.80	27.720	1,848,000,000	−3.650
247	19.12.2013	105.30	518.50	27.650	1,848,000,000	0.000
248	20.12.2013	105.30	514.00	27.655	1,843,550,000	4.450
249	23.12.2013	105.30	515.00	27.575	1,843,750,000	−0.200
250	27.12.2013	105.30	515.60	27.440	1,843,000,000	0.750
251	30.12.2013	105.30	517.00	27.445	1,844,450,000	−1.450
252	31.12.2013	105.30	517.00	27.425	1,844,250,000	0.200

Source: calculated by EViews

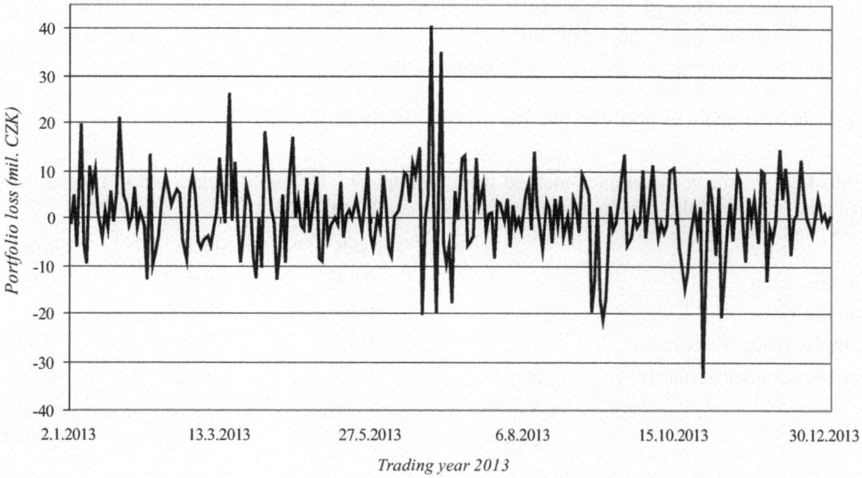

Fig. 11.7 Daily portfolio loss in the year 2013 from Example 11.1 (negative values mean profits)

$$\mathrm{P}(X > x) = \frac{1}{T}\sum_{t=1}^{T} I_{[x_t > x]}, \tag{11.25}$$

where x_1, x_2, …, x_T are observed losses during a period of length T (e.g., T trading days).

Example 11.2 (*Calculation of VaR by method of historical simulation*). Let us consider the portfolio from Example 11.1 composed of government bonds, stocks of ČEZ, and euro deposit. Table 11.5 presents 15 highest daily losses. The value at risk of this portfolio with confidence level 95% can be found according to (11.25): as 12/252 = 4.76% (the twelfth highest daily loss is 13.450 million CZK) and 13/252 = 5.16% (the thirteenth highest daily loss is 13.250 million CZK), hence it follows approximately by means of interpolation (according to Table 11.5 with the

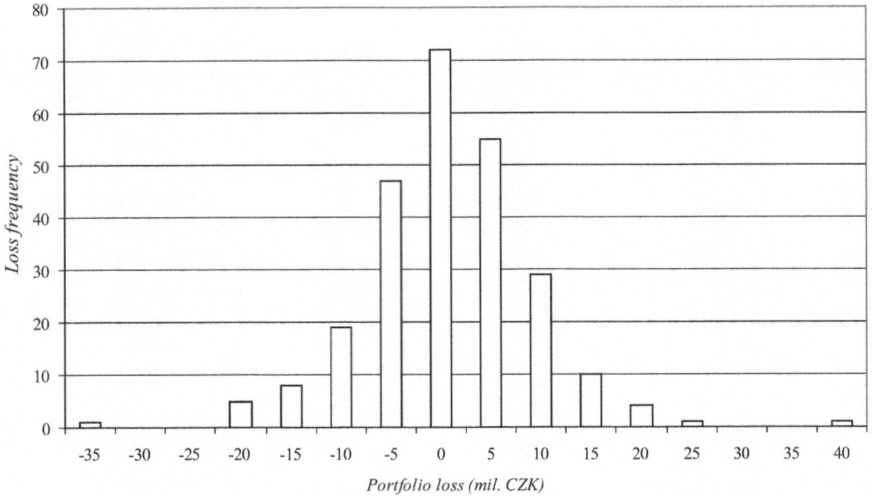

Fig. 11.8 Histogram of portfolio loss in the year 2013 from Example 11.1

Table 11.4 Variance-covariance method from Example 11.1: sample means and sample covariance matrix of portfolio components (bonds, stocks, and euro deposit)

		Bonds	Stocks	Euros
Sample mean		0.113	0.647	−0.091
Sample variance		0.634	75.831	0.919
Sample standard deviation		0.796	8.708	0.959
Sample covariance matrix:	Bonds	0.634	−1.358	−0.054
	Stocks	−1.358	75.831	−0.342
	Euros	−0.054	−0.342	0.919

daily losses during 2013 ordered from the highest 40.700 million CZK to the lowest one)

$$VaR_{95\%} \approx 13.4 \text{ million CZK.}$$

This value at risk is significantly lower than the value 14.8 million CZK from Example 11.1.

◇

3. Various Modification of Methods of Historical Simulation

The method of historical simulation described above in its basic form can be modified in various ways:

(a) *Method simulating previous development*:

This method simulates additional data respecting the development of previous ratios among neighboring observations (see Example 11.3):

Table 11.5 Method of historical simulation from Example 11.2: daily portfolio losses ordered downward

Trading days ordered according to losses	Date	Loss (million CZK)
1	21.6.2013	40.700
2	15.3.2013	26.400
3	24.1.2013	21.050
4	25.6.2013	20.000
5	7.1.2013	19.700
6	3.4.2013	18.000
7	16.4.2013	16.950
8	17.6.2013	14.850
9	3.12.2013	14.450
10	9.8.2013	14.000
11	20.9.2013	13.600
12	7.2.2013	13.450
13	9.7.2013	13.250
14	12.3.2013	12.650
15	8.7.2013	12.550
⋮	⋮	⋮

Example 11.3 (*Calculation of VaR by method simulating previous development*). We again make use of portfolio from Example 11.1 (for simplicity, we ignore the coupon and dividend income now). The task is to predict *VaR* for the first trading day of the year 2014 using data of year 2013 (see Table 11.6). We shall proceed by the method simulating previous development:

The reference row is chosen as the values observed on December 12, 2013 (see Table 11.6), and we generate gradually data respecting the development of previous ratios in such a way that in the first simulation (see the row of Table 11.6) the prices of bond, stock, and euro are

$$\frac{108.16}{108.16} \cdot 105.30 = 105.30 \text{ CZK}, \qquad \frac{680.20}{680.00} \cdot 517.00 = 517.15 \text{ CZK},$$
$$\frac{25.225}{25.140} \cdot 27.425 = 27.518 \text{ CZK},$$

i.e., the prices from the reference row of 31.12.2013 are multiplied by growth rates between neighboring trading days December12, 2012, and January 2, 2013; in the second simulation, prices from the reference row of December 31, 2013, are multiplied by growth rates between neighboring trading days January 2, 2013, and January 3, 2013, presenting further possibility of change of the reference row to the neighboring date January 2, 2014 (it is the first trading day of year 2014, for which *VaR* is predicted in this example), and so on.

In this way, one obtains 252 simulations in Table 11.6 including corresponding losses. For example, the loss (i.e., the profit with negative sign) generated in the first simulation is

Table 11.6 Method from Example 11.3 simulating previous development: 252 simulated losses (the first row contains values for reference date)

Order number of simulation	Price of bond 3.40/15 (%)	Price of stock ČEZ (CZK)	Exchange rate EUR/CZK (CZK)	Price of portfolio (CZK)	Loss (million CZK)
31.12.2013	105.30	517.00	27.425	1,844,250,000	
1	105.30	517.15	27.518	1,845,329,316	−1.079
2	105.30	513.05	27.463	1,840,678,158	3.572
3	105.30	520.83	27.528	1,849,111,053	−4.861
4	105.30	500.65	27.620	1,829,850,630	14.399
5	105.30	520.93	27.473	1,848,658,896	−4.409
6	105.30	524.79	27.371	1,851,505,949	−7.256
7	105.30	507.79	27.532	1,836,112,645	8.137
8	105.30	512.00	27.409	1,839,087,530	5.162
9	105.30	508.95	27.425	1,836,200,237	8.050
10	105.30	516.28	27.420	1,843,474,960	0.775
11	105.30	520.61	27.393	1,847,541,316	−3.291
12	104.90	519.55	27.382	1,842,380,681	1.869
13	105.30	517.95	27.522	1,846,168,397	−1.918
14	105.30	512.41	27.420	1,839,603,758	4.646
15	105.30	517.72	27.409	1,844,808,518	−0.559
16	105.30	513.01	27.414	1,840,153,715	4.096
17	105.30	500.12	27.420	1,827,311,521	16.938
18	105.30	512.84	27.436	1,840,201,201	4.049
19	105.30	513.65	27.527	1,841,915,824	2.334
20	105.30	518.69	27.382	1,845,509,938	−1.260
21	105.30	517.00	27.425	1,844,250,000	0.000
22	104.98	514.56	27.382	1,838,159,635	6.090
⋮	⋮	⋮	⋮	⋮	⋮
230	105.30	507.42	27.485	1,835,270,637	8.979
231	105.30	529.35	27.435	1,856,698,616	−12.449
232	105.30	518.87	27.435	1,846,220,112	−1.970
233	105.30	520.73	27.465	1,848,377,223	−4.127
234	105.30	517.92	27.440	1,845,325,058	−1.075
235	105.30	503.15	27.480	1,830,952,187	13.298
236	105.30	513.02	27.420	1,840,215,844	4.034
237	105.30	506.68	27.420	1,833,875,350	10.375
238	105.30	512.61	27.465	1,840,260,013	3.990
239	105.30	524.38	27.435	1,851,728,451	−7.478
240	105.30	517.68	27.375	1,844,430,351	−0.180
241	105.30	516.23	27.410	1,843,325,169	0.925
242	105.30	504.48	27.470	1,832,184,730	12.065
243	105.30	511.43	27.480	1,839,231,207	5.019
244	105.30	516.50	27.485	1,844,345,076	−0.095
245	105.30	517.91	27.485	1,845,751,733	−1.502

(continued)

Table 11.6 (continued)

Order number of simulation	Price of bond 3.40/15 (%)	Price of stock ČEZ (CZK)	Exchange rate EUR/CZK (CZK)	Price of portfolio (CZK)	Loss (million CZK)
246	105.30	520.01	27.489	1,847,907,415	−3.657
247	105.30	517.70	27.356	1,844,256,368	−0.006
248	105.30	512.51	27.430	1,839,812,611	4.437
249	105.30	518.01	27.346	1,844,462,490	−0.212
250	105.30	517.60	27.291	1,843,509,674	0.740
251	105.30	518.40	27.430	1,845,703,774	−1.454
252	105.30	517.00	27.405	1,844,050,146	0.200

$$\begin{aligned}-(1,845,329,316 - 1,844,250,000) &= -1,079,316 \text{ CZK}\\ &= -1.079 \text{ million CZK},\end{aligned}$$

and similarly for further simulations. Table 11.7 contains such daily losses for each of 252 simulations ordered downward. Hence in the same way as in Table 11.5, one can find the value at risk approximately as

$$VaR_{95\%} \approx 13.4 \text{ million CZK}.$$

This value at risk is significantly lower than the value 14.8 million CZK from Example 11.1, and it is nearly the same as in Example 11.2.

◇

(b) *Method of historical simulation based on principle EWMA*:

It is the classical method of historical simulation from Example 11.1 supplemented by the principle EWMA (*exponentially weighted moving average*). This principle, which weighs time data by means of weights decreasing exponentially to the past, is frequently applied for financial time series (see Sects. 3.3.1 or 8.3.1). When we constructed the corresponding quantile $q_{0.95} = VaR_{95\%}$ of losses in Example 11.1, we looked for such a loss among T losses ordered downward that its order number i fulfills as the first one the inequality $i/T \geq 0.05$. If using the principle EWMA, we assign to the ith loss in their descending arrangement the weight

$$\lambda^{i-1}\frac{1-\lambda}{1-\lambda^T}, \tag{11.26}$$

so that now we look for such a loss among T losses ordered downwards that its order number i fulfills as the first one the inequality

Table 11.7 Method from Example 11.3 simulating previous development: simulated daily portfolio losses ordered downward

Trading days ordered according to losses	Loss (million CZK)
1	40.854
2	22.757
3	20.885
4	16.938
5	15.839
6	15.672
7	14.997
8	14.905
9	14.399
10	14.280
11	13.959
12	13.437
13	13.298
14	13.092
15	12.215
⋮	⋮

$$\frac{1-\lambda}{1-\lambda^T}+\lambda\frac{1-\lambda}{1-\lambda^T}+\cdots+\lambda^{i-1}\frac{1-\lambda}{1-\lambda^T}=\frac{1-\lambda^i}{1-\lambda^T}\geq 0.05, \tag{11.27}$$

where T is the number of losses (it can be compared with the original inequality $i/T \geq 0.05$; see above). The coefficient λ must be chosen a priori controlling the impact of time arrangement of losses: the closer to 1 this coefficient λ is, the less important is the time allocation of losses so that losses more remote in the past may have impact on *VaR* (the weights (11.26) converge to $1/T$ for λ going to 1, so that the method converts to the classical calculation of *VaR* by means of historical simulation, where the time arrangement does not play any role). For values λ usual in practice, Table 11.8 indicates the order number of such a loss among losses ordered downward which determines the corresponding $VaR_{95\%}$ (e.g., for $\lambda = 0.99$, the position of the asterisk indicates the loss order $i = 5$).

Example 11.4 (*Method of historical simulation based on principle EWMA*). We shall again demonstrate this method by means of portfolio from Example 11.1. Let us choose, e.g., $\lambda = 0.99$, so that Table 11.8 indicates the value at risk $VaR_{0.95}$ as the fifth loss in the descending arrangement of losses in Table 11.5, i.e.,

$$VaR_{95\%} \approx 19.7 \ \text{million CZK}.$$

This value at risk is by far the highest one in comparison with all previous results so that the time allocation of losses has a significant impact for construction of *VaR*.

Table 11.8 Method of historical simulation based on principle EWMA from Example 11.4: order number of loss in descending arrangement for construction of $VaR_{95\%}$

Order number of loss i	$\lambda = 0.9$	$\lambda = 0.95$	$\lambda = 0.99$	$\lambda = 0.995$	$\lambda = 0.999$
1	0.1000*	0.0500*	0.0109	0.0070	0.0045
2	0.1900	0.0975	0.0216	0.0139	0.0090
3	0.2710	0.1426	0.0323	0.0208	0.0134
4	0.3439	0.1855	0.0428	0.0277	0.0179
5	0.4095	0.2262	0.0532*	0.0345	0.0224
6	0.4686	0.2649	0.0636	0.0413	0.0269
7	0.5217	0.3017	0.0738	0.0481*	0.0313
8	0.5695	0.3366	0.0839	0.0548	0.0358
9	0.6126	0.3698	0.0939	0.0615	0.0402
10	0.6513	0.4013	0.1039	0.0682	0.0447
11	0.6862	0.4312	0.1137	0.0748	0.0491*
12	0.7176	0.4596	0.1234	0.0814	0.0536
13	0.7458	0.4867	0.1330	0.0880	0.0580
14	0.7712	0.5123	0.1426	0.0945	0.0624
15	0.7941	0.5367	0.1520	0.1010	0.0668
⋮	⋮	⋮	⋮	⋮	⋮

◇

4. Method of Simulation Monte Carlo

This method usually combines parametric and nonparametric approaches:

At first one estimates parametrically the probability distribution of losses. There are various alternatives how to do it: (1) to estimate separately the marginal distributions of particular loss components and their correlation (or copula structure), (2) to estimate directly the multivariate distribution of loss vector, and (3) to estimate the loss dynamically as a multivariate time series (e.g., by means of a multivariate GARCH model; see Sect. 13.3).

In the second step, one realizes Monte Carlo simulations based on calibrated (estimated) model. It results in a set of mutually independent loss values referred to the time moment of constructed *VaR*. These loss realizations enable us to calculate the corresponding value at risk in the same (nonparametric) way as in the previous methods described in this section. The simulation technique denoted as *bootstrap* is preferred in this context (then the resulting *VaR* is sometimes called *resampled value at risk*).

The advantage of this Monte Carlo simulation method consists mainly in the fact that the volume of simulated losses can be much larger than the volume of observed losses (e.g., in the case of credit portfolio, the volume of observed losses is relatively limited). On the other hand, there are also drawbacks of this method, namely the calculation complexity (particularly for portfolios with financial derivatives, which must be newly priced for each simulation) and high demands on the quality of simulation models.

11.3 Extreme Value Theory

This section presents basic facts on quantitative approach to extreme values. Even though the corresponding theory denoted explicitly as EVT (*Extreme Value Theory*; see Embrechts et al. (1997), McNeil et al. (2005), and others) comprises very complex and nontrivial results, its applications are very broad including time series data not only in economy (particularly in finance and insurance, e.g., financial losses or insured claims) but also in technical and environmental disciplines (e.g., river flows in hydrology, wind forces in climatology, exhaust concentrations in environmental control) and others. As the risk measures based on the value at risk principle have some extreme properties, the theme of EVT is included in this chapter.

The EVT makes use mainly of parametric methods because extreme values are rare (i.e., with small probabilities that can be quantified only in a parametric way). As the extreme value methodology is concerned, the following two approaches are preferred in practical data analysis:

- *Block maxima* (or *minima*): this approach segments particular data to blocks and then uses maximum (or minimum) values of particular blocks.
- *Threshold excesses*: this approach uses data exceeding a given threshold only.

11.3.1 Block Maxima

The model of *block maxima* is a traditional model in extreme problems: it is the model of maxima in particular blocks, which originate by segmentation of original (large) sample of independent identically distributed observations (the assumption of independence can be weakened). The most frequent application in finance are daily maxima of log returns of a given investment asset (e.g., stocks) over a specific period (see Remark 6.20).

Let us consider a block generated by independent identically distributed random variables (*iid*) with distribution function $F(x) = \mathrm{P}(Y \le x)$ that can be denoted for simplicity as $Y_1, \ldots, Y_n$. We denote the corresponding maximum (i.e., maximum random variable) as

$$M_n = \max(Y_1, \ldots, Y_n). \tag{11.28}$$

The distribution function of this maximum obviously fulfills

$$\mathrm{P}(M_n \le x) = (F(x))^n \tag{11.29}$$

(we shall write simply $F^n(x)$).

1. Generalized Extreme Value Distribution

The *generalized extreme value distribution* GEV has for EVT a similar meaning as the normal distribution for CLT (*Central Limit Theorem*). Its distribution function has the form

$$H_{\xi}(x) = \begin{cases} \exp\left(-(1+\xi x)^{-1/\xi}\right), & \xi \neq 0, \\ \exp\left(-\mathrm{e}^{-x}\right), & \xi = 0, \end{cases} \tag{11.30}$$

where ξ is a real parameter (it is so-called shape parameter) and $1 + \xi x > 0$. Generally, one can add a parameter of location μ and a positive parameter of variability σ so that one has three-parametric GEV with distribution function $H_{\xi,\mu,\sigma}(x) = H_{\xi}((x-\mu)/\sigma)$. Here so-called *Fisher–Tippett Theorem* plays the role of Central Limit Theorem: If there exist sequences of real constants $c_n > 0$ and d_n such that

$$\lim_{n\to\infty} \mathrm{P}\left(\frac{M_n - d_n}{c_n} \leq x\right) = \lim_{n\to\infty} F^n(c_n x + d_n) = H(x) \tag{11.31}$$

for a non-degenerated (i.e., non-concentrated in a single point) distribution function H, then this limit function is the distribution function of generalized extreme value distribution (11.30).

Remark 11.7 If (11.31) holds, then one says that F is in so-called *maximum domain of attraction* of H and writes $F \in \mathrm{MDA}(H)$. The parameter ξ of the generalized extreme value distribution $H_{\xi,\mu,\sigma}$ originating as the limit distribution in (11.31) is determined unambiguously, while the parameters of location and variability μ and σ depend on chosen sequences c_n and d_n. Since for a suitable choice of these constants one can achieve that the limit distribution is directly H_{ξ} according to (11.30) (i.e., $H_{\xi,0,1}$ with the zero parameter of location and the unit parameter of variability), one can confine oneself to this one-parametric GEV only.

◇

Example 11.5 Let independent random variables Y_i have the exponential distribution with distribution function

$$F(x) = 1 - \exp(-\lambda \cdot x)\ , \quad x \geq 0 \tag{11.32}$$

(λ is a positive parameter). Then for choice of $c_n = 1/\lambda$ and $d_n = (\ln n)/\lambda$ it holds

$$\mathrm{P}\left(\frac{M_n - d_n}{c_n} \leq x\right) = F^n(c_n x + d_n) = \left(1 - \frac{1}{n}\exp(-x)\right)^n, \quad x \geq \ln n,$$

and hence

$$\lim_{n\to\infty} \mathrm{P}\left(\frac{M_n - d_n}{c_n} \le x\right) = \lim_{n\to\infty} F^n(c_n x + d_n) = \exp\left(-\mathrm{e}^{-x}\right), \quad x \in \mathrm{R},$$

i.e., $F \in \mathrm{MDA}(H_0)$ (the distribution corresponding to the distribution function H_0 is called *Gumbel distribution*).

For instance, if $T_1, T_2, \ldots$ are the periods between mortgage defaults in a bank credit portfolio with the numbers of defaults modeled as Poisson process (see Fig. 2.9), then the random variables $T_1, T_2, \ldots$ are *iid* with exponential distribution (see Sect. 2.4.1). Hence the maximum length among particular defaults has asymptotically Gumbel distribution (this result can be applied in an analogical way as the asymptotic normality following from CLT).

◊

The generalized extreme value distribution H_ξ (see (11.30)) includes three possible cases in dependence on the sign of the parameter ξ:

- For $\xi > 0$: Fréchet distribution.
- For $\xi = 0$: Gumbel distribution.
- For $\xi < 0$: Weibull distribution.

We shall discuss shortly these possibilities (all with zero parameter of location and unit parameter of variability) from the point of view of EVT (see also Fig. 11.9):

Fréchet distribution (see (11.30) for $\xi > 0$ with support $(-1/\xi, \infty)$) has in its maximum domain of attraction the distributions with so-called (right) *heavy tail* fulfilling

$$\mathrm{P}(Y > x) = 1 - F(x) \approx x^{-1/\xi}, \quad x \to \infty \tag{11.33}$$

(a more exact description would necessitate to introduce the concept of so-called *slowly varying function*). Such distributions are denoted as the distributions with *tail index* $1/\xi$ and are frequent in financial applications since they possess not only heavy tails but also infinite higher moments: one can even show that each nonnegative random variable Y with distribution lying in the maximum domain of attraction of Fréchet distribution with tail index $1/\xi$ ($\xi > 0$) has $\mathrm{E}(Y^k) = \infty$ for $k > 1/\xi$. The maximum domain of attraction of Fréchet distribution contains some common distributions, e.g., F-, (generalized) Pareto, t-, and other types of distributions.

Gumbel distribution (see (11.30) for $\xi = 0$ with support $(-\infty, \infty)$) has in its maximum domain of attraction the distributions both with bounded and unbounded support. The heavy tails in the case of unbounded support can decrease quickly (see, e.g., exponential distribution) or slowly (see, e.g., log-normal distribution), so that the maximum domain of attraction of Gumbel distribution can be also useful for financial applications. Each nonnegative random variable Y with distribution lying in the maximum domain of attraction of Gumbel distribution has all moments finite, i.e., $\mathrm{E}(Y^k) < \infty$ for $k > 0$. The maximum domain of attraction of Gumbel distribution contains, e.g., exponential, gamma, chi-, log-normal, normal, and other distributions.

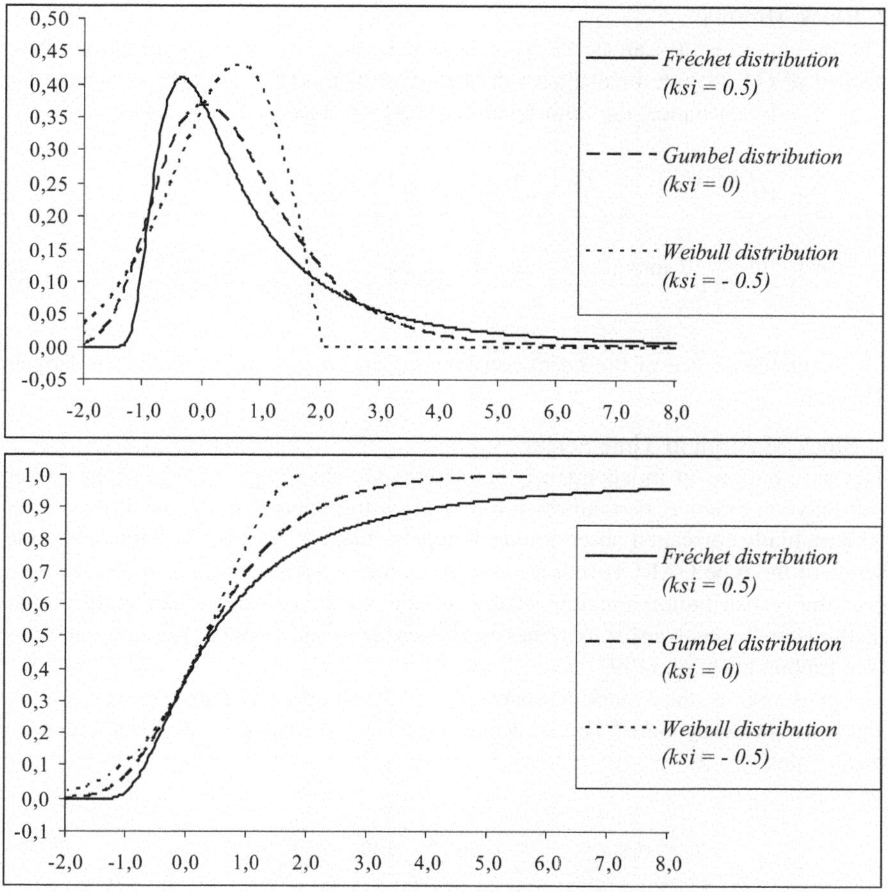

Fig. 11.9 Probability density (upper figure) and distribution function (bottom figure) of Fréchet distribution ($\xi = 0.5$), Gumbel distribution ($\xi = 0$), and Weibull distribution ($\xi = -0.5$) (see (11.30))

Weibull distribution (see (11.30) for $\xi < 0$ with support $(-\infty, -1/\xi)$) contains in its maximum domain of attraction the distributions that are mostly uninteresting in financial applications since their support is bounded from the right-hand side:

$$x_F = \sup\{x : \ F(x) < 1\} < \infty \tag{11.34}$$

(see Fig. 11.9, where the support of Weibull distribution with $\xi = -0.5$ is bounded from the right-hand side by the point $x_F = -1/\xi = 2$). The maximum domain of attraction of Weibull distribution contains, e.g., beta and uniform distribution.

2. Block Minima

The previous results can be easily extended to the case of block minima where instead of (11.28) one investigates the behavior of $\min(Y_1, \ldots, Y_n) = -\max(-Y_1, \ldots, -Y_n)$. For instance, the limit relation (11.31) implies

$$\lim_{n\to\infty} \mathrm{P}\left(\frac{\min(Y_1, \ldots, Y_n) - b_n}{a_n} \le x\right)$$

$$= 1 - \lim_{n\to\infty} \mathrm{P}\left(\frac{\max(-Y_1, \ldots, -Y_n) + b_n}{a_n} \le -x\right) = 1 - H(-x) \qquad (11.35)$$

for a suitable choice of the norm constants a_n and b_n according to Fisher–Tippett Theorem.

3. Block Maxima in Time Series

The assumption on independence of random variables $Y_1, \ldots, Y_n$ can be rather restrictive in practice, particularly if one deals with extremes in financial time series with mutually correlated observations. It appears that for the models of financial time series of the type GARCH (but also for so-called *strictly stationary processes* with probability distribution remaining stable in time; see Sect. 6.1), one can apply similar methods as for analysis of extremes in sequences of independent random variables (see Embrechts et al. (1997)).

Let us assume that a random process $Y_1, \ldots, Y_n$ observed as a time series $y_1, \ldots, y_n$ can be modeled by GARCH(r, s) according to (8.55) written in the form with zero mean value

$$y_t = \sigma_t \varepsilon_t, \qquad \sigma_t^2 = \alpha_0 + \sum_{i=1}^{r} \alpha_i y_{t-i}^2 + \sum_{j=1}^{s} \beta_j \sigma_{t-j}^2, \qquad (11.36)$$

where $\{\varepsilon_t\}$ are *iid* random variables with zero mean value and unit variance, and the parameters of model fulfill

$$\alpha_0 > 0, \quad \alpha_i \ge 0, \quad \beta_j \ge 0, \quad \alpha_1 + \cdots + \alpha_r + \beta_1 + \cdots + \beta_s < 1. \qquad (11.37)$$

Then Fisher–Tippett Theorem (see above) can be reformulated to the form, in which so-called *extreme index* θ ($0 < \theta < 1$); see, e.g., Table 11.9 for the model ARCH (1) from (8.33). Its existence is guaranteed for each process GARCH. Then instead of (11.31) it holds

$$\lim_{n\to\infty} \mathrm{P}\left(\frac{M_n - d_n}{c_n} \le x\right) = \lim_{n\to\infty} F^{n\theta}(c_n x + d_n) = H^{\theta}(x) \qquad (11.38)$$

(the extreme index should not be confused with the tail index $1/\xi$, see above). Therefore, instead of n dependent observations, it is possible to investigate the

Table 11.9 Extreme index θ for selected values of parameter α_1 in the model ARCH(1)

α_1	0.1	0.3	0.5	0.7	0.9
θ	0.999	0.939	0.835	0.721	0.612

maximum of $n\theta$ independent observations with the same distribution function (if the number n of observations is higher): one may imagine that $n\theta$ is the number of mutually independent clusters in the sequence of n dependent observations. From the practical point of view, it means that for higher n one can approximate the distribution of the maximum of process GARCH by

$$P(M_n \le z) \approx H^{\theta}\left(\frac{z - d_n}{c_n}\right). \tag{11.39}$$

The processes with extreme index $\theta = 1$ (e.g., ARMA processes with normally distributed white noise) do not show tendency to cluster high values, and their extremes behave as in the case of independent random variables.

4. Statistical Analysis of Block Maxima

The statistical analysis of block maxima demands a data sample of observed maxima. Therefore, we usually apply the design where data $y_1, y_2, \ldots$ are divided into m blocks of size n and maxima in particular blocks are denoted as $m_{n1}, \ldots, m_{nm}$. These are, e.g., daily maxima of log returns of a stock index during one calendar year. If the data are generated from the same distribution with a known distribution function F and are mutually independent (or possibly of the type GARCH), then according to the theory described above it suffices (for higher n) to approximate the distribution of block maxima by the three-parametric distribution $H_{\xi,\mu,\sigma}(x) = H_{\xi}((x-\mu)/\sigma)$ (see its standardized form H_{ξ} in (11.30)). If $h_{\xi,\mu,\sigma}$ denotes the corresponding probability density, then the unknown parameters ξ, μ, and σ identifying the distribution of block maxima can be estimated using the maximum likelihood method by maximizing over these parameters the log likelihood function of the form

$$l(\xi, \mu, \sigma; m_{n1}, \ldots, m_{nm}) = \sum_{m=1}^{m} \ln h_{\xi,\mu,\sigma}(m_{ni}) =$$
$$= -m \ln \sigma - \left(1 + \frac{1}{\xi}\right) \sum_{i=1}^{m} \ln\left(1 + \xi \frac{m_{ni} - \mu}{\sigma}\right) - \sum_{i=1}^{m} \ln\left(1 + \xi \frac{m_{ni} - \mu}{\sigma}\right)^{-1/\xi} \tag{11.40}$$

under the conditions $\sigma > 0$ and $1 + \xi\,(m_{ni} - \mu)/\sigma > 0$ for all i. These estimates have convenient properties of maximum likelihood estimates even though the range of their feasible values may depend on the observed data (in the case of $\xi > -0.5$).

Remark 11.8 There exists a conflict of interests between the number and the size of blocks. It is convenient from the point of view of estimation if the number of blocks

m (i.e., the number of observations to estimates the parameters) is higher. On the other hand, it means that in the case of fixed total number of observations one must reduce the size of blocks n (or $n\theta$ in the case of dependent observations), which worsens the approximation of maxima distribution by the limit distribution GEV.

◇

In practice, the previous results are often used to find return level and return period (see, e.g., McNeil et al. (2005)):

- *Return level* relates to the problem how to find the size of an extreme event which will occur on average with a frequency given a priori. A more exact formulation is as follows: if H is the distribution function of block maxima in blocks of size n, then for a given k the corresponding return level is defined as

$$r_{n,k} = q_{1-1/k}(H), \tag{11.41}$$

where $q_u = \inf\{x\colon H(x) \geq u\} = H^{-1}(u)$ is the u-quantile of distribution H. The value $r_{n,k}$ may be obviously interpreted as the level, which is exceeded by the block maxima in each k-tuple of blocks of size n on average just once: e.g., $r_{260,10}$ is the level which is exceeded by the yearly maximum of daily log returns of a given stock index on average just once in 10 years (260 is an average number of trading days in one calendar year). After substituting the estimated distribution function H, one obtains the estimated return level in the form

$$\widehat{r}_{n,k} = \widehat{\mu} + \frac{\widehat{\sigma}}{\widehat{\xi}}\left(\left(-\ln\left(1 - 1/k\right)\right)^{-\widehat{\xi}} - 1\right). \tag{11.42}$$

- *Return period* relates to the problem of how to find the average frequency of occurrence of an extreme event over a given level. A more exact formulation is as follows: if H is the distribution function of block maxima in blocks of size n, then for a given u the corresponding return period is defined as

$$k_{n,u} = \frac{1}{1 - H(u)}. \tag{11.43}$$

The value $k_{n,u}$ may be obviously interpreted in such a way that in each $k_{n,u}$-tuple of blocks of size n we can on average expect the occurrence of just one block, in which the level u will be exceeded; e.g., $k_{260,\,u}$ is the number of years, in which the yearly maximum of daily log returns just once exceeds the level u. After substituting the estimated distribution function H, one obtains the estimated return period in the form

$$\widehat{k}_{n,u} = \frac{1}{1 - H_{\widehat{\xi,\mu,\sigma}}(u)}. \tag{11.44}$$

Example 11.6 McNeil et al. (2005) apply the theory of block extremes to the time series of daily drops (in percent) of stock index S&P 500 (this index is used globally as a barometer of stock markets) for the period 1960 to Friday, October 16, 1987, when during one day the given index dropped by 5.25% (as a forerunner of Black Monday, October 19, 1987, with the catastrophic fall of this index by 20.5%). Therefore, one analyzed the yearly and semiannual block maxima of daily drops (recorded in absolute values), i.e., 28 and 56 observed values of block maxima, respectively. First one constructed the maximum likelihood estimates according to (11.40) obtaining

- For yearly block maxima: very unstable estimates $\widehat{\xi} = 0.27$, $\widehat{\mu} = 2.04$ and $\widehat{\sigma} = 0.72$ with high standard deviations 0.21, 0.16, and 0.14 (the limit Fréchet distribution shows a very heavy right tail and infinite fourth moment, since $4 > 1/0.27$).
- For semiannual block maxima: more stable estimates $\widehat{\xi} = 0.36$, $\widehat{\mu} = 1.65$, and $\widehat{\sigma} = 0.54$ with more reasonable standard deviations 0.15, 0.09, and 0.08.

Further one estimated the return level according to (11.42), namely

- Ten-year return level: $\widehat{r}_{260,10} = 4.3\%$ with estimated 95% confidence interval (3.4%; 7.1%).
- Twenty-year return level: $\widehat{r}_{130,20} = 4.5\%$ with estimated 95% confidence interval (3.5%; 7.4%).

The drop by 20.5% during Black Monday, October 19, 1987 (i.e., just on the beginning of future period from the point of view of performed analysis) missed significantly previous confidence intervals both for the 10-year and for 20-year return level.

◊

11.3.2 Threshold Excesses

This approach explores the observations exceeding a given level (or threshold). Its main advantage consists in the fact that it does not "waste" data as the method of block maxima from Sect. 11.3.1 which exploits only maxima of (large) blocks and throws away remaining information. The data which we handle in this method are extreme in the sense that they exceed a given (usually high) level so that they may be denoted as excesses (e.g., in the framework of reinsurance of commercial insurance companies one can confine oneself to such parts of losses that lie in the layer that has origin in a designated level).

1. Generalized Pareto Distribution

The *generalized Pareto distribution* GPD plays for excesses a similar role as the generalized extreme value distribution GEV for block maxima (remind that GDP itself belongs to the domain of attraction of GEV; see Sect. 11.3.1). The distribution function of GDP can be parameterized as

$$G_{\xi,\beta}(x) = \begin{cases} 1 - (1 + \xi x/\beta)^{-1/\xi}, & \xi \neq 0, \\ 1 - \exp\left(-\dfrac{x}{\beta}\right), & \xi = 0, \end{cases} \tag{11.45}$$

where β and ξ are scale and shape real parameters ($\beta > 0$). One has $x \geq 0$ for $\xi \geq 0$, while $0 \leq x \leq -\beta/\xi$ for $\xi < 0$. The mean value of GPD is equal to $\beta/(1 - \xi)$ for $\xi < 1$, and $\mathrm{E}(Y^k) = \infty$ for $k \geq 1/\xi$, so that, e.g., the variance of GDP is infinite for $\xi = 0.5$ as a consequence of heavy tails. Similarly as GEV, the generalized Pareto distribution includes three possible cases in dependence on the sign of the parameter ξ (see Fig. 11.10):

- For $\xi > 0$: Pareto distribution.
- For $\xi = 0$: exponential distribution.
- For $\xi < 0$: Pareto type II distribution (it has the bounded support $(0, -\beta/\xi)$).

2. Distribution of Excesses

Let Y be a random variable with distribution function F. Then the excess $Y - u$ over a given level (or threshold) u has so-called *excess distribution function* $F_u(x)$ of the form

$$F_u(x) = \mathrm{P}(Y - u \leq x | Y > u) = \frac{F(x + u) - F(u)}{1 - F(u)} \tag{11.46}$$

for $0 \leq x \leq x_F - u$, where x_F is the right endpoint of support of F (mostly $x_F = \infty$). Moreover, one defines the *mean excess function* $e(u)$ as

$$e(u) = \mathrm{E}(Y - u | Y > u). \tag{11.47}$$

Example 11.7 Let us assume in addition that Y has the generalized Pareto distribution (11.45). Then according to (11.46) it holds

$$F_u(x) = 1 - (1 + \xi x/(\beta + \xi u))^{-1/\xi} = G_{\xi,\beta+\xi\cdot u}(x), \tag{11.48}$$

where $x \geq 0$ for $\xi \geq 0$, while $0 \leq x \leq -\beta/\xi - u$ for $\xi < 0$. It means that the excess distribution remains of the type GPD. Particularly for the exponential distribution $F(x) = G_{0,\beta}(x)$, it stays $F_u(x) = G_{0,\beta}(x) = F(x)$, which confirms the characteristic "loss of memory" of exponential distribution (i.e., the excess distribution of exponential distribution remains the identical exponential distribution regardless of the size of level u).

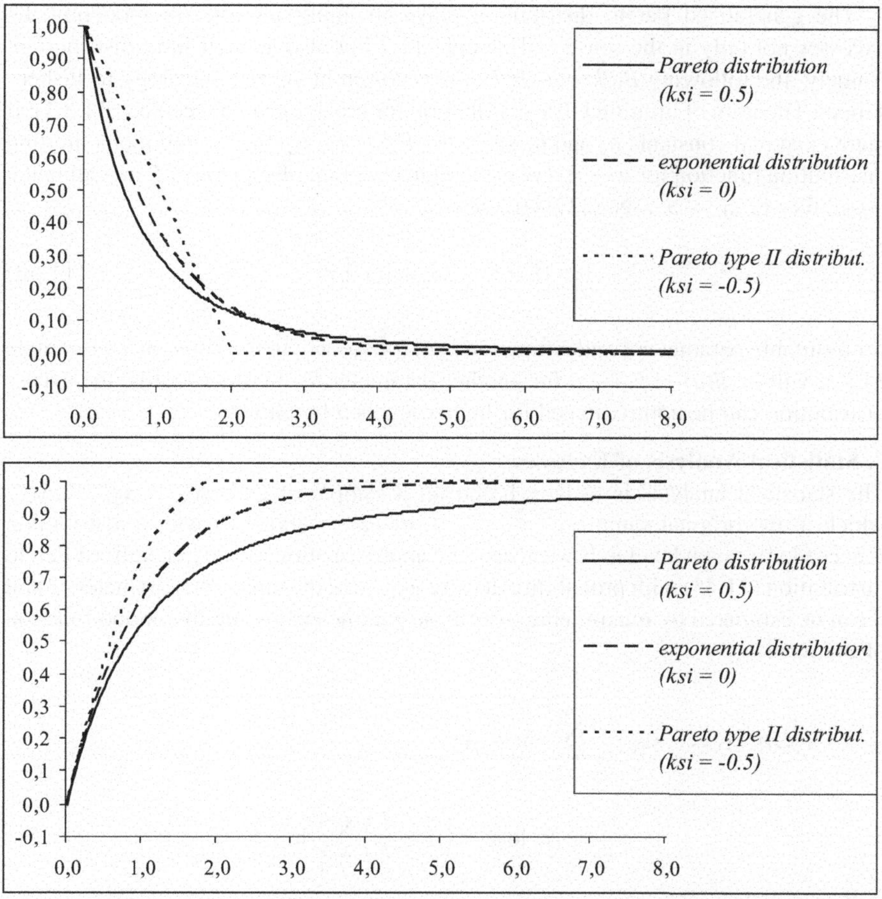

Fig. 11.10 Probability density (upper figure) and distribution function (bottom figure) of Pareto distribution ($\xi = 0.5$), exponential distribution ($\xi = 0$), and Pareto type II distribution ($\xi = -0.5$) (see (11.45) with $\beta = 1$)

Furthermore, the mean excess function of generalized Pareto distribution has the form

$$e(u) = \frac{\beta + \xi u}{1 - \xi}, \tag{11.49}$$

where $u \geq 0$ for $0 \leq \xi < 1$, while $0 \leq u \leq -\beta/\xi$ for $\xi < 0$. Hence the mean excess of GPD is the linear function of u, which is a useful property in some related statistical procedures (e.g., for the identification of distribution GDP).

◇

The generalized Pareto distribution plays an important role for modeling the excesses not only in the sense of Example 11.7 but also as their limit distribution. Namely, the following *Balkema–de Haan Theorem* holds as an analogy to Fisher–Tippett Theorem of limiting GEV distribution for block maxima (see Sect. 11.3.1): if there exist real constants a_u and b_u such that $F_u(a_u x + b_u)$ has a continuous limiting distribution function for $u \to x_F$ (x_F is the right endpoint of support of F including the possibility of $x_F = \infty$; see (11.34)), then

$$\lim_{u \to x_F} \left| F_u(x) - G_{\xi,\beta(u)}(x) \right| = 0 \tag{11.50}$$

for a suitable parameter ξ and a function $\beta(u)$ (obviously in the situation of Example 11.7 it will be $\beta(u) = \beta + \xi \cdot u$ for another parameter β). In other words, the excess distribution can be approximated for higher levels u by GPD.

3. Statistical Analysis of Excesses

The statistical analysis is usually based on a sample of excesses $x_1, x_2, \ldots, x_{N_u}$, which in the original sample $y_1, y_2, \ldots, y_n$ of *iid* observations with a distribution F exceeded a given level u. If we accept the approximation by the generalized Pareto distribution (11.45) with probability density $g_{\xi,\beta}$, then the unknown parameters ξ and β can be estimated by maximizing over these parameters the log likelihood function of the form

$$\begin{aligned} l(\xi, \beta; x_1, \ldots, x_{N_u}) &= \sum_{j=1}^{N_u} \ln g_{\xi,\beta}(x_j) \\ &= -N_u \ln \beta - \left(1 + \frac{1}{\xi}\right) \sum_{j=1}^{N_u} \ln\left(1 + \xi \frac{x_j}{\beta}\right) \end{aligned} \tag{11.51}$$

under conditions $\beta > 0$ and $1 + \xi\, x_j / \beta > 0$ for all j. Similarly as in the case of block maxima, it is again possible to generalize this procedure by means of extreme index to time series of correlated observations (e.g., for the models GARCH; see Embrechts et al. (1997)).

Moreover, the model constructed for a given level u can be transformed to models with excesses over any higher levels $v \geq u$ since one can easily show that it holds

$$F_v(x) = G_{\xi,\beta+\xi\cdot(v-u)}(x) \tag{11.52}$$

and similarly

$$e(v) = \frac{\beta + \xi(v - u)}{1 - \xi} = \frac{\xi v}{1 - \xi} + \frac{\beta - \xi u}{1 - \xi}, \tag{11.53}$$

where $u \leq v < \infty$ for $0 \leq \xi < 1$, while $u \leq v \leq u - \beta/\xi$ for $\xi < 0$. The linearity of the mean excess function (11.53) of argument v (for fixed u) is helpful when looking for such a level u that the excesses over u can be modeled by means of GDP (see Example 11.8).

For loss observations $y_1, y_2, \ldots, y_n$, the following *sample mean excess* (11.54) can be used as a statistical estimate of the mean excess over level u:

$$e_n(u) = \frac{\sum_{i=1}^{n}(y_i - u)I_{[y_i\ >\ u]}}{\sum_{i=1}^{n} I_{[y_i\ >\ u]}}. \tag{11.54}$$

The sample excesses are often used to identify graphically the GDP of excesses in real data. Data $y_1, y_2, \ldots, y_n$ are ordered by their size in ascending order to the form $y_{(1)} \leq y_{(2)} \leq \ldots \leq y_{(n)}$ (so-called ordered statistics) and then plotted in a plane graph as points with coordinates $(y_{(i)}, e_n(y_{(i)}))$ for $i = 2, \ldots, n$, where $e_n(\cdot)$ is the sample mean excess according to (11.54). If these points lie on a line approximately starting with some ordered statistics, then according to (11.53) the approximation of excess distribution by the generalized Pareto distribution is proper starting again with the level corresponding to this ordered statistics. Moreover, in such a case the slope of identified line corresponds to the size of parameter ξ in the given GPD (see (11.53)).

Example 11.8 In the context of the excess modeling, the example of Danish fire insurance data is well known (see, e.g., McNeil et al. (2005)): Table 11.10 and Fig. 11.11 present time series of losses over 1 million DKK (Danish crowns) harmed by fires in Denmark in the period 1980–1990.

Figure 11.12 shows the sample mean excess (11.54) as a function of level u (more exactly, the graph plots points with coordinates $(y_{(i)}, e_n(y_{(i)}))$; see the discussion above. If one ignores the points with high levels u, where the sample estimates of mean excesses (11.54) are unreliable due to small number of data, then in the graph starting approximately with the level of 10 million DKK one can identify an increasing line. Therefore starting with this level it is possible to approximate the excess distribution by the generalized Pareto distribution with a positive parameter ξ, i.e., by the classical ("non-generalized") Pareto distribution. The maximum likelihood estimates of parameters maximizing the log likelihood (11.51) are then $\widehat{\beta} = 7.0$ with standard deviation 1.1 and $\widehat{\xi} = 0.50$ with standard deviation 0.14. The GPD for various levels u can be obtained by means of the simple transformation (11.48).

◊

Table 11.10 Losses over 1 million DKK harmed by fires in Denmark in Example 11.8

Date	Losses over 1 million DKK		Date	Losses over 1 million DKK		Date	Losses over 1 million DKK
		⋮	⋮	⋮	⋮	⋮	⋮
01/03/1980	1.683748	⋮	12/02/1984	1.256545	⋮	11/27/1990	1.134488
01/04/1980	2.093704	⋮	12/03/1984	1.103048	⋮	11/29/1990	3.407591
01/05/1980	1.732581	⋮	12/03/1984	1.204188	⋮	11/29/1990	1.072607
01/07/1980	1.779754	⋮	12/08/1984	1.151832	⋮	11/30/1990	1.167492
01/07/1980	4.612006	⋮	12/08/1984	1.884817	⋮	11/30/1990	1.072607
01/10/1980	8.725274	⋮	12/10/1984	7.539267	⋮	12/04/1990	1.270627
01/10/1980	7.898975	⋮	12/11/1984	1.099476	⋮	12/05/1990	1.472772
01/16/1980	2.208045	⋮	12/12/1984	1.570681	⋮	12/06/1990	1.036304
01/16/1980	1.486091	⋮	12/13/1984	2.670157	⋮	12/07/1990	1.650165
01/19/1980	2.796171	⋮	12/17/1984	1.151832	⋮	12/08/1990	1.678218
01/21/1980	7.320644	⋮	12/19/1984	3.874346	⋮	12/09/1990	2.640264
01/21/1980	3.367496	⋮	12/22/1984	5.026178	⋮	12/09/1990	1.601485
01/24/1980	1.464129	⋮	12/28/1984	1.780105	⋮	12/10/1990	17.739274
01/25/1980	1.722223	⋮	12/29/1984	4.764398	⋮	12/14/1990	4.372937
01/26/1980	11.374817	⋮	12/31/1984	1.151832	⋮	12/15/1990	1.361386
01/26/1980	2.482739	⋮	01/01/1985	1.500000	⋮	12/16/1990	1.183993
01/28/1980	26.214641	⋮	01/03/1985	1.251000	⋮	12/17/1990	2.970297
02/03/1980	2.002430	⋮	01/04/1985	1.030000	⋮	12/19/1990	1.023102
02/05/1980	4.530015	⋮	01/05/1985	1.050000	⋮	12/20/1990	1.130363
02/07/1980	1.841753	⋮	01/05/1985	1.900000	⋮	12/21/1990	3.011551
02/10/1980	3.806735	⋮	01/05/1985	1.100000	⋮	12/21/1990	1.402640
02/13/1980	14.122076	⋮	01/06/1985	1.881750	⋮	12/22/1990	2.322607
02/16/1980	5.424253	⋮	01/07/1985	1.007000	⋮	12/23/1990	1.115512
02/19/1980	11.713031	⋮	01/07/1985	1.630000	⋮	12/23/1990	1.691419
02/20/1980	1.515373	⋮	01/07/1985	1.025000	⋮	12/24/1990	1.237624
02/21/1980	2.538589	⋮	01/08/1985	1.007274	⋮	12/27/1990	1.114686
02/22/1980	2.049780	⋮	01/08/1985	3.500000	⋮	12/30/1990	1.402640
02/23/1980	12.465593	⋮	01/10/1985	2.900000	⋮	12/30/1990	4.867987
02/25/1980	1.735445	⋮	01/11/1985	2.463137	⋮	12/30/1990	1.072607
02/27/1980	1.683748	⋮	01/11/1985	4.625000	⋮	12/31/1990	4.125413
⋮	⋮	⋮	⋮	⋮	⋮		

Source: Copenhagen Reinsurance

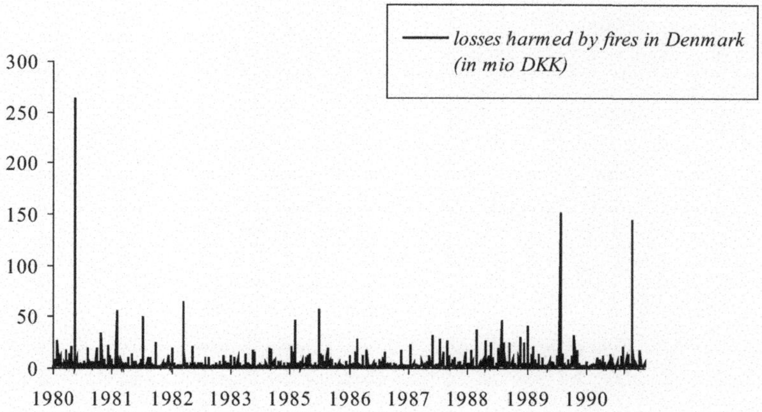

Fig. 11.11 Losses over 1 million DKK harmed by fires in Denmark in Example 11.8. Source: Copenhagen Reinsurance

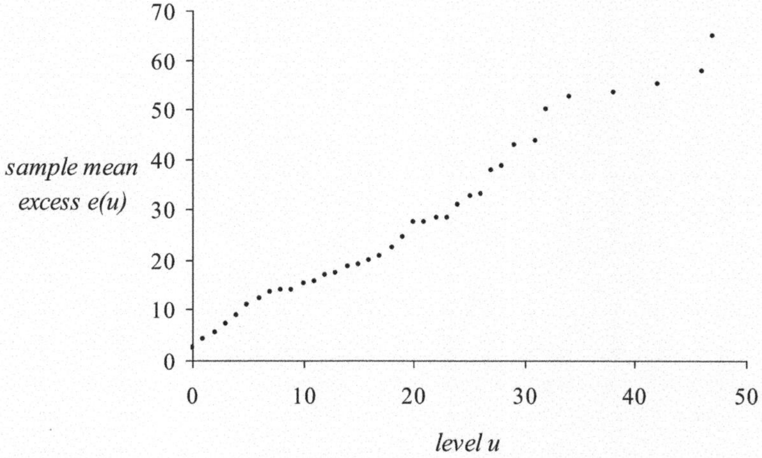

Fig. 11.12 Sample mean excesses as function of level u for losses over 1 million DKK harmed by fires in Denmark in Example 11.8. Source: calculated by EViews

11.4 Exercises

Exercise 11.1 Repeat the calculation of *VaR* from Examples 11.1–11.4 (calculation of *VaR* for daily losses in given investment portfolio in the year 2013), but only for the last month of December 2013.

Part V
Multivariate Time Series

Chapter 12
Methods for Multivariate Time Series

12.1 Generalization of Methods for Univariate Time Series

Most procedures for univariate time series from previous chapters can be generalized for multivariate time series, where instead of scalar values y_t we observe m-variate vector values $\mathbf{y}_t = (y_{1t}, \ldots, y_{mt})'$ in time as realizations of a vector random process (see Sect. 2.1). The transfer from univariate to multivariate dimension mostly means only higher formal and numerical complexity of methods described in previous parts of this text (decomposition methods, methods for linear and nonlinear processes, and the like), which will be demonstrated briefly in this section by means of examples of stationary multivariate time series. Later we shall see that such a parallel description of several scalar processes brings to the analysis further elements that have exclusively the multivariate character (examples are the routine methodology VAR for multivariate time series, the cointegration among particular univariate components, and others).

(*Weak*) *stationarity* of multivariate time series $\{\mathbf{y}_t\}$ means again that the corresponding process is invariant to time shifts of the first and second moments, i.e.,

$$\mathrm{E}(\mathbf{y}_t) = \boldsymbol{\mu} = const, \tag{12.1}$$

$$\mathrm{cov}(\mathbf{y}_s, \mathbf{y}_t) = \mathrm{E}(\mathbf{y}_s - \boldsymbol{\mu})(\mathbf{y}_t - \boldsymbol{\mu})' = \mathrm{cov}\left(\mathbf{y}_{s+h}, \mathbf{y}_{t+h}\right) \quad \text{for arbitrary } h, \tag{12.2}$$

i.e., particularly also

$$\mathrm{var}(\mathbf{y}_t) = \boldsymbol{\Sigma}_{\mathbf{yy}} = const. \tag{12.3}$$

For stationary multivariate time series, one can define analogously as in the univariate case the (matrix) *autocovariance function*

T. Cipra, *Time Series in Economics and Finance*,
https://doi.org/10.1007/978-3-030-46347-2_12

$$\boldsymbol{\Gamma}_k = \text{cov}(\mathbf{y}_t, \mathbf{y}_{t-k}) = \text{E}(\mathbf{y}_t - \boldsymbol{\mu})(\mathbf{y}_{t-k} - \boldsymbol{\mu})', \qquad k = \ldots, -1, 0, 1, \ldots \tag{12.4}$$

and the (matrix) *autocorrelation function*

$$\boldsymbol{\rho}_k = \mathbf{D}^{-1/2}\boldsymbol{\Gamma}_k\mathbf{D}^{-1/2}, \qquad k = \ldots, -1, 0, 1, \ldots, \tag{12.5}$$

where $\mathbf{D} = \mathbf{diag}\{\boldsymbol{\Sigma}_{\mathbf{yy}}\} = \mathbf{diag}\{\text{var}(y_{1t}), \ldots, \text{var}(y_{mt})\}$. The element $\gamma_{ij}(k)$ in position (i, j) of matrix $\boldsymbol{\Gamma}_k = (\gamma_{ij}(k))$ is called the *mutual covariance function* of time series $\{y_{it}\}$ and $\{y_{jt}\}$. Quite analogously, the term *mutual correlation function* $\rho_{ij}(k)$ is used. Apparently it holds

$$\boldsymbol{\Gamma}_k = \boldsymbol{\Gamma}'_{-k}, \qquad \boldsymbol{\rho}_k = \boldsymbol{\rho}'_{-k}, \tag{12.6}$$

i.e., $\gamma_{ij}(k) = \gamma_{ji}(-k)$ and $\rho_{ij}(k) = \rho_{ji}(-k)$. Then the *estimated (matrix) autocovariance function* (estimated by means of $\mathbf{y}_1, \ldots, \mathbf{y}_n$) is simply

$$\mathbf{C}_k = \frac{1}{n}\sum_{t=k+1}^{n}(\mathbf{y}_t - \bar{\mathbf{y}})(\mathbf{y}_{t-k} - \bar{\mathbf{y}})', \qquad k = 0, 1, \ldots, n-1 \tag{12.7}$$

and the *estimated (matrix) autocorrelation function* is

$$\mathbf{R}_k = \widehat{\mathbf{D}}^{-1/2}\mathbf{C}_k\widehat{\mathbf{D}}^{-1/2}, \qquad k = 0, 1, \ldots, n-1, \tag{12.8}$$

where the diagonal matrix $\widehat{\mathbf{D}}$ has the main diagonal identical with the main diagonal of $\mathbf{C}_0$.

Remark 12.1 Let us consider components $\{y_{it}\}$ and $\{y_{jt}\}$ of multivariate time series $\{\mathbf{y}_t\}$. Then the mutual correlation function $\rho_{ij}(k)$ describes (linear) dependence between the series $\{y_{it}\}$ and $\{y_{jt}\}$ in time, e.g.,

- If $\rho_{ij}(k) = \rho_{ji}(k) = 0$ for all $k \geq 0$, then $\{y_{it}\}$ and $\{y_{jt}\}$ are *mutually uncorrelated* (i.e., there exists no stochastic linear dependence between them).
- If $\rho_{ij}(0) = 0$, then $\{y_{it}\}$ and $\{y_{jt}\}$ are *simultaneously uncorrelated* (in the opposite case they are *simultaneously correlated*).
- If $\rho_{ij}(k) = \rho_{ji}(k) = 0$ for all $k > 0$, then $\{y_{it}\}$ and $\{y_{jt}\}$ are *uncoupled.*
- If $\rho_{ij}(k) = 0$ for all $k > 0$, but $\rho_{ji}(l) \neq 0$ for some $l > 0$, then there exists a *unidirectional dependency relationship* of $\{y_{jt}\}$ on $\{y_{it}\}$ (i.e., y_{it} depends on no past value y_{jt}, but y_{jt} depends on some past value y_{it}).
- If $\rho_{ij}(k) \neq 0$ for some $k > 0$ and $\rho_{ji}(l) \neq 0$ for some $l > 0$, then there exists a *feedback* between $\{y_{it}\}$ and $\{y_{jt}\}$.

◇

Example 12.1 Table 12.1 and Fig. 12.1 present the first differences of monthly yields to maturity YTM for 3-month T-bills (so-called *short-term interest rates*

Table 12.1 Monthly data in Example 12.1 (the first differences of monthly yields to maturity for 3-month T-bills and corporate bonds AAA in USA in % p.a.); see also Table 12.16

Month	DTB3	DAAA	Obs	DTB3	DAAA	Obs	DTB3	DAAA
1985M01	−0.40	−0.05	1988M05	0.35	0.23	1991M09	−0.14	−0.14
1985M02	0.46	0.05	1988M06	0.23	−0.04	1991M10	−0.22	−0.06
1985M03	0.35	0.43	1988M07	0.23	0.10	1991M11	−0.43	−0.07
1985M04	−0.57	−0.33	1988M08	0.29	0.15	1991M12	−0.48	−0.17
1985M05	−0.44	−0.51	1988M09	0.21	−0.29	1992M01	−0.28	−0.11
1985M06	−0.55	−0.78	1988M10	0.11	−0.31	1992M02	0.00	0.09
1985M07	0.04	0.03	1988M11	0.34	−0.06	1992M03	0.21	0.06
1985M08	0.13	0.08	1988M12	0.41	0.12	1992M04	−0.24	−0.02
1985M09	−0.10	0.02	1989M01	0.20	0.05	1992M05	−0.15	−0.05
1985M10	0.09	−0.05	1989M02	0.19	0.01	1992M06	0.04	−0.06
1985M11	0.03	−0.47	1989M03	0.35	0.17	1992M07	−0.42	−0.15
1985M12	−0.13	−0.39	1989M04	−0.13	−0.01	1992M08	−0.14	−0.12
1986M01	−0.03	−0.11	1989M05	−0.30	−0.22	1992M09	−0.17	−0.03
1986M02	−0.01	−0.38	1989M06	−0.18	−0.47	1992M10	−0.13	0.07
1986M03	−0.44	−0.67	1989M07	−0.30	−0.17	1992M11	0.30	0.11
1986M04	−0.53	−0.21	1989M08	−0.01	0.03	1992M12	0.11	−0.12
1986M05	0.06	0.30	1989M09	−0.19	0.05	1993M01	−0.19	−0.07
1986M06	0.09	0.04	1989M10	−0.13	−0.09	1993M02	−0.11	−0.20
1986M07	−0.37	−0.25	1989M11	0.08	−0.03	1993M03	0.02	−0.13
1986M08	−0.27	−0.16	1989M12	−0.03	−0.03	1993M04	−0.08	−0.12
1986M09	−0.38	0.17	1990M01	0.00	0.13	1993M05	0.07	−0.03
1986M10	−0.01	−0.03	1990M02	0.12	0.23	1993M06	0.14	−0.10
1986M11	0.17	−0.18	1990M03	0.11	0.15	1993M07	−0.05	−0.16
1986M12	0.14	−0.19	1990M04	−0.09	0.09	1993M08	0.00	−0.32
1987M01	−0.04	−0.13	1990M05	0.00	0.01	1993M09	−0.09	−0.19
1987M02	0.14	0.02	1990M06	−0.04	−0.21	1993M10	0.08	0.01
1987M03	−0.03	−0.02	1990M07	−0.08	−0.02	1993M11	0.08	0.26
1987M04	0.20	0.49	1990M08	−0.22	0.17	1993M12	−0.04	0.00
1987M05	−0.01	0.48	1990M09	−0.06	0.15	1994M01	−0.06	−0.01
1987M06	−0.06	−0.01	1990M10	−0.19	−0.03	1994M02	0.19	0.16
1987M07	0.09	0.10	1990M11	−0.12	−0.23	1994M03	0.31	0.40
1987M08	0.22	0.25	1990M12	−0.26	−0.25	1994M04	0.22	0.40
1987M09	0.32	0.51	1991M01	−0.51	−0.01	1994M05	0.45	0.11
1987M10	0.08	0.34	1991M02	−0.35	−0.21	1994M06	−0.01	−0.02
1987M11	−0.59	−0.51	1991M03	−0.04	0.10	1994M07	0.21	0.14
1987M12	−0.01	0.10	1991M04	−0.24	−0.07	1994M08	0.11	−0.04
1988M01	0.10	−0.23	1991M05	−0.16	0.00	1994M09	0.14	0.27
1988M02	−0.21	−0.48	1991M06	0.09	0.15	1994M10	0.32	0.23
1988M03	0.00	−0.01	1991M07	−0.02	−0.01	1994M11	0.29	0.11
1988M04	0.23	0.28	1991M08	−0.19	−0.25	1994M12	0.39	−0.22

Source: calculated by EViews
https://fred.stlouisfed.org/graph/?id=TB3MA, https://fred.stlouisfed.org/graph/?id=AAA

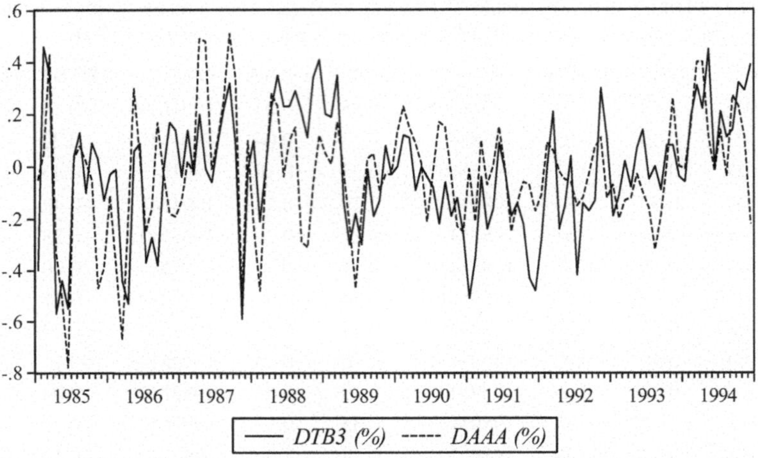

Fig. 12.1 Monthly data in Example 12.1 (the first differences of monthly yields to maturity for 3-month T-bills and corporate bonds AAA in the USA in % p.a.)

denoted as *DTB*3 in % p.a.) and for corporate bonds of the highest rating AAA by S&P (denoted as *DAAA* in % p.a.) during 10-year period 1985–1994 in the USA. Graphs of both time series of lengths 120 in Fig. 12.1 can be regarded as stationary (it is just the reason why the first differences are analyzed; see non-differenced time series in Table 12.16 and Fig. 12.10).

Evidently, there is a relatively strong positive correlation between these time series, which is confirmed by the estimated correlation coefficient of size 0.563 and scatterplot in Fig. 12.2. Due to the estimated (matrix) autocorrelation function in Table 12.2 and the partial correlograms (not shown here) one could identify for individual time series *DTB*3 and *DAAA* models AR(1) (or AR(3)) and AR(2),

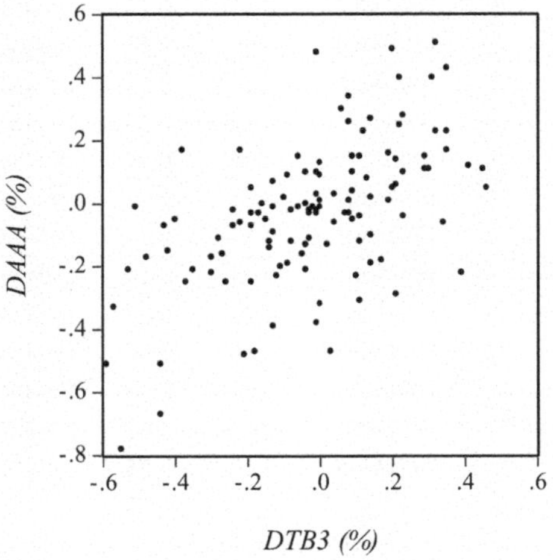

Fig. 12.2 Scatterplot for data from Example 12.1

Table 12.2 Estimated (matrix) autocorrelation function for time series *DTB3* and *DAAA* from Example 12.1

k	$\rho_{11}(k)$	$\rho_{22}(k)$	$\rho_{12}(k)$	$\rho_{12}(-k)$
0	1.000	1.000	0.563	0.563
1	0.443	0.409	0.301	0.234
2	0.151	−0.033	0.007	0.001
3	0.214	−0.058	0.020	−0.025
4	0.150	0.085	0.096	−0.032
5	0.181	0.071	0.047	0.036
6	0.129	−0.018	−0.046	0.034
7	0.096	0.080	0.011	0.078
8	0.148	0.122	0.063	0.072
9	0.140	−0.022	0.051	−0.021
10	0.100	0.024	0.106	0.077

respectively. The relationship between these time series has again the form of feedback (in both directions till the lag one; see Remark 12.1).

◇

Example 12.2 Table 12.3 and Fig. 12.3 present the first differences of logarithms of annual gross domestic products (i.e., log returns; see (8.1)) during the period 1951–1992 in seven countries (France, Germany, Italy, the UK, Japan, the USA, and Canada) denoted as *RGDP_FRA*, *RGDP_GER*, *RGDP_ITA*, *RGDP_UK*, *RGDP_JAP*, *RGDP_US*, *RGDP_CAN*, respectively. The graphs of these seven time series of lengths 42 in Fig. 12.3 can be again regarded as stationary.

There are again strong positive correlations among these time series, which is confirmed by the estimated correlation matrix in Table 12.4 and scatterplots in Fig. 12.4. Due to the estimated correlograms and partial correlograms (not shown here), one could identify for each individual time series models AR(1) (or white noise). There exist unidirectional dependency relationships from Remark 12.1 between some pairs of these time series, e.g., between France and Germany (see Table 12.5).

◇

Besides the mutual correlation function $\rho_{ij}(k)$, one applies also *partial mutual correlation function* denoted as $\rho_{ij}(k,k)$ and defined as the partial correlation coefficient between y_{it} and $y_{j,t-k}$ under fixed values $\mathbf{y}_{t-k+1}, \ldots, \mathbf{y}_{t-1}$. Its estimate $r_{ij}(k,k)$ can be obtained as the estimated parameter $\left(\widehat{\boldsymbol{\Phi}}_{kk}\right)_{ij}$ in the model

$$\mathbf{y}_t = \boldsymbol{\Phi}_{11}\mathbf{y}_{t-1} + \cdots + \boldsymbol{\Phi}_{kk}\mathbf{y}_{t-k} + \boldsymbol{\varepsilon}_t, \tag{12.9}$$

where the *multivariate white noise* $\{\boldsymbol{\varepsilon}_t\}$ is quite analogical to the univariate white noise, i.e., particular components of vectors $\boldsymbol{\varepsilon}_t$ have zero means and are mutually uncorrelated in different times, but are simultaneously correlated with a constant positive definite variance matrix $\boldsymbol{\Sigma}$

Table 12.3 Annual data in Example 12.2 (log returns of annual gross domestic products for France, Germany, Italy, the UK, Japan, the USA, and Canada)

Year	RGDP_FRA	RGDP_GER	RGDP_ITA	RGDP_UK	RGDP_JPN	RGDP_US	RGDP_CAN
1951	0.041	0.073	0.069	0.030	0.112	0.051	0.028
1952	0.041	0.089	0.035	0.006	0.080	0.009	0.056
1953	0.019	0.064	0.068	0.048	0.053	0.025	0.013
1954	0.026	0.068	0.033	0.031	0.046	−0.029	−0.051
1955	0.041	0.115	0.062	0.031	0.067	0.061	0.062
1956	0.066	0.048	0.034	0.013	0.065	0.000	0.062
1957	0.037	0.040	0.037	0.013	0.062	−0.003	−0.019
1958	0.014	0.031	0.039	0.003	0.045	−0.026	0.021
1959	0.027	0.066	0.055	0.035	0.080	0.045	0.017
1960	0.059	0.071	0.065	0.036	0.117	0.006	−0.007
1961	0.044	0.041	0.076	0.023	0.123	0.007	0.001
1962	0.048	0.034	0.058	0.003	0.060	0.041	0.046
1963	0.040	0.014	0.056	0.032	0.089	0.026	0.033
1964	0.062	0.059	0.015	0.047	0.113	0.038	0.046
1965	0.036	0.045	0.019	0.015	0.039	0.051	0.052
1966	0.045	0.012	0.053	0.016	0.093	0.045	0.049
1967	0.040	−0.010	0.067	0.025	0.100	0.014	0.014
1968	0.038	0.059	0.055	0.031	0.117	0.033	0.034
1969	0.067	0.070	0.060	0.007	0.095	0.020	0.038
1970	0.048	0.056	0.056	0.028	0.093	−0.000	0.016
1971	0.038	0.022	0.005	0.015	0.036	0.019	0.042
1972	0.035	0.034	0.018	0.038	0.069	0.037	0.050
1973	0.052	0.039	0.054	0.053	0.060	0.046	0.074
1974	−0.007	−0.022	0.030	−0.044	−0.047	−0.032	0.046
1975	−0.018	−0.008	−0.041	0.001	0.003	−0.023	−0.001
1976	0.041	0.056	0.059	0.022	0.033	0.042	0.054

1977	0.023	0.030	0.025	0.026	0.037	0.035	0.011
1978	0.031	0.038	0.031	0.045	0.049	0.038	0.025
1979	0.027	0.037	0.053	0.033	0.026	0.010	0.048
1980	−0.007	−0.009	0.042	−0.019	0.006	−0.030	0.015
1981	−0.009	−0.021	−0.009	−0.013	0.031	0.014	0.032
1982	0.019	−0.010	0.004	0.020	0.020	−0.033	−0.063
1983	−0.000	0.027	0.009	0.039	0.020	0.029	0.028
1984	0.007	0.025	0.025	0.023	0.039	0.060	0.056
1985	0.018	0.019	0.024	0.033	0.046	0.021	0.038
1986	0.042	0.050	0.046	0.034	0.046	0.016	0.022
1987	0.020	0.023	0.032	0.047	0.036	0.017	0.043
1988	0.040	0.034	0.039	0.057	0.058	0.031	0.045
1989	0.026	0.024	0.025	0.024	0.038	0.021	0.021
1990	0.020	0.039	0.025	0.000	0.037	−0.003	−0.026
1991	0.000	0.028	0.022	−0.027	0.044	−0.020	−0.051
1992	0.007	0.006	0.004	−0.004	0.015	0.020	−0.002

Source: calculated by EViews

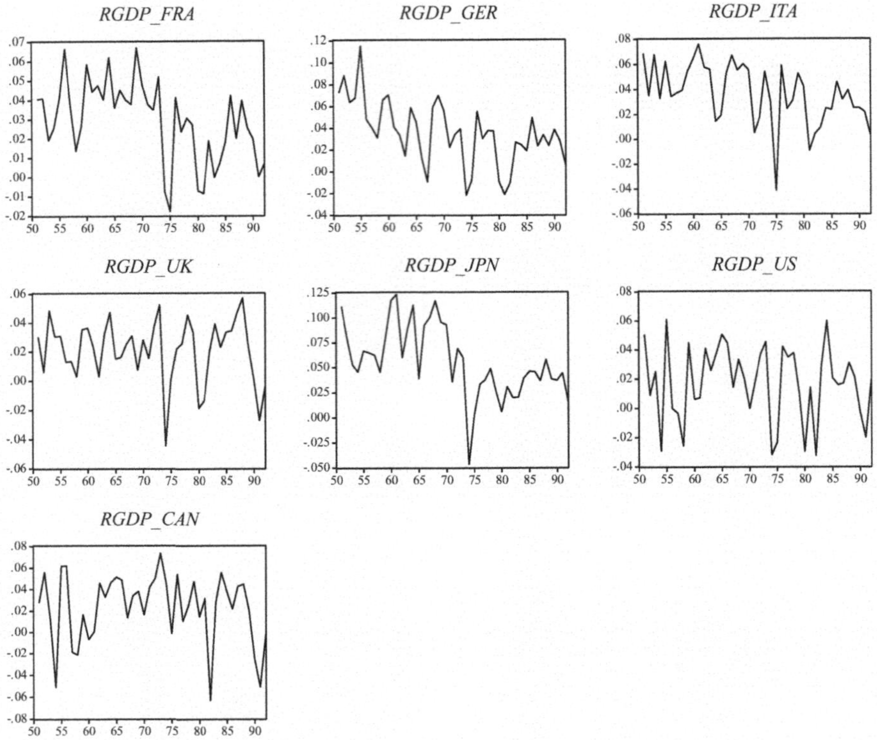

Fig. 12.3 Annual data in Example 12.2 (log returns of annual gross domestic products for France, Germany, Italy, the UK, Japan, the USA, and Canada). Source: OECD (https://data.oecd.org/gdp/gross-domestic-product-gdp.htm)

$$\mathrm{E}(\boldsymbol{\varepsilon}_t) = \mathbf{0}, \qquad \mathrm{E}\left(\boldsymbol{\varepsilon}_s\boldsymbol{\varepsilon}_t'\right) = \delta_{st}\boldsymbol{\Sigma}. \tag{12.10}$$

The model (12.9) is an example of so-called *multivariate linear process* that is analogical to the univariate case (6.17). Its general form is

$$\mathbf{y}_t = \boldsymbol{\varepsilon}_t + \boldsymbol{\Psi}_1\boldsymbol{\varepsilon}_{t-1} + \boldsymbol{\Psi}_2\boldsymbol{\varepsilon}_{t-2} + \cdots = \left(1 + \boldsymbol{\Psi}_1 B + \boldsymbol{\Psi}_2 B^2 + \cdots\right)\boldsymbol{\varepsilon}_t = \boldsymbol{\Psi}(B)\boldsymbol{\varepsilon}_t \tag{12.11}$$

under certain conditions for matrix parameters $\boldsymbol{\Psi}_i$ to achieve stationarity and invertibility. In particular, similarly as in the univariate case one defines *vector mixed process* VARMA(p, q)

$$\mathbf{y}_t = \boldsymbol{\Phi}_1\mathbf{y}_{t-1} + \cdots + \boldsymbol{\Phi}_p\mathbf{y}_{t-p} + \boldsymbol{\varepsilon}_t + \boldsymbol{\Theta}_1\boldsymbol{\varepsilon}_{t-1} + \cdots + \boldsymbol{\Theta}_q\boldsymbol{\varepsilon}_{t-q}, \text{ i.e., } \boldsymbol{\Phi}(B)\mathbf{y}_t = \boldsymbol{\Theta}(B)\boldsymbol{\varepsilon}_t. \tag{12.12}$$

Table 12.4 Estimated correlation matrix for seven time series from Example 12.2

	RGDP_FRA	RGDP_GER	RGDP_ITA	RGDP_UK	RGDP_JPN	RGDP_US	RGDP_CAN
RGDP_FRA	1.000	0.610	0.591	0.489	0.748	0.409	0.345
RGDP_GER	0.610	1.000	0.510	0.445	0.553	0.400	0.177
RGDP_ITA	0.591	0.510	1.000	0.303	0.591	0.284	0.189
RGDP_UK	0.489	0.445	0.303	1.000	0.468	0.543	0.250
RGDP_JPN	0.748	0.553	0.591	0.468	1.000	0.388	0.104
RGDP_US	0.409	0.400	0.284	0.543	0.388	1.000	0.667
RGDP_CAN	0.345	0.177	0.189	0.250	0.104	0.667	1.000

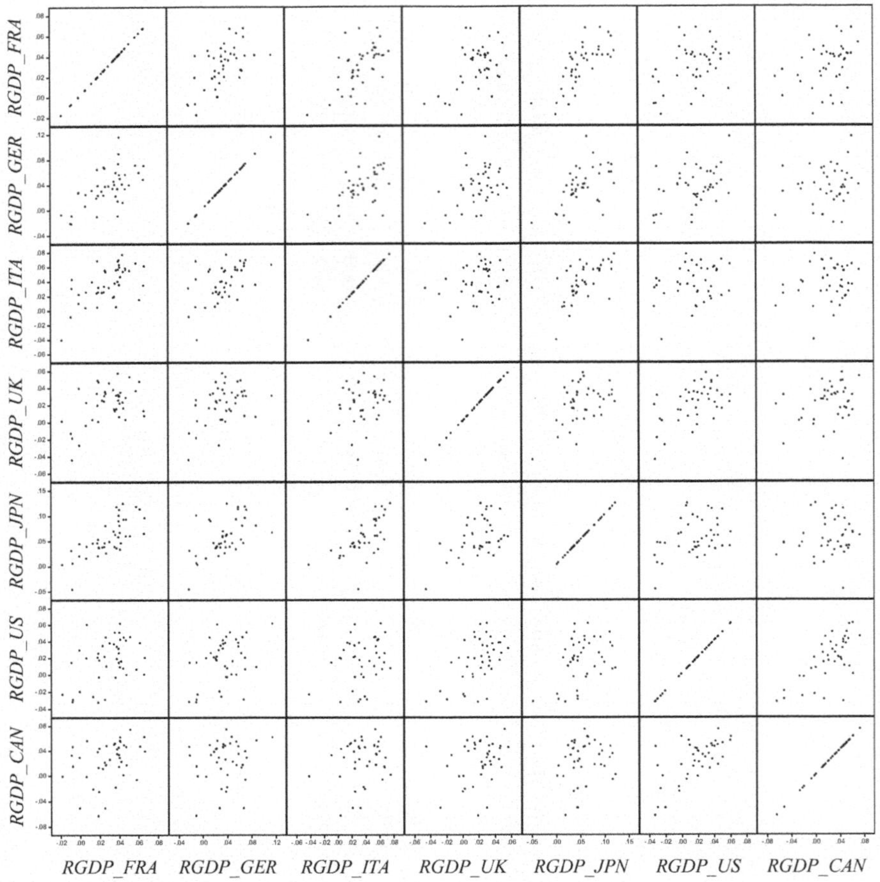

Fig. 12.4 Scatterplots for data from Example 12.2

Table 12.5 Estimated mutual correlation function $\rho_{12}(\cdot)$ (France versus Germany) and $\rho_{15}(\cdot)$ (France versus Japan) from Example 12.2

k	$\rho_{12}(k)$	$\rho_{12}(-k)$	$\rho_{15}(k)$	$\rho_{15}(-k)$
0	0.6097	0.6097	0.7475	0.7475
1	0.4218	0.1649	0.6257	0.3150
2	0.1619	−0.0103	0.2335	0.2033
3	0.1829	0.0855	0.3920	0.2320
4	0.1749	0.1371	0.3416	0.2791
5	0.1879	0.0208	0.3356	0.1555
6	0.0713	0.0252	0.2571	0.1132
7	0.2083	0.0617	0.1292	0.0691
8	0.3618	0.0936	0.1251	0.1216
9	0.3873	0.0249	0.3186	−0.0157
10	0.1296	−0.1749	0.0846	−0.1678

In Sect. 12.2, we will present in more detail a special case of VARMA, namely the *vector autoregressive process* VAR(p)

$$\mathbf{y}_t = \boldsymbol{\Phi}_1 \mathbf{y}_{t-1} + \cdots + \boldsymbol{\Phi}_p \mathbf{y}_{t-p} + \varepsilon_t, \text{ i.e., } \boldsymbol{\Phi}(B)\mathbf{y}_t = \boldsymbol{\varepsilon}_t, \tag{12.13}$$

since nowadays this model is broadly applied just for dynamic economic data.

12.2 Vector Autoregression VAR

The *vector autoregression* (12.13) (see, e.g., Lütkepohl (2005)) is a natural extension of the univariate autoregressive process. In econometrics, it represents a useful instrument in the context of *simultaneous equation models* SEM (see, e.g., Greene (2012) or Heij et al. (2004)).

The VAR have several pros and cons in the framework of practical analysis of economic and financial time series:

+ It is not necessary to distinguish between *exogenous variables* (they originate outside the model) and *endogenous variables* (they originate as outputs of the given model).
+ Models VAR has a richer structure than univariate processes AR since each variable can depend on further variables (and not only on its lagged values with added white noise).
+ The classical OLS estimate has usually acceptable properties in VAR models.
+ Empirical experiences show that predictions by means of VAR are sufficient for routine situations in practice.
– The application of VAR is sometimes "too technical" without deeper arguments justifying the given model (in practice, this approach is popular in the context of *data mining*).
– The number of parameters which must be estimated can be large (particularly for higher dimensions m and orders p of VAR). Moreover, one must solve the problem of an adequate choice of p in practice).
– One must stationarize the modeled data before the VAR is constructed. However, the necessary adjustments and transformations to achieve stationarity (mainly differencing) may imply a substantial loss of information contained originally in the data.

At first let us consider the following model VAR(1) (the description is simpler than for the general VAR(p) and the results derived for VAR(1) can be extended easily to the general order; see Remark 12.3)

$$\mathbf{y}_t = \boldsymbol{\varphi}_0 + \boldsymbol{\Phi}\mathbf{y}_{t-1} + \boldsymbol{\varepsilon}_t, \tag{12.14}$$

where $\boldsymbol{\varepsilon}_t$ is m-variate white noise (see (12.10)). In comparison with (12.13), the relation (12.14) contains in addition an m-variate intercept $\boldsymbol{\varphi}_0$. For example, if $m = 2$, then VAR(1) is formed by two equations which can be written explicitly as

$$\begin{aligned} y_{1t} &= \varphi_{10} + \varphi_{11} y_{1,t-1} + \varphi_{12} y_{2,t-1} + \varepsilon_{1t}, \\ y_{2t} &= \varphi_{20} + \varphi_{21} y_{1,t-1} + \varphi_{22} y_{2,t-1} + \varepsilon_{2t}. \end{aligned} \tag{12.15}$$

Remark 12.2 The explicit form (12.15) demonstrates how the model parameters influence relations between series $\{y_{1t}\}$ and $\{y_{2t}\}$ in time (see Remark 12.1). If in addition the covariance matrix $\boldsymbol{\Sigma}$ of white noise $\{\boldsymbol{\varepsilon}_t\}$ is diagonal (i.e., its components are mutually uncorrelated), then it holds:

- If $\varphi_{12} = \varphi_{21} = 0$, then $\{y_{1t}\}$ and $\{y_{2t}\}$ are uncoupled.
- If $\varphi_{12} = 0$ and $\varphi_{21} \neq 0$, then there exists a unidirectional dependency relationship of $\{y_{2t}\}$ on $\{y_{1t}\}$.
- If $\varphi_{12} \neq 0$ and $\varphi_{21} \neq 0$, then there exists a feedback between $\{y_{1t}\}$ and $\{y_{2t}\}$.

◇

The formula (12.14) is called *reduced form* of the model VAR. If we consider the ith equation in (12.14) (or more generally in (12.13)), then only the variable y_i is present in the current form (i.e., without lag). Moreover, in the reduced form, the simultaneous correlation between $\{y_{it}\}$ and $\{y_{jt}\}$ is represented only by means of the element σ_{ij} of the covariance matrix of white noise $\boldsymbol{\Sigma}$. However, sometimes one needs to express the simultaneous relation between $\{y_{it}\}$ and $\{y_{jt}\}$ more explicitly. In such a case, one can use so-called *structural form* of the model VAR, namely by means of *Cholesky decomposition* from the matrix theory: as the matrix $\boldsymbol{\Sigma}$ is positive definite, then there is a lower triangular matrix $\mathbf{L}$ with units on the main diagonal and a diagonal matrix $\mathbf{D}$ such that

$$\boldsymbol{\Sigma} = \mathbf{LDL}', \quad \text{i.e.} \quad \mathbf{L}^{-1}\boldsymbol{\Sigma}(\mathbf{L}')^{-1} = \mathbf{D}. \tag{12.16}$$

The original model (12.14) is then transferred to the structural form

$$\mathbf{L}^{-1}\mathbf{y}_t = \boldsymbol{\varphi}_0^* + \boldsymbol{\Phi}^*\mathbf{y}_{t-1} + \mathbf{u}_t, \tag{12.17}$$

where

$$\boldsymbol{\varphi}_0^* = \mathbf{L}^{-1}\boldsymbol{\varphi}_0, \; \boldsymbol{\Phi}^* = \mathbf{L}^{-1}\boldsymbol{\Phi}, \; \mathbf{u}_t = \mathbf{L}^{-1}\boldsymbol{\varepsilon}_t, \; \mathrm{E}(\mathbf{u}_t) = \mathbf{0}, \; \mathrm{var}(\mathbf{u}_t) = \mathbf{L}^{-1}\boldsymbol{\Sigma}(\mathbf{L}')^{-1} = \mathbf{D}. \tag{12.18}$$

Particularly, $\{\mathbf{u}_t\}$ is an m-variate white noise with diagonal covariance matrix (i.e., components of $\{\mathbf{u}_t\}$ are simultaneously uncorrelated). If we denote the last row of the inverted matrix $\mathbf{L}^{-1}$ as $(\lambda_{m1}, \ldots, \lambda_{m,m\text{-}1}, 1)$, then the mth equation in (12.17) is

$$y_{mt} + \sum_{i=1}^{m-1} \lambda_{mi} y_{it} = \varphi_{m0}^{*} + \sum_{i=1}^{m} \varphi_{mi}^{*} y_{i,\ t-1} + u_{mt}. \qquad (12.19)$$

This structural form presents explicitly the simultaneous (i.e., at time t) linear dependence of y_{mt} on y_{it} for $i = 1, \ldots, m - 1$ (since u_{mt} is uncorrelated with y_{it} which follows from the facts that u_{mt} is uncorrelated with u_{it} and $\mathbf{L}^{-1}$ in (12.17) is a lower triangular matrix with units on the main diagonal similarly as $\mathbf{L}$). As the components of vector $\mathbf{y}_t$ can be rearranged in an arbitrary way, one obtains the same conclusion as for y_{mt} also for other components y_{jt} ($j = 1, \ldots, m - 1$) of vector $\mathbf{y}_t$.

In practice, one prefers the reduced form of the model VAR since

- The estimation of the reduced form is relatively easy.
- The predictions in the structural form are complicated due to the links of predicted variable with the simultaneous values of further variables (see above).

The conditions of (weak) stationarity (see Sect. 12.1) and the first and second moments of VAR(1) can be found analogically as for the scalar (univariate) autoregressive process (see Sect. 6.2.3):

A sufficient condition of stationarity of the model VAR(1) written in the form (12.14) (this condition also allows to express VAR in the form (12.22)) usually demands that all m eigenvalues of the matrix $\mathbf{\Phi}$ lie inside the unit circle in complex plane (i.e., their absolute values are lower than one). The eigenvalues of matrix $\mathbf{\Phi}$ are the roots of polynomial equation $\det(\lambda\,\mathbf{I} - \mathbf{\Phi}) = 0$ (or equivalently the inverted roots of polynomial equation $\det(\mathbf{I} - \mathbf{\Phi}\cdot z) = 0$). Therefore, the condition of stationarity can be formulated also in such a way that all m roots of autoregressive (matrix) polynomial $\mathbf{\Phi}(z) = \mathbf{I} - \mathbf{\Phi}\cdot z$ lie outside the unit circle in complex plane (or equivalently all m inverted roots of this polynomial lie inside the unit circle in complex plane).

Under the condition of stationarity (see above), the matrix $\mathbf{I} - \mathbf{\Phi}$ is regular so that (12.14) can be rewritten in the form

$$\mathbf{y}_t - \boldsymbol{\mu} = \mathbf{\Phi}(\mathbf{y}_{t-1} - \boldsymbol{\mu}) + \boldsymbol{\varepsilon}_t, \qquad (12.20)$$

where

$$\boldsymbol{\mu} = (\mathbf{I} - \mathbf{\Phi})^{-1} \boldsymbol{\varphi}_0 = (\mathbf{\Phi}(1))^{-1} \boldsymbol{\varphi}_0 \qquad (12.21)$$

is the mean vector of the stationary process $\{\mathbf{y}_t\}$ (obviously, $\mathbf{\Phi}(1) = \mathbf{I} - \mathbf{\Phi}$). Moreover, this process can be then also written in the form of m-variate linear process

$$\mathbf{y}_t = \boldsymbol{\mu} + \boldsymbol{\varepsilon}_t + \mathbf{\Phi}\ \boldsymbol{\varepsilon}_{t-1} + \mathbf{\Phi}^2 \boldsymbol{\varepsilon}_{t-2} + \mathbf{\Phi}^3 \boldsymbol{\varepsilon}_{t-3} + \ldots \qquad (12.22)$$

(compare with (6.17)) so that its covariance matrix is

$$\mathbf{\Gamma}_0 = \text{var}(\mathbf{y}_t) = \mathbf{\Sigma} + \mathbf{\Phi}\mathbf{\Sigma}\mathbf{\Phi}' + \mathbf{\Phi}^2\mathbf{\Sigma}(\mathbf{\Phi}^2)' + \ldots \tag{12.23}$$

and its autocovariance function (12.4) fulfills

$$\mathbf{\Gamma}_k = \mathbf{\Phi}^k\mathbf{\Gamma}_0. \tag{12.24}$$

Remark 12.3 All previous formulas can be extended to the model VAR(p)

$$\mathbf{y}_t = \boldsymbol{\varphi}_0 + \mathbf{\Phi}_1\mathbf{y}_{t-1} + \cdots + \mathbf{\Phi}_p\mathbf{y}_{t-p} + \boldsymbol{\varepsilon}_t. \tag{12.25}$$

The sufficient condition of stationarity and linearity of this process is that

$$\text{all roots of } \ \mathbf{\Phi}(z) = \mathbf{I} - \mathbf{\Phi}_1 z - \cdots - \mathbf{\Phi}_p z^p \ \text{ lie outside the unit circle} \tag{12.26}$$

(in complex plane). Then one can rewrite (12.25) in the form

$$\mathbf{y}_t - \boldsymbol{\mu} = \mathbf{\Phi}_1(\mathbf{y}_{t-1} - \boldsymbol{\mu}) + \cdots + \mathbf{\Phi}_p(\mathbf{y}_{t-p} - \boldsymbol{\mu}) + \boldsymbol{\varepsilon}_t, \tag{12.27}$$

where

$$\boldsymbol{\mu} = (\mathbf{I} - \mathbf{\Phi}_1 - \cdots - \mathbf{\Phi}_p)^{-1}\boldsymbol{\varphi}_0 = (\mathbf{\Phi}(1))^{-1}\boldsymbol{\varphi}_0 \tag{12.28}$$

is the mean vector of the stationary process $\{\mathbf{y}_t\}$. Its autocovariance function (12.4) fulfills the multivariate version of system of Yule–Walker equations (6.35)

$$\mathbf{\Gamma}_k = \mathbf{\Phi}_1\mathbf{\Gamma}_{k-1} + \cdots + \mathbf{\Phi}_p\mathbf{\Gamma}_{k-p} \quad \text{for } k > 0 \tag{12.29}$$

and

$$\mathbf{\Gamma}_0 = \mathbf{\Phi}_1\mathbf{\Gamma}_1 + \cdots + \mathbf{\Phi}_p\mathbf{\Gamma}_p + \mathbf{\Sigma}. \tag{12.30}$$

◇

The construction of VAR(p) based on observations $\mathbf{y}_1, \ldots, \mathbf{y}_n$ is entirely analogical to the univariate AR(p):

1. Identification of VAR Order

The order p could be identified by generalizing partial correlograms for multivariate case (see Sect. 6.3.1.1), but such an approach is rather elaborate. Therefore, in practice one makes use of identification procedures based either on statistical tests or on information criteria.

The *likelihood ratio test* (*LR test*) modified in a sequential way is used typically for the VAR order determination (see Lütkepohl (2005) or Tables 12.7 and 12.11 by EViews). This test makes use of critical regions with confidence level α of the form

Table 12.6 Wald test for the identification of model VAR in Example 12.3 (*DTB3*$_t$ *and DAAA*$_t$) calculated by means of EViews

VAR lag exclusion Wald tests			
Included observations: 117			
Chi-squared test statistics for lag exclusion			
Numbers in [ ] are *p*-values			
	DTB3	DAAA	Joint
Lag 1	31.18664	31.30155	54.13323
	[1.69e-07]	[1.60e-07]	[4.94e-11]
Lag 2	4.323035	7.686472	8.446665
	[0.115150]	[0.021424]	[0.076520]
Lag 3	6.233328	0.484196	2.614466
	[0.044305]	[0.784979]	[0.624263]
df	2	2	4

$$LR = n\left(\ln|\widehat{\boldsymbol{\Sigma}}_R| - \ln|\widehat{\boldsymbol{\Sigma}}_U|\right) > \chi^2_{1-\alpha}(qm^2). \tag{12.31}$$

In more detail, one tests the null hypothesis that the last q lags of an original VAR model with a high number of lags (regarded as an upper bound for the VAR order) have zero parameters (i.e., zero matrices $\boldsymbol{\Phi}_i$ for the last q lags). The symbols $\widehat{\boldsymbol{\Sigma}}_R$ and $\widehat{\boldsymbol{\Sigma}}_U$ in (12.31) denote the estimated covariance matrix of estimated residuals in the *restricted* model VAR (i.e., under the restrictions of null hypothesis) and the *unrestricted* model VAR (i.e., without such restrictions), respectively.

Another test recommended in this context is *Wald test* (see Lütkepohl (2005) or Table 12.6 by EViews) that is similar to the classical F-test in linear regression models, but it is based on χ^2-distribution.

As the information criteria are concerned, one applies them in the same way as for the order determination of univariate time series models (see Sect. 6.3.1.2). For example, the m-variate version of *AIC criterion* is

$$AIC(k) = \ln|\widehat{\boldsymbol{\Sigma}}_k| + \frac{2k^*}{n}, \tag{12.32}$$

where $\widehat{\boldsymbol{\Sigma}}_k$ is the estimated covariance matrix of the estimated residuals in the model VAR(k) and $k^* = m(km + 1)$ is the number of parameters, which must be estimated in the m-variate model VAR(k) with nonzero mean vector (see Tables 12.7 and 12.11 by EViews).

2. Estimation of Model VAR

The model VAR is usually estimated by means of the ML method (i.e., by maximizing the (log) likelihood function under the assumption of normal distribution of white noise) even though the reduced form of model VAR (see (12.25)) may be also estimated by means of the classical OLS method. Under routine conditions, both approaches are asymptotically equivalent, and the estimates have asymptotically the normal distribution.

Table 12.7 LR test and information criteria *AIC*, *BIC*, and *HQ* for the identification of model VAR in Example 12.3 ($DTB3_t$ *and* $DAAA_t$) calculated by means of EViews

VAR lag order selection criteria					
Included observations: 112					
Lag	LogL	LR	AIC	BIC	HQ
0	42.94176	NA	−0.731103	−0.682558	−0.711407
1	72.27577	57.09657	−1.183496	−1.037862[a]	−1.124408[a]
2	76.80968	8.663012	−1.193030	−0.950307	−1.094550
3	82.06239	9.848829[a]	−1.215400[a]	−0.875588	−1.077527
4	82.58925	0.969044	−1.153380	−0.716478	−0.976115
5	83.27286	1.232937	−1.094158	−0.560167	−0.877501

[a]Lag order selected by the criterion
LR sequential modified LR test statistic (each test at 5% level), *AIC* Akaike information criterion, *BIC* Schwarz information criterion, *HQ* Hannan–Quinn information criterion

3. Diagnostic of Model VAR

The diagnostic of VAR checks various properties of the constructed model. Primarily, the condition of stationarity should be confirmed, i.e., the inverted roots of the estimated autoregressive polynomial should lie inside the unit circle in complex plane.

Further diagnostic procedures check the serial uncorrelatedness of the estimated white noise), namely

- By means of *Bartlett's approximation* (see also (6.65)), where the critical bound of size $2\sqrt{(1/n)}$ (applying the significance level of 5%) is used for particular estimated autocorrelations and mutual correlations of estimated residuals with nonzero delays (e.g., EViews software offers graphical outputs for all elements of the estimated matrix autocorrelation function with plotted critical bounds; see Fig. 12.6).
- By means of m-variate version of *Q-test* (or *portmanteau test*) with the critical region (see, e.g., Lütkepohl (2005) or (6.67) or Tables 12.9 and 12.13 by EViews)

$$Q_m = n^2 \sum_{k=1}^{K} \frac{1}{n-k} \operatorname{tr}\left(\widehat{\boldsymbol{\Gamma}}_k' \widehat{\boldsymbol{\Gamma}}_0^{-1} \widehat{\boldsymbol{\Gamma}}_k \widehat{\boldsymbol{\Gamma}}_0^{-1}\right) \geq \chi^2_{1-\alpha}\left(m^2(K-p)\right) \tag{12.33}$$

  for various choices of K (one recommends to try primarily $K \approx \sqrt{n}$).

- By means of *Lagrange multiplier test* (*LM test*; see Lütkepohl (2005) or Table 12.10 by EViews) which is applied in such a way that we run an auxiliary regression of the estimated residual at time t on its lagged value at time $t-h$ and on the original right-hand-side lagged regressors $\mathbf{y}_{t-1}, \ldots, \mathbf{y}_{t-p}$: under the null hypothesis of no serial correlation of order h, the corresponding LM statistics is asymptotically distributed as $\chi^2(m^2)$.

- By means of tests of normality of the estimated residuals (see, e.g., the *test Jarque–Bera* in Example 8.1 or Table 12.14 by EViews).

4. Predictions in Model VAR

The construction of predictions in the model VAR is quite analogical to the univariate model AR. Thus, again one makes recursively use of the relation

$$\widehat{\mathbf{y}}_{t+k}(t) = \boldsymbol{\varphi}_0 + \boldsymbol{\Phi}_1 \widehat{\mathbf{y}}_{t+k-1}(t) + \cdots + \boldsymbol{\Phi}_p \widehat{\mathbf{y}}_{t+k-p}(t), \tag{12.34}$$

where

$$\widehat{\mathbf{y}}_{t+j}(t) = \mathbf{y}_{t+j} \quad \text{for } j \leq 0. \tag{12.35}$$

Remark 12.4 As the models VMA and VARMA are concerned, the application of the OLS method is not so straightforward, and one prefers the ML method in such models. For example in the model VMA(1)

$$\mathbf{y}_t = \boldsymbol{\vartheta}_0 + \boldsymbol{\varepsilon}_t + \boldsymbol{\Theta}_1 \boldsymbol{\varepsilon}_{t-1}, \qquad \boldsymbol{\varepsilon}_t \sim N(\mathbf{0}, \boldsymbol{\Sigma}) \tag{12.36}$$

the (conditional) likelihood function has the form

$$L(\boldsymbol{\vartheta}_0, \boldsymbol{\Theta}_1, \boldsymbol{\Sigma}) = \prod_{t=1}^{n} \frac{1}{(2\pi)^{m/2} |\boldsymbol{\Sigma}|^{1/2}} \exp\left(-\frac{1}{2} \boldsymbol{\varepsilon}_t' \boldsymbol{\Sigma}^{-1} \boldsymbol{\varepsilon}_t\right), \tag{12.37}$$

where the vectors of white noise are calculated recursively in time

$$\boldsymbol{\varepsilon}_1 = \mathbf{y}_1 - \boldsymbol{\vartheta}_0, \quad \boldsymbol{\varepsilon}_2 = \mathbf{y}_2 - \boldsymbol{\vartheta}_0 - \boldsymbol{\Theta}_1 \boldsymbol{\varepsilon}_1, \quad \boldsymbol{\varepsilon}_3 = \mathbf{y}_3 - \boldsymbol{\vartheta}_0 - \boldsymbol{\Theta}_1 \boldsymbol{\varepsilon}_2, \ldots . \tag{12.38}$$

◊

Remark 12.5 Particular components of models VARMA have the form of univariate models ARMA: in the case of m-variate model VARMA(p, q) are these *marginal models* of the type ARMA(mp, $(m-1)p + q$). For example, the bivariate model VAR(1) in (12.15) with zero mean vector can be written as

$$\begin{pmatrix} 1 - \varphi_{11}B & -\varphi_{12}B \\ -\varphi_{21}B & 1 - \varphi_{22}B \end{pmatrix} \begin{pmatrix} y_{1t} \\ y_{2t} \end{pmatrix} = \begin{pmatrix} \varepsilon_{1t} \\ \varepsilon_{2t} \end{pmatrix}. \tag{12.39}$$

If we multiply the relation (12.39) from the left by the matrix

$$\begin{pmatrix} 1-\varphi_{22}B & \varphi_{12}B \\ \varphi_{21}B & 1-\varphi_{11}B \end{pmatrix},$$

then we can write due to diagonality of the matrix product on the left-hand side of (12.20)

$$\left((1-\varphi_{11}B)(1-\varphi_{22}B)-\varphi_{12}\varphi_{21}B^2\right)\begin{pmatrix} y_{1t} \\ y_{2t} \end{pmatrix}$$
$$=\begin{pmatrix} 1-\varphi_{22}B & \varphi_{12}B \\ \varphi_{21}B & 1-\varphi_{11}B \end{pmatrix}\begin{pmatrix} \varepsilon_{1t} \\ \varepsilon_{2t} \end{pmatrix}, \tag{12.40}$$

so that particular components are modeled as ARMA(2,1) (the order $q = 1$ of moving averages follows from the fact that the autocorrelation functions of both univariate processes of moving averages on the right-hand side of (12.40) have truncation points $k_0 = 1$).

◇

Example 12.3 We shall construct the model VAR for the data from Example 12.1 (the bivariate time series of length 120 with the first differences of monthly yields to maturity (YTM) for three-month T-bills denoted as $DTB3_t$ in the first component and for corporate bonds denoted as $DAAA_t$ in the second component):

1. *Identification*:
   - Wald test in Table 12.6 identifies the given bivariate time series overall as VAR(1) (with significance level of 5%); however, the second component *DAAA* individually demands the order of 2 on this significance level so that a reasonable model seems to be VAR(2).
   - The application of LR test and information criteria *AIC*, *BIC*, and *HQ* in Table 12.7 results in the model VAR(1) or VAR(3).

2. *Estimation*:

   The model VAR(2) was finally chosen for the given data (also taking into account the diagnostic results; see 3). The estimation of this model is realized in Table 12.8. Obviously, it would be possible to omit some lagged regressors in the estimated model by applying estimated standard deviations or *t*-ratio presented in this table.

3. *Diagnostics*:
   - Figure 12.5 verifies that the estimated model VAR(2) is stationary (all four inverted roots of autoregressive polynomial lie inside the unit circle in complex plane).
   - The estimated matrix autocorrelation function of the estimated residuals with plotted critical bounds in Fig. 12.6 confirms the uncorrelatedness of estimated residuals.

Table 12.8 Estimation of model VAR(2) in Example 12.3 ($DTB3_t$ and $DAAA_t$) by means of EViews

Vector autoregression estimates		
Included observations: 120		
Standard errors in ( ) and t-statistics in [ ]		
	DTB3	DAAA
DTB3(−1)	0.454168	0.002001
	(0.11239)	(0.10863)
	[4.04089]	[0.01842]
DTB3(−2)	−0.017927	0.015799
	(0.10254)	(0.09911)
	[−0.17483]	[0.15940]
DAAA(−1)	0.121705	0.514193
	(0.11407)	(0.11026)
	[1.06690]	[4.66359]
DAAA(−2)	−0.149127	−0.258736
	(0.11277)	(0.10900)
	[−1.32238]	[−2.37376]
C	−0.009735	−0.023150
	(0.01951)	(0.01885)
	[−0.49906]	[−1.22789]
S.E. equation	0.211679	0.204597

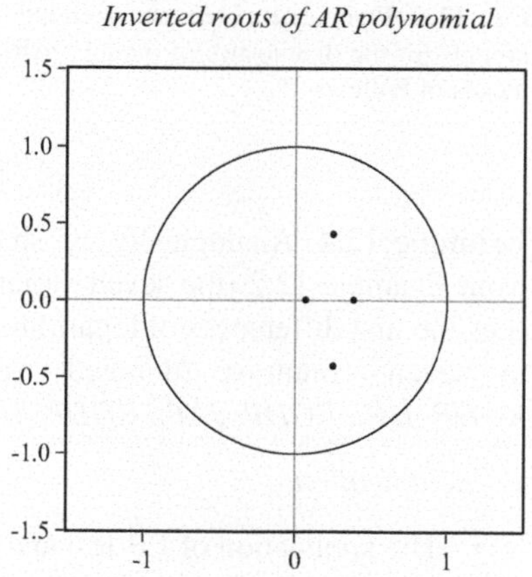

Fig. 12.5 Inverted roots of estimated autoregressive polynomial in Example 12.3 ($DTB3_t$ and $DAAA_t$) calculated by means of EViews

- *Q*-test applied to the estimated residuals in Table 12.9 confirms the uncorrelatedness of estimated residuals.
- LM test applied to the estimated residuals in Table 12.10 confirms the uncorrelatedness of estimated residuals as well.

◇

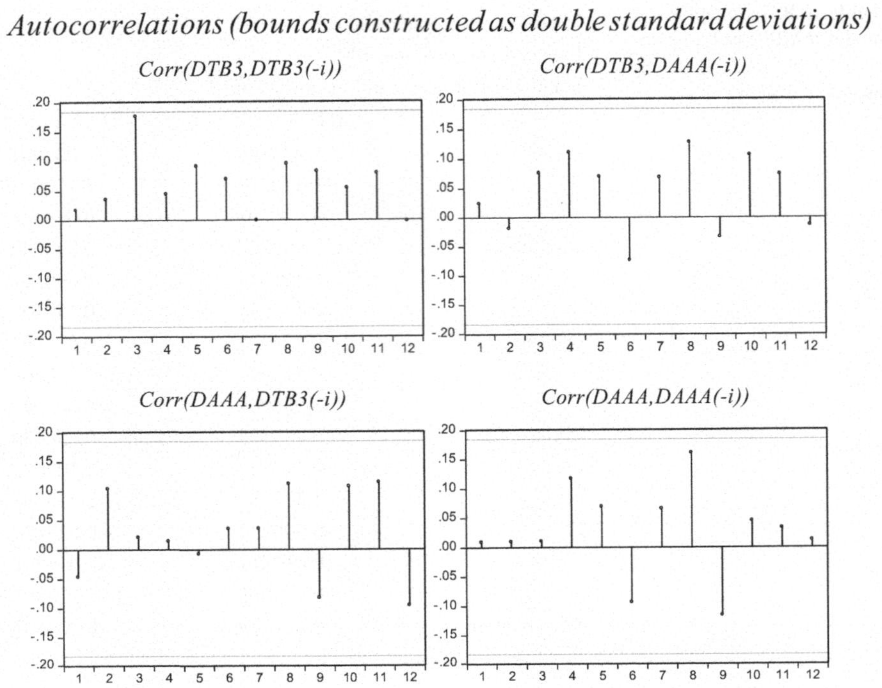

Fig. 12.6 Estimated matrix autocorrelation function of estimated residuals with plotted critical bounds for the diagnostics of model VAR(2) in Example 12.3 ($DTB3_t$ *and* $DAAA_t$) calculated by means of EViews

Example 12.4 Analogically we shall construct the model VAR also for the data from Example 12.2 (the seven-variate time series of length 42 with the log returns (i.e., the first differences of logarithms) of the annual gross domestic products (GDP) in seven countries (denoted as $RGDP\_FRA_t$, $RGDP\_GER_t$, $RGDP\_ITA_t$, $RGDP\_UK_t$, $RGDP\_JAP_t$, $RGDP\_USA_t$, $RGDP\_CAN_t$):

1. *Identification*:

   - The application of LR test and information criteria in Table 12.11 identify the given time series as VAR(1).

2. *Estimation*:

The estimation of the model VAR(1) is realized by means of EViews in Table 12.12. Analogically as in Example 12.3 one could omit some lagged regressors in the estimated model by applying estimated standard deviations or *t*-ratio

Table 12.9 *Q*-test applied to estimated residuals for the diagnostics of model VAR(2) in Example 12.3 ($DTB3_t$ *and* $DAAA_t$) calculated by means of EViews

VAR residual Portmanteau tests for autocorrelations
H0: no residual autocorrelations up to lag h
Included observations: 118

Lags	Q-Stat	Prob.	Adj Q-Stat	Prob.	df
1	0.759100	NA*	0.765588	NA*	NA*
2	2.509240	NA*	2.545903	NA*	NA*
3	7.221494	0.1246	7.381086	0.1171	4
4	9.651343	0.2904	9.896192	0.2724	8
5	12.22388	0.4279	12.58255	0.4001	12
6	15.53721	0.4857	16.07339	0.4479	16
7	16.56259	0.6812	17.16344	0.6423	20
8	20.01529	0.6959	20.86724	0.6465	24
9	24.81018	0.6381	26.05804	0.5699	28
10	27.85802	0.6764	29.38809	0.5994	32

*The test is valid only for lags larger than the VAR lag order
df is degrees of freedom for (approximate) chi-square distribution

Table 12.10 LM test applied to estimated residuals for the diagnostics of model VAR(2) in Example 12.3 ($DTB3_t$ *and* $DAAA_t$) calculated by means of EViews

VAR residual serial correlation LM tests
H0: no serial correlation at lag order h
Included observations: 118

Lags	LM-Stat	Prob
1	7.522854	0.1107
2	7.499448	0.1117
3	5.172111	0.2701
4	2.696262	0.6099
5	2.692773	0.6105
6	3.632837	0.4580
7	1.420712	0.8406
8	3.703578	0.4476
9	5.453784	0.2438
10	3.642071	0.4566

Probs from chi-square with 4 df

(even the most of lagged regressors since the given time series has large dimension of 7 but small length of 42, which causes relatively broad confidence intervals).

3. *Diagnostics*:

- The estimated model VAR(1) is stationary according to Fig. 12.7.
- *Q*-test applied to the estimated residuals in Table 12.13 confirms the uncorrelatedness of estimated residuals.
- The test Jarque–Bera applied to the estimated residuals in Table 12.14 confirms the normality of these residuals.

◇

Table 12.11 LR test and information criteria *AIC*, *BIC*, and *HQ* for the identification of model VAR in Example 12.4 ($RGDP\_FRA_t$, ...) calculated by means of EViews

VAR lag order selection criteria					
Included observations: 39					
Lag	LogL	LR	AIC	BIC	HQ
0	677.8285	NA	−34.40146	−34.10287*	−34.29433
1	744.7419	106.3751*	−35.32010*	−32.93139	−34.46305*
2	770.1225	31.23772	−34.10885	−29.63003	−32.50188
3	817.3099	41.13769	−34.01589	−27.44695	−31.65901

*Lag order selected by the criterion
LR sequential modified LR test statistic (each test at 5% level), *FPE* Final prediction error, *AIC* Akaike information criterion, *BIC* Schwarz information criterion, *HQ* Hannan–Quinn information criterion

12.3 Tests of Causality

An important aspect of the multivariate time series analysis is the investigation of *causality* among various blocks of modeled variables. A general approach to the causality relates to predicting: if a time series influences causally another time series, then it should improve prediction results of the affected time series (see, e.g., Granger (1969), Sims (1972)).

In econometric practice, the conclusions on causality can be easily realized just in the models VAR, where they can be obtained in a very objective and computationally easy way. The so-called *Granger causality* means that there exists a correlation between the current value of a variable and past values of other variables (see also Remark 12.1). The structure of models VAR enables in this context to transfer such a causality to testing whether blocks of certain parameters in the estimated model VAR are zero (particularly, it can be done simply by applying F- or Wald tests under the assumption of stationarity).

More specifically, the following terminology is used in this context (similarly to the one in Remark 12.1):

- If lagged values of a variable y_i in the VAR equation explaining a variable y_j are statistically significant (globally, e.g., in the sense of F-test), then the variable y_i *causes* (or *G-causes*) the variable y_j.
- If y_i causes y_j, but y_j does not cause y_i, then there exists a *unidirectional relationship from y_i to y_j*. In such a case, one denotes the variable y_i in the VAR equation explaining the variable y_j as *strongly exogenous.*
- If y_i causes y_j and also y_j causes y_i, then there exists a *feedback* between y_i and y_j.
- If y_i does not cause y_j and also y_j does not cause y_i, then y_i and y_j are *G-independent.*

Let us consider, e.g., the following bivariate model VAR(1):

Table 12.12 Estimation of model VAR(1) in Example 12.4 ($RGDP\_FRA_t, \ldots$) by means of EViews

Vector autoregression estimates

Included observations: 41 after adjustments

Standard errors in ( ) and t-statistics in [ ]

	RGDP_FRA	RGDP_GER	RGDP_ITA	RGDP_UK	RGDP_JPN	RGDP_US	RGDP_CAN
RGDP_FRA(−1)	−0.153093	−0.432343	−0.042512	0.028843	0.041562	−0.001648	−0.055642
	(0.22493)	(0.32673)	(0.28877)	(0.31717)	(0.42006)	(0.35775)	(0.46020)
	[−0.68062]	[−1.32326]	[−0.14721]	[0.09094]	[0.09894]	[−0.00461]	[−0.12091]
RGDP_GER(−1)	0.120615	0.516444	0.186485	−0.034213	−0.085963	−0.019305	0.024199
	(0.11702)	(0.16998)	(0.15024)	(0.16501)	(0.21854)	(0.18612)	(0.23942)
	[1.03070]	[3.03824]	[1.24127]	[−0.20734]	[−0.39335]	[−0.10372]	[0.10107]
RGDP_ITA(−1)	−0.207291	−0.206014	−0.167585	−0.239290	0.113849	−0.443415	−0.232438
	(0.13888)	(0.20172)	(0.17829)	(0.19582)	(0.25935)	(0.22088)	(0.28413)
	[−1.49264]	[−1.02127]	[−0.93995]	[−1.22197]	[0.43898]	[−2.00749]	[−0.81807]
RGDP_UK(−1)	−0.001346	0.344634	0.189435	0.218979	−0.429568	0.210135	0.527973
	(0.15675)	(0.22769)	(0.20124)	(0.22103)	(0.29274)	(0.24931)	(0.32071)
	[−0.00859]	[1.51360]	[0.94132]	[0.99072]	[−1.46742]	[0.84285]	[1.64629]
RGDP_JPN(−1)	0.429235	0.363092	0.316353	0.070350	0.558444	0.228788	0.131389
	(0.12418)	(0.18038)	(0.15943)	(0.17510)	(0.23191)	(0.19751)	(0.25407)
	[3.45650]	[2.01292]	[1.98429]	[0.40176]	[2.40802]	[1.15835]	[0.51714]
RGDP_US(−1)	0.138609	−0.247262	0.017365	0.080385	0.400317	0.010285	−0.130052
	(0.17285)	(0.25108)	(0.22192)	(0.24374)	(0.32281)	(0.27492)	(0.35365)
	[0.80188]	[−0.98479]	[0.07825]	[0.32980]	[1.24011]	[0.03741]	[−0.36774]
RGDP_CAN(−1)	0.031245	−0.149500	−0.053745	−0.041571	−0.240112	−0.118187	0.203071
	(0.12512)	(0.18175)	(0.16064)	(0.17643)	(0.23367)	(0.19901)	(0.25599)
	[0.24971]	[−0.82257]	[−0.33458]	[−0.23562]	[−1.02759]	[−0.59388]	[0.79327]

(continued)

Table 12.12 (continued)

Vector autoregression estimates

Included observations: 41 after adjustments

Standard errors in ( ) and t-statistics in [ ]

	RGDP_FRA	RGDP_GER	RGDP_ITA	RGDP_UK	RGDP_JPN	RGDP_US	RGDP_CAN
C	0.009702	0.016474	0.015393	0.020895	0.028488	0.018920	0.011073
	(0.00574)	(0.00834)	(0.00738)	(0.00810)	(0.01073)	(0.00914)	(0.01175)
	[1.68886]	[1.97409]	[2.08701]	[2.57942]	[2.65534]	[2.07066]	[0.94212]
S.E. equation	0.016176	0.023496	0.020767	0.022809	0.030208	0.025727	0.033095

Fig. 12.7 Inverted roots of estimated autoregressive polynomial in Example 12.4 ($RGDP\_FRA_t, \ldots$) calculated by means of EViews

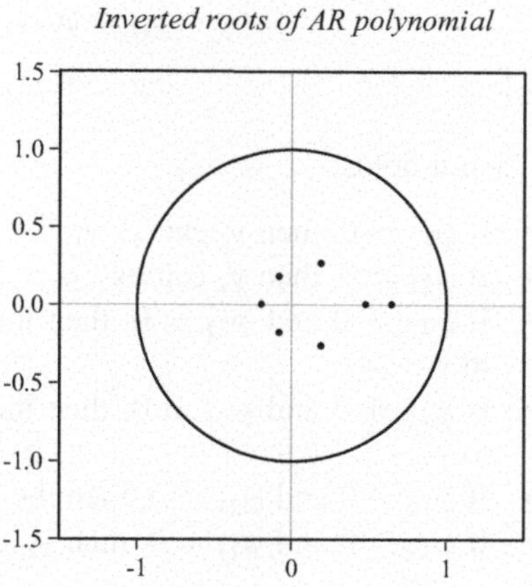

Table 12.13 Q-test applied to estimated residuals for the diagnostics of model VAR(1) in Example 12.4 ($RGDP\_FRA_t, \ldots$) calculated by means of EViews

VAR residual Portmanteau tests for autocorrelations
H0: no residual autocorrelations up to lag h
Included observations: 41

Lags	Q-Stat	Prob.	Adj Q-Stat	Prob.	df
1	11.39627	NA*	11.68118	NA*	NA*
2	37.89006	0.8753	39.53363	0.8308	49
3	83.33112	0.8547	88.56214	0.7418	98
4	121.3687	0.9397	130.7119	0.8285	147
5	164.3181	0.9517	179.6265	0.7931	196
6	197.2777	0.9888	218.2363	0.8898	245
7	231.0652	0.9973	258.9800	0.9303	294

*The test is valid only for lags larger than the VAR lag order
df is degrees of freedom for (approximate) chi-square distribution

Table 12.14 Test Jarque–Bera applied to estimated residuals for the diagnostics of model VAR(1) in Example 12.4 ($RGDP\_FRA_t, \ldots$) calculated by means of EViews

VAR residual normality tests
Orthogonalization: Cholesky (Lutkepohl)
H0: residuals are multivariate normal
Included observations: 41

Component	Jarque-Bera	df	Prob.
1	3.418095	2	0.1810
2	2.690175	2	0.2605
3	2.919029	2	0.2323
4	1.465934	2	0.4805
5	2.644147	2	0.2666
6	1.757411	2	0.4153
7	2.742968	2	0.2537
Joint	17.63776	14	0.2238

$$y_{1t} = \varphi_{11} y_{1,t-1} + \varphi_{12} y_{2,t-1} + \varepsilon_{1t},$$
$$y_{2t} = \varphi_{21} y_{1,t-1} + \varphi_{22} y_{2,t-1} + \varepsilon_{2t}. \tag{12.41}$$

Then it holds:

- If $\varphi_{12} \neq 0$, then y_2 causes y_1.
- If $\varphi_{21} \neq 0$, then y_1 causes y_2.
- If $\varphi_{12} \neq 0$ and $\varphi_{21} = 0$, then there exists a unidirectional relationship from y_2 to y_1.
- If $\varphi_{12} = 0$ and $\varphi_{21} \neq 0$, then there exists a unidirectional relationship from y_1 to y_2.
- If $\varphi_{12} \neq 0$ and $\varphi_{21} \neq 0$, then there exists a feedback between y_1 and y_2.
- If $\varphi_{12} = 0$ and $\varphi_{21} = 0$, then y_1 and y_2 are G-independent.

Remark 12.6 The presented approach to the problem of causality concerns not only the causality relations between two scalar variables but also between blocks of more variables of given model VAR. However, it has sense only in the case that the given model VAR is unambiguously identified. If it is not the case, then various transformations of such a model may exist delivering different causality results.

◇

Example 12.5 Let us consider the model VAR(1) estimated in Example 12.4 (the seven-variate time series of length 42 with the log returns of annual gross domestic products in seven countries). Table 12.15 presents the causality analysis based on the corresponding p-values delivered by EViews. One can see (applying the significance level of 5%) that:

- There exists a unidirectional relationship from *RGDP_JPN* to *RGDP_FRA*.
- There exists a unidirectional relationship from *RGDP_JPN* to *RGDP_GER*.
- There exists a unidirectional relationship from *RGDP_JPN* to *RGDP_ITA*.
- There exists a unidirectional relationship from *RGDP_ITA* to *RGDP_US*.
- *RGDP_FRA* is influenced casually by all remaining six variables.
- *RGDP_GER* is influenced casually by all remaining six variables.
- *RGDP_ITA* is influenced casually by all remaining six variables.

◇

12.4 Impulse Response and Variance Decomposition

The causality analysis in Sect. 12.3 based on F- or analogical tests does not answer questions, which is the sign of a causality relation or how long the effect of various one-shot changes will survive. Such information can be obtained by means of procedures denoted as impulse response and variance decomposition.

Table 12.15 Causality analysis of model VAR(1) in Example 12.5 (*RGDP_FRA*$_t$, . . .) calculated by means of EViews

VAR Granger Causality/Block Exogeneity Wald tests			
Included observations: 41			
Excluded	Chi-sq	df	Prob.
Dependent variable: RGDP_FRA			
RGDP_GER	1.062342	1	0.3027
RGDP_ITA	2.227988	1	0.1355
RGDP_UK	7.37E-05	1	0.9932
RGDP_JPN	11.94738	1	0.0005
RGDP_US	0.643009	1	0.4226
RGDP_CAN	0.062357	1	0.8028
All	20.79029	6	0.0020
Dependent variable: RGDP_GER			
RGDP_FRA	1.751010	1	0.1857
RGDP_ITA	1.042991	1	0.3071
RGDP_UK	2.290986	1	0.1301
RGDP_JPN	4.051830	1	0.0441
RGDP_US	0.969813	1	0.3247
RGDP_CAN	0.676629	1	0.4107
All	14.54094	6	0.0241
Dependent variable: RGDP_ITA			
RGDP_FRA	0.021672	1	0.8830
RGDP_GER	1.540756	1	0.2145
RGDP_UK	0.886088	1	0.3465
RGDP_JPN	3.937413	1	0.0472
RGDP_US	0.006123	1	0.9376
RGDP_CAN	0.111941	1	0.7379
All	17.32903	6	0.0081
Dependent variable: RGDP_UK			
RGDP_FRA	0.008270	1	0.9275
RGDP_GER	0.042991	1	0.8357
RGDP_ITA	1.493219	1	0.2217
RGDP_JPN	0.161412	1	0.6879
RGDP_US	0.108771	1	0.7415
RGDP_CAN	0.055517	1	0.8137
All	1.936258	6	0.9255
Dependent variable: RGDP_JPN			
RGDP_FRA	0.009790	1	0.9212
RGDP_GER	0.154727	1	0.6941
RGDP_ITA	0.192703	1	0.6607
RGDP_UK	2.153336	1	0.1423
RGDP_US	1.537884	1	0.2149
RGDP_CAN	1.055934	1	0.3041
All	3.346580	6	0.7643

(continued)

Table 12.15 (continued)

VAR Granger Causality/Block Exogeneity Wald tests			
Included observations: 41			
Excluded	Chi-sq	df	Prob.
Dependent variable: RGDP_US			
RGDP_FRA	2.12E-05	1	0.9963
RGDP_GER	0.010758	1	0.9174
RGDP_ITA	4.030007	1	0.0447
RGDP_UK	0.710396	1	0.3993
RGDP_JPN	1.341782	1	0.2467
RGDP_CAN	0.352699	1	0.5526
All	7.351121	6	0.2896
Dependent variable: RGDP_CAN			
RGDP_FRA	0.014619	1	0.9038
RGDP_GER	0.010215	1	0.9195
RGDP_ITA	0.669240	1	0.4133
RGDP_UK	2.710257	1	0.0997
RGDP_JPN	0.267433	1	0.6051
RGDP_US	0.135235	1	0.7131
All	4.685880	6	0.5847

1. Impulse Response

Impulse response (see also Sect. 6.3.3.1) investigates the reaction of a chosen dependent variable of the given model VAR to an impulse (innovation shock) generated in a chosen row of this model. Obviously, a shock generated in the ith row of the model affects not only the variable y_i, but it is also transmitted to other variables through the dynamic lag structure of VAR. Thus, in the estimated m-variate model VAR one can investigate in time (starting at the moment of impulse) altogether m^2 responses (namely m responses for each of m dependent variables y_{1t}, …, y_{mt} on the left-hand sides of particular rows in time t). Under the assumption of stationarity of this model, the impacts of impulses in all (i.e., m^2) response situations gradually dampen (even though with different intensities, which is often useful to investigate).

For example in the bivariate model VAR(1) with zero mean vector of the form

$$\begin{pmatrix} y_{1t} \\ y_{2t} \end{pmatrix} = \begin{pmatrix} 0.6 & 0.2 \\ 0 & 0.3 \end{pmatrix} \begin{pmatrix} y_{1,t-1} \\ y_{2,t-1} \end{pmatrix} + \begin{pmatrix} \varepsilon_{1t} \\ \varepsilon_{2t} \end{pmatrix}$$

one obtains gradually in time the following responses to a (deterministic) unit impulse generated at time $t = 0$ in the first row (i.e., the innovation shock is $\varepsilon_{1,0} = 1$, while the second one remains at the zero level):

$$\begin{pmatrix} y_{1,0} \\ y_{2,0} \end{pmatrix} = \begin{pmatrix} \varepsilon_{1,0} \\ \varepsilon_{2,0} \end{pmatrix} = \begin{pmatrix} 1 \\ 0 \end{pmatrix}, \begin{pmatrix} y_{1,1} \\ y_{2,1} \end{pmatrix} = \begin{pmatrix} 0.6 & 0.2 \\ 0 & 0.3 \end{pmatrix} \begin{pmatrix} 1 \\ 0 \end{pmatrix} = \begin{pmatrix} 0.6 \\ 0 \end{pmatrix},$$

$$\begin{pmatrix} y_{1,2} \\ y_{2,2} \end{pmatrix} = \begin{pmatrix} 0.6 & 0.2 \\ 0 & 0.3 \end{pmatrix} \begin{pmatrix} 0.6 \\ 0 \end{pmatrix} = \begin{pmatrix} 0.36 \\ 0 \end{pmatrix}, \quad \ldots$$

and the following responses to a unit impulse generated at time $t = 0$ in the second row (i.e., the innovation shock is $\varepsilon_{2,0} = 1$, while the first one remains at the zero level):

$$\begin{pmatrix} y_{1,0} \\ y_{2,0} \end{pmatrix} = \begin{pmatrix} \varepsilon_{1,0} \\ \varepsilon_{2,0} \end{pmatrix} = \begin{pmatrix} 0 \\ 1 \end{pmatrix}, \begin{pmatrix} y_{1,1} \\ y_{2,1} \end{pmatrix} = \begin{pmatrix} 0.6 & 0.2 \\ 0 & 0.3 \end{pmatrix} \begin{pmatrix} 0 \\ 1 \end{pmatrix} = \begin{pmatrix} 0.2 \\ 0.3 \end{pmatrix},$$

$$\begin{pmatrix} y_{1,2} \\ y_{2,2} \end{pmatrix} = \begin{pmatrix} 0.6 & 0.2 \\ 0 & 0.3 \end{pmatrix} \begin{pmatrix} 0.2 \\ 0.3 \end{pmatrix} = \begin{pmatrix} 0.18 \\ 0.09 \end{pmatrix}, \quad \ldots$$

(one can see that the both responses dampen in time and that the response of y_2 to the impulse generated in the first row is zero in all times since $\varphi_{21} = 0$).

Remark 12.7 If the model VAR can be written as the linear process of the form

$$\mathbf{y}_t = \boldsymbol{\varepsilon}_t + \boldsymbol{\Psi}_1 \boldsymbol{\varepsilon}_{t-1} + \boldsymbol{\Psi}_2 \boldsymbol{\varepsilon}_{t-2} + \cdots \tag{12.42}$$

(see also (12.22)), then obviously the elements of the ith column of matrix $\boldsymbol{\Psi}_k$ represent the responses of particular dependent variables to the unit innovation shock generated in the ith row at time $t - k$ (while the other values of the multivariate white noise remain at the zero level).

◇

Remark 12.8 Several technical problems must be solved when the impulse response analysis is applied (see, e.g., EViews):

- Sometimes it is reasonable to investigate the response to such an impulse that does not occur in one shot, but repeatedly starting at a given time moment; in this case, the response in the given stationary model VAR does not dampen, but after some time stabilizes to a (nonzero) level.
- The impulses are usually generated randomly being set to one standard deviation of the estimated residuals (or to multiples of these standard deviations). In particular, such a standardization is reasonable when different variables are measured in different scales.
- Standard deviations for particular responses can be constructed (analytically or by means of simulations; see Fig. 12.8).
- The impulses can be also orthogonalized in advance (e.g., using Cholesky decomposition similarly as in the transformation (12.17)) to guarantee the mutual

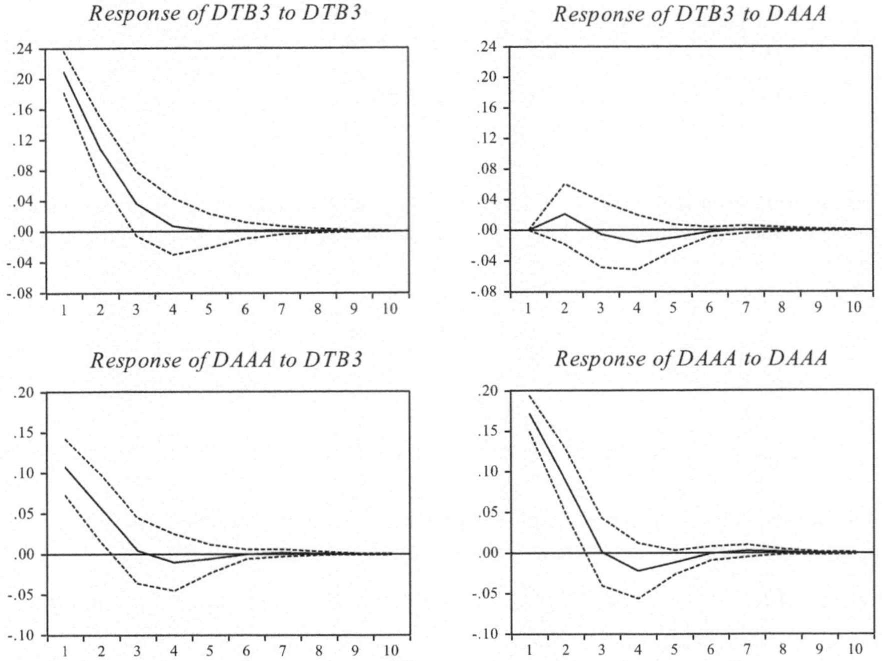

Fig. 12.8 Impulse response analysis of model VAR(2) in Example 12.6 (*DTB3$_t$ and DAAA$_t$*) calculated by means of EViews

uncorrelatedness of (random) impulses generated in different rows of the model VAR.

- The ordering of particular variables included in the given model VAR is very important and should be consistent with the economic theory: by applying Cholesky decomposition (12.16), one can rewrite the linear process (12.42) in the form

$$\mathbf{y}_t = \mathbf{L}\mathbf{L}^{-1}\boldsymbol{\varepsilon}_t + \boldsymbol{\Psi}_1\mathbf{L}\mathbf{L}^{-1}\boldsymbol{\varepsilon}_{t-1} + \cdots = \mathbf{L}\mathbf{u}_t + \boldsymbol{\Psi}_1\mathbf{L}\mathbf{u}_{t-1} + \cdots, \tag{12.43}$$

where $\{\mathbf{u}_t\} = \{\mathbf{L}^{-1}\boldsymbol{\varepsilon}_t\}$ is the orthogonalized white noise with simultaneously uncorrelated components and the matrix $\mathbf{L}$ applied to the current value of white noise in $\mathbf{u}_t$ in (12.43) is lower triangular. Hence one should choose (*i*) the first variable y_1 such that the shock u_{1t} in the first row may affect immediately at time t each of variables $y_{1t}, \ldots, y_{mt}$, (*ii*) similarly the second variable y_2 such that the shock u_{2t} in the second row may affect immediately at time t each of variables $y_{2t}, \ldots, y_{mt}$ but not the variable y_{1t}, etc.

◊

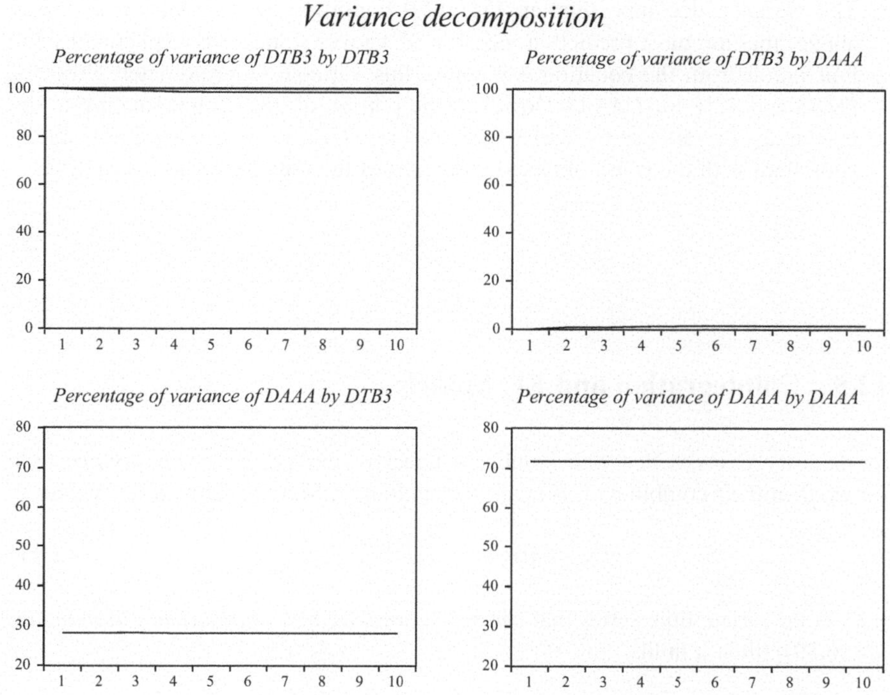

Fig. 12.9 Variance decomposition analysis of model VAR(2) in Example 12.6 ($DTB3_t$ and $DAAA_t$) calculated by means of EViews

2. Variance Decomposition

While the response to impulse captures the impact of an impulse generated in a chosen row of VAR for a chosen dependent variable, the *variance decomposition* provides information on such a portion of the variance of prediction error (when predicting a chosen dependent variable) that is explained by innovations from particular rows of VAR. In practice, the main portion of this variance for y_{it} is usually explained by the innovation from the *i*th row of the model.

Example 12.6 Figures 12.8 and 12.9 present the results of the analysis based on impulse response and variance decomposition for the model VAR(2) estimated in Example 12.3 (the bivariate time series of length 120 with the first differences of monthly yields to maturity for 3-month T-bills $DTB3_t$ in the first component and for corporate bonds $DAAA_t$ in the second component):

- The response to (random) impulses set to one standard deviation of the estimated residuals in Fig. 12.8 is performed after the orthogonalization by Cholesky decomposition (the standard deviations for estimated responses are also plotted; see Remark 12.8). Obviously, each of four responses gradually dampens so that the given model VAR is stable (the model stationarity is also confirmed in this way). Note the nonnegligible response of *DAAA* to the impulse generated in the first equation for *DTB3* (see the lower graph on the left).

- The variance decomposition in Fig. 12.9 confirms the well-known fact (see above) that the most prediction variance of a given variable is explained by the innovation from the equation explaining this variable (approximately 99% for *DTB*3 and 72% for *DAAA*). However, the portion of 28% corresponding to the percentage of variance of *DAAA* explained by *DTB3* is not nonnegligible and is consistent with the result obtained by means of the impulse response analysis.

◇

12.5 Cointegration and EC Models

In majority cases, when one combines linearly (univariate) *nonstationary* time series, then their combination is again nonstationary. More specifically, if we have

$$y_{it} \sim \mathrm{I}(d_i), \qquad i = 1, \ldots, m \tag{12.44}$$

(i.e., m univariate time series that can be stationarized by appropriate differencing; see (6.89)), then it holds

$$\sum_{i=1}^{m} \alpha_i y_{it} \sim \mathrm{I}\left(\max_{i=1, \ \ldots, \ m} d_i \right) \tag{12.45}$$

for arbitrary nontrivial linear combination of considered time series. Hence, particularly, any linear combination of time series with (stochastic) linear trend $y_{it} \sim \mathrm{I}(1)$ usually includes a (stochastic) linear trend as well.

On the other hand, economic and financial time series can be sometimes combined in such a way that the resulting linear combination of nonstationary time series becomes stationary. Such a phenomenon is denoted as *cointegration* and can be interpreted as relationship of a *long-run equilibrium* among economic variables: particular time series are nonstationary, but their ("cointegrated") movement in time tends (as a consequence of various market forces) to a balanced state of equilibrium (even though in *short-run* segments deviations from such a long-run balance persist in time). Particularly, in finance there exist various examples of cointegration, e.g.:

- Among spot and futures prices of various assets (commodities, securities, and the like).
- Among price ratios in different countries (i.e., the ratios of prices of the same goods) and corresponding currency rates.
- Among market prices of stocks and volumes of dividends.

In given (and other) examples, the absence of long-run equilibrium would give birth to arbitrage opportunities. Therefore, the principle of cointegration including so-called EC models (see below) becomes one of the main econometric topics nowadays (see, e.g., the seminal work by Engle and Granger (1987)).

Both the theoretical and the practical analyses of cointegration become simpler in the framework of the models VAR, which is demonstrated by the following Example 12.7.

Example 12.7 Let us consider two time series $\{y_{1t}\}$ and $\{y_{2t}\}$ that can be modeled simultaneously as the bivariate model VAR(1) of the form

$$\begin{pmatrix} y_{1t} \\ y_{2t} \end{pmatrix} = \boldsymbol{\Phi} \cdot \begin{pmatrix} y_{1,t-1} \\ y_{2,t-1} \end{pmatrix} + \begin{pmatrix} \varepsilon_{1t} \\ \varepsilon_{2t} \end{pmatrix} = \begin{pmatrix} 0.5 & -0.25 \\ -1 & 0.5 \end{pmatrix} \begin{pmatrix} y_{1,t-1} \\ y_{2,t-1} \end{pmatrix} + \begin{pmatrix} \varepsilon_{1t} \\ \varepsilon_{2t} \end{pmatrix}, \tag{12.46}$$

where the bivariate white noise $\{\boldsymbol{\varepsilon}_t\}$ has a general covariance matrix $\boldsymbol{\Sigma} > 0$. However, this model is not stationary, since the eigenvalues of matrix $\boldsymbol{\Phi}$ in (12.46), i.e., the roots of polynomial equation $\det(\lambda\,\mathbf{I} - \boldsymbol{\Phi}) = 0$), are 0 and 1, so that one of them does not lie inside the unit circle in complex plane. Each of marginal time series $\{y_{1t}\}$ and $\{y_{2t}\}$ is nonstationary as well: if the model (12.46) is multiplied from the left-hand side as

$$\begin{pmatrix} 1-0.5B & -0.25B \\ -B & 1-0.5B \end{pmatrix} \cdot \left| \begin{pmatrix} 1-0.5B & 0.25B \\ B & 1-0.5B \end{pmatrix} \begin{pmatrix} y_{1t} \\ y_{2t} \end{pmatrix} = \begin{pmatrix} \varepsilon_{1t} \\ \varepsilon_{2t} \end{pmatrix}, \right. \tag{12.47}$$

then one obtains an equivalent model VAR(1)

$$\begin{pmatrix} 1-B & 0 \\ 0 & 1-B \end{pmatrix} \begin{pmatrix} y_{1t} \\ y_{2t} \end{pmatrix} = \begin{pmatrix} 1-0.5B & -0.25B \\ -B & 1-0.5B \end{pmatrix} \begin{pmatrix} \varepsilon_{1t} \\ \varepsilon_{2t} \end{pmatrix}. \tag{12.48}$$

It explicitly demonstrates that both marginal time series $\{y_{1t}\}$ and $\{y_{2t}\}$ can be modeled as models ARIMA(0,1,1); therefore, these univariate time series include stochastic linear trends and are nonstationary.

Now let us transform the time series $\{y_{1t}\}$ and $\{y_{2t}\}$ to time series $\{z_{1t}\}$ and $\{z_{2t}\}$ and the white noise $\{\boldsymbol{\varepsilon}_t\}$ to another white noise $\{\mathbf{u}_t\}$ by means of transformation

$$\begin{gathered} \begin{pmatrix} z_{1t} \\ z_{2t} \end{pmatrix} = \mathbf{P} \cdot \begin{pmatrix} y_{1t} \\ y_{2t} \end{pmatrix} = \begin{pmatrix} 1 & 0.5 \\ -2 & 1 \end{pmatrix} \begin{pmatrix} y_{1t} \\ y_{2t} \end{pmatrix}, \\ \begin{pmatrix} u_{1t} \\ u_{2t} \end{pmatrix} = \mathbf{P} \cdot \begin{pmatrix} \varepsilon_{1t} \\ \varepsilon_{2t} \end{pmatrix} = \begin{pmatrix} 1 & 0.5 \\ -2 & 1 \end{pmatrix} \begin{pmatrix} \varepsilon_{1t} \\ \varepsilon_{2t} \end{pmatrix}. \end{gathered} \tag{12.49}$$

Since (12.46) can be rewritten as

$$\mathbf{P}\cdot\begin{pmatrix} y_{1t} \\ y_{2t} \end{pmatrix} = \mathbf{P\Phi P}^{-1}\mathbf{P}\cdot\begin{pmatrix} y_{1,t-1} \\ y_{2,t-1} \end{pmatrix} + \mathbf{P}\cdot\begin{pmatrix} \varepsilon_{1t} \\ \varepsilon_{2t} \end{pmatrix}, \tag{12.50}$$

one obtains finally

$$\begin{pmatrix} z_{1t} \\ z_{2t} \end{pmatrix} = \begin{pmatrix} 0 & 0 \\ 0 & 1 \end{pmatrix}\begin{pmatrix} z_{1,t-1} \\ z_{2,t-1} \end{pmatrix} + \begin{pmatrix} u_{1t} \\ u_{2t} \end{pmatrix}, \tag{12.51}$$

i.e., for individual equations

$$\begin{aligned} z_{1t} &= u_{1t}, \\ z_{2t} &= z_{2,t-1} + u_{2t} \end{aligned} \tag{12.52}$$

(apparently, the time series $\{z_{1t}\}$ is directly a white noise and the time series $\{z_{2t}\}$ is a random walk I(1)).

We arrived at an evident paradox: both original time series $\{y_{1t}\}$ and $\{y_{2t}\}$ are nonstationary with nonstationarity caused according to (12.48) for each of time series always by one unit root (i.e., as if two unit roots figured seemingly in the system), while after the transformation only one time series $\{z_{2t}\}$ is nonstationary in the system (it confirms the previous conclusion that the system considered as a whole possesses only one unit root). This paradox may be explained just by the existence of the cointegration relation

$$y_{1t} + 0.5y_{2t} = z_{1t} = u_{1t} \tag{12.53}$$

(see (12.49) and (12.52)): both univariate time series $\{y_{1t}\}$ and $\{y_{2t}\}$ are nonstationary, while their linear combination (12.53) is stationary.

◇

The cointegration can be defined exactly by two equivalent ways (in both cases, we confine ourselves to a special case that appears in practice most frequently). Let $\{y_{1t}\}, \ldots, \{y_{mt}\}$ be nonstationary time series with the nonstationarity caused always just by one unit root of the corresponding autoregressive polynomial (particularly it may be $y_{1t} \sim \mathrm{I}(1), \ldots, y_{mt} \sim \mathrm{I}(1)$). Then the time series $\{y_{1t}\}, \ldots, \{y_{mt}\}$ are *cointegrated* if

(1) there exists their nontrivial (i.e., nonzero) linear combination that is stationary;
(2) equivalently: the corresponding model VAR of multivariate time series $(y_{1t}, \ldots, y_{mt})'$ has $m-r$ unit roots, where $r (0 < r < m)$ presents the number of cointegration relations of the type (12.53).

For instance in Example 12.7: (1) both time series $\{y_{1t}\}$ and $\{y_{2t}\}$ are of the type I(1) and their stationary linear combination exists (see (12.53)), or equivalently (2) there exists their bivariate model VAR with one unit root (see (12.46)), i.e.,

$m = 2, r = 1$. Therefore, $\{y_{1t}\}$ and $\{y_{2t}\}$ are cointegrated with the single cointegration relation (12.53) (if we ignore its scalar multiples).

Remark 12.9 More generally, one can define *cointegration of order* (d, b) ($b > 0$, $d > 0$), where $y_{1t} \sim \mathrm{I}(d), \ldots, y_{mt} \sim \mathrm{I}(d)$, and there exists a nontrivial linear combination of given time series that is of the type $\mathrm{I}(d-b)$. Then one writes $\{\mathbf{y}_t\} \sim \mathrm{CI}(d, b)$ (i.e., CoIntegrated).

◊

1. EC Model (ECM)

When analyzing a univariate nonstationary time series, then the usual recommendation consists in differencing this time series at first (e.g., if $\{y_t\} \sim \mathrm{I}(1)$, then one transfers it to the first differences $\{\Delta y_t\}$; see Sect. 6.4.2). However, when we deal with more nonstationary variables observed in time and are interested in their mutual link in time, then transferring to differences may be correct statistically, but the model constructed for differenced variables may not recover relations of long-run equilibrium among original (non-differenced) variables, which is an important feature just in the case of cointegration.

For example, let us consider two time series $\{x_t\}$ and $\{y_t\}$, which are both nonstationary of the type I(1). There is a conjecture that the time series $\{x_t\}$ influences $\{y_t\}$. Since both time series are nonstationary, this conjecture could be possibly investigated by means of the model

$$\Delta y_t = \gamma \cdot \Delta x_t + \varepsilon_t. \tag{12.54}$$

However, we are interested in the relation between variables x and y after its balancing to a long-run equilibrium, when the accruals of variables within time units are (nearly) zero. Therefore, the relation (12.54) has no informative value from this point of view. The situation is different if the time series $\{x_t\}$ and $\{y_t\}$ seem to be cointegrated in long-term horizon: then the model (12.54) can be *corrected* to the form

$$\Delta y_t = \gamma \cdot \Delta x_t + \alpha \cdot (y_{t-1} - \beta \cdot x_{t-1}) + \varepsilon_t \tag{12.55}$$

by including a correction term that is based on level (and not on differenced) values of given variables at previous time $t - 1$. The model (12.55) describes not only the short-run relation between accruals Δx_t and Δy_t, but simultaneously it guarantees corrections in the case when short-run changes of both variables deviate the levels of these variables from their long-run equilibrium state. Let us stress that the correction of changes of variables x and y from time $t - 1$ to time t is based on the correction term constructed at time $t - 1$, since its value at time t is not known yet when corrections for this time are constructed. If the time series $\{x_t\}$ and $\{y_t\}$ are really cointegrated and the correction term in (12.55) is chosen as the cointegration relation providing a stationary time series, then all terms in the model (12.55) are stationary. Thus, the situation when one uses both stationary and nonstationary terms simultaneously in one model is avoided (it could cause problems when constructing such a model).

As the terminology is concerned, the model of the type (12.55) is mostly called *EC model* (*error correction* or *equilibrium correction*). Sometimes one calls it also *VEC model* (*vector error correction*) to stress the VAR context. The terms of type $y_{t-1} - \beta \cdot x_{t-1}$ are called *error correction terms*. The parameters of type β describe long-run cointegration relations among variables and they are usually ordered to so-called *cointegration vectors* of type $(1, -\beta)'$. The parameters of type γ describe short-run cointegration relations among variables. Finally the parameters of type α control the rate of adjustment to the equilibrium state. Moreover, there may be intercepts or linear trends in the model (including the error correction terms), e.g.,

$$\Delta y_t = \gamma_1 + \gamma_2 \cdot \Delta x_t + \alpha \cdot (y_{t-1} - \beta_1 - \beta_2 \cdot x_{t-1}) + \varepsilon_t \tag{12.56}$$

(obviously, the parameter γ_1 means the intercept from the point of view of differential variables, but it means the deterministic linear trend from the point of view of level variables).

2. EC Model Formulated as VAR

The theory of EC models (but also their practical testing and constructing) is the most elaborate in the context of vector autoregressive models VAR. For instance, let us consider the bivariate model VAR(1)

$$\begin{pmatrix} y_{1t} \\ y_{2t} \end{pmatrix} = \mathbf{\Phi} \cdot \begin{pmatrix} y_{1,t-1} \\ y_{2,t-1} \end{pmatrix} + \begin{pmatrix} \varepsilon_{1t} \\ \varepsilon_{2t} \end{pmatrix} \tag{12.57}$$

which can be rewritten in the form

$$\begin{pmatrix} \Delta y_{1t} \\ \Delta y_{2t} \end{pmatrix} = \mathbf{\Pi} \cdot \begin{pmatrix} y_{1,t-1} \\ y_{2,t-1} \end{pmatrix} + \begin{pmatrix} \varepsilon_{1t} \\ \varepsilon_{2t} \end{pmatrix}, \quad \text{where} \quad \mathbf{\Pi} = \mathbf{\Phi} - \mathbf{I}. \tag{12.58}$$

The key role for classification of (12.58) as an EC model plays the rank $r = r(\mathbf{\Pi})$ of matrix $\mathbf{\Pi}$ (in our case of the type 2×2), which is closely related to the form of eigenvalues of matrix $\mathbf{\Phi}$ or equivalently to the form of roots of autoregressive polynomial $\mathbf{\Phi}(z) = \mathbf{I} - \mathbf{\Phi}{\cdot}z$:

1. $r(\mathbf{\Pi}) = 0$: In this case $\mathbf{\Pi} = \mathbf{0}$ so that according to (12.58) both time series $\{y_{1t}\}$ and $\{y_{2t}\}$ are nonstationary of the type I(1), and no cointegration relation exists between them.
2. $r(\mathbf{\Pi}) = 2$: In this case, $\mathbf{\Pi}$ has the full rank so that both eigenvalues of this matrix are nonzero, and hence no root of polynomial $\mathbf{\Phi}(z)$ is unit. Moreover, if we assume that both roots of $\mathbf{\Phi}(z)$ lie outside the unit circle in complex plane (i.e., both inverted roots of $\mathbf{\Phi}(z)$, which are simultaneously the eigenvalues of matrix $\mathbf{\Phi}$, lie inside the unit circle), then the model VAR model (12.57) is stationary, and it makes no sense to transfer it by differencing of the type (12.58) to the EC model.

3. $r(\mathbf{\Pi}) = 1$: This case is from the point of view of cointegration and EC methodology the most interesting. Just one of both eigenvalues of $\mathbf{\Pi}$ is nonzero, i.e., just one of both roots of polynomial $\mathbf{\Phi}(z)$ is unit. If we again assume that the remaining root of $\mathbf{\Phi}(z)$ lies outside the unit circle, then one can show that both univariate time series $\{\Delta y_{1t}\}$ and $\{\Delta y_{2t}\}$ are stationary (more specifically, each of non-differenced time series $\{y_{1t}\}$ and $\{y_{2t}\}$ is of the type ARIMA(1,1,1)). Moreover, (12.58) can be rewritten as

$$\begin{pmatrix} \Delta y_{1t} \\ \Delta y_{2t} \end{pmatrix} = \boldsymbol{\alpha}\boldsymbol{\beta}' \begin{pmatrix} y_{1,t-1} \\ y_{2,t-1} \end{pmatrix} + \begin{pmatrix} \varepsilon_{1t} \\ \varepsilon_{2t} \end{pmatrix} = \begin{pmatrix} \alpha_1 \\ \alpha_2 \end{pmatrix} (\beta_1 y_{1,t-1} + \beta_2 y_{2,t-1}) + \begin{pmatrix} \varepsilon_{1t} \\ \varepsilon_{2t} \end{pmatrix} \tag{12.59}$$

(the existence of bivariate column vectors $\boldsymbol{\alpha}$ and $\boldsymbol{\beta}$ follows from the unit rank $r(\mathbf{\Pi}) = 1$). The series $\{\Delta y_{1t}\}$ and $\{\Delta y_{2t}\}$ are stationary; hence also the time series $\{\beta_1 y_{1,t-1} + \beta_2 y_{2,t-1}\}$ must be stationary (otherwise, (12.59) would equate stationary and nonstationary terms) representing the cointegration relation between $\{y_{1t}\}$ and $\{y_{2t}\}$. The construction of EC model (12.59) serves as an example of application of *Granger's representation theorem* (see, e.g., Engle and Granger (1987)). Generally, it holds that the rank r of matrix $\mathbf{\Pi}$ equals the number of cointegration relations (if we ignore scalar multiples of these relations) in the corresponding EC model, while $m - r$ is the number of unit roots in the considered m-variate model VAR.

Remark 12.10 There is a direct analogy to DF test (Dickey and Fuller (1979); see (6.81)), where the validity of null hypothesis

$$\mathrm{H}_0 : \Delta y_t = \psi y_{t-1} + \varepsilon_t \quad \text{for } \psi = 0 \tag{12.60}$$

means the existence of unit root in the tested time series $\{y_t\}$ (obviously, $\psi = \varphi_1 - 1 = \pi$).

◇

Example 12.8 Let us consider two time series $\{y_{1t}\}$ and $\{y_{2t}\}$ from Example 12.7 modeled simultaneously as the bivariate model VAR(1) of the form (12.46). This model can be easily transferred to the form (12.59), namely

$$\begin{aligned} \begin{pmatrix} \Delta y_{1t} \\ \Delta y_{2t} \end{pmatrix} &= \begin{pmatrix} -0.5 & -0.25 \\ -1 & -0.5 \end{pmatrix} \begin{pmatrix} y_{1,t-1} \\ y_{2,t-1} \end{pmatrix} + \begin{pmatrix} \varepsilon_{1t} \\ \varepsilon_{2t} \end{pmatrix} \\ &= \begin{pmatrix} -0.5 \\ -1 \end{pmatrix} \cdot (1 \;\; 0.5) \cdot \begin{pmatrix} y_{1,t-1} \\ y_{2,t-1} \end{pmatrix} + \begin{pmatrix} \varepsilon_{1t} \\ \varepsilon_{2t} \end{pmatrix} \end{aligned} \tag{12.61}$$

(here the matrix $\mathbf{\Pi}$ has the rank $r = 1$). The corresponding cointegration vector (just one if we ignore its scalar multiples) is

$$\boldsymbol{\beta} = (1, 0.5)' \tag{12.62}$$

(see also (12.53)). The bivariate model has really just one unit root (see Example 12.7). The first and second components of vector $\boldsymbol{\alpha} = (-0.5, -1)'$ control the rate of adjustment to the equilibrium state in the first and second equations, respectively.

The motivation for the model (12.61) can be presented also in another way: The time series $\{y_{1t}\}$ and $\{y_{2t}\}$ considered individually ignoring their cointegration relations possess altogether two unit roots (one for each of them; see (12.48)), i.e., more than one unit root of the bivariate VAR model (12.46). It implies that by differencing each component to achieve stationarity we would overdifference the given system (so-called *overdifferencing*), which has some negative consequences: the invertibility may be damaged due to lagged terms of the type $\varepsilon_{1,t-1}$, estimation and prediction problems may appear, and the like. On the other hand, if we *correct* the model by subtracting the vector $\mathbf{y}_{t-1}$ from both sides of (12.46) (as it is the case of (12.61)), then the MA structure of this model does not change (of course, there must be a compensation allowed for maintenance of invertibility, namely the presence of the level variable $\mathbf{y}_{t-1}$ in the model).

◇

In general, the m-variate model VAR(p) of the form

$$\mathbf{y}_t = \boldsymbol{\Phi}_1 \mathbf{y}_{t-1} + \cdots + \boldsymbol{\Phi}_p \mathbf{y}_{t-p} + \boldsymbol{\varepsilon}_t \tag{12.63}$$

constructed for components $y_{1t} \sim \mathrm{I}(1), \ldots, y_{mt} \sim \mathrm{I}(1)$ has the following EC representation:

$$\Delta \mathbf{y}_t = \boldsymbol{\Pi} \mathbf{y}_{t-1} + \boldsymbol{\Gamma}_1 \Delta \mathbf{y}_{t-1} + \cdots + \boldsymbol{\Gamma}_{p-1} \Delta \mathbf{y}_{t-p+1} + \boldsymbol{\varepsilon}_t, \tag{12.64}$$

where

$$\begin{gathered} \boldsymbol{\Pi} = \boldsymbol{\Phi}_1 + \cdots + \boldsymbol{\Phi}_p - \mathbf{I} = -\boldsymbol{\Phi}(1), \\ \boldsymbol{\Gamma}_1 = -\boldsymbol{\Phi}_2 - \ldots - \boldsymbol{\Phi}_p, \ldots, \boldsymbol{\Gamma}_{p-2} = -\boldsymbol{\Phi}_{p-1} - \boldsymbol{\Phi}_p, \boldsymbol{\Gamma}_{p-1} = -\boldsymbol{\Phi}_p. \end{gathered} \tag{12.65}$$

If the rank r of matrix $\boldsymbol{\Pi}$ fulfills $0 < r < m$ (the boundary cases $r = 0$ and $r = m$ are discussed in the commentary below (12.58)), then (again according to Granger's theorem) there exist matrices $\boldsymbol{\alpha}$ and $\boldsymbol{\beta}$ (both are rectangular $m \times r$ and have the full column rank r) so that $\boldsymbol{\Pi} = \boldsymbol{\alpha}\boldsymbol{\beta}'$ and, moreover, each component of the vector $\boldsymbol{\beta}'\mathbf{y}_t$ can be modeled as I(0). In other words, there exists an EC representation of the form

$$\Delta \mathbf{y}_t = \boldsymbol{\alpha}\boldsymbol{\beta}' \mathbf{y}_{t-1} + \boldsymbol{\Gamma}_1 \Delta \mathbf{y}_{t-1} + \cdots + \boldsymbol{\Gamma}_{p-1} \Delta \mathbf{y}_{t-p+1} + \boldsymbol{\varepsilon}_t \tag{12.66}$$

with r cointegration relations (each column of the matrix $\boldsymbol{\beta}$ corresponds to one of these cointegration relations).

◇

Remark 12.11 The matrices $\boldsymbol{\alpha}$ and $\boldsymbol{\beta}$ in (12.66) are not constructed unambiguously: if $\boldsymbol{\gamma}$ is an arbitrary regular matrix of the type $r \times r$, then the matrices $\boldsymbol{\alpha\gamma}^{-1}$ and $\boldsymbol{\beta\gamma}'$ present further possible decomposition of the matrix $\boldsymbol{\Pi}$. Therefore, various adjustments are recommended in practice, e.g., normalizations of parametric matrices $\boldsymbol{\alpha}$ and $\boldsymbol{\beta}$ or application of prescribed a priori constraints.

It is also usual in practice that various exogenous variables figure on the right-hand side of the EC model (12.66): intercepts, polynomial trends, dummy variables (e.g., in seasonal models, the dummy variables should be centered in a suitable way; see Johansen (1995)) or exogenous variables from other systems. As it has significant consequences for tests and construction of cointegrated models, the corresponding software systems usually enable to classify various types of EC models:

- The cointegration relations contain intercepts, e.g., instead of (12.59) it holds

$$\begin{pmatrix} \Delta y_{1t} \\ \Delta y_{2t} \end{pmatrix} = \begin{pmatrix} \alpha_1 \\ \alpha_2 \end{pmatrix} (\delta_0 + \beta_1 y_{1,t-1} + \beta_2 y_{2,t-1}) + \begin{pmatrix} \varepsilon_{1t} \\ \varepsilon_{2t} \end{pmatrix}. \tag{12.67}$$

- The EC model contains intercepts outside of the cointegration relations, e.g., instead of (12.59) it holds

$$\begin{pmatrix} \Delta y_{1t} \\ \Delta y_{2t} \end{pmatrix} = \begin{pmatrix} \varphi_{10} \\ \varphi_{20} \end{pmatrix} + \begin{pmatrix} \alpha_1 \\ \alpha_2 \end{pmatrix} (\beta_1 y_{1,t-1} + \beta_2 y_{2,t-1}) + \begin{pmatrix} \varepsilon_{1t} \\ \varepsilon_{2t} \end{pmatrix} \tag{12.68}$$

  (as a matter of fact, the parameters φ_{10} and φ_{20} mean the intercepts from the point of view of differential variables, but they mean the deterministic linear trends from the point of view of level variables y_1 and y_2).

- The intercepts are included both in the cointegration relations and outside of them: in such a case, one must distinguish strictly among "inner" and "outer" intercepts (otherwise, there may be problems with the identification of these models).

- The cointegration relations contain linear trends, e.g., instead of (12.59) it holds

$$\begin{pmatrix} \Delta y_{1t} \\ \Delta y_{2t} \end{pmatrix} = \begin{pmatrix} \alpha_1 \\ \alpha_2 \end{pmatrix} (\delta_0 + \delta_1 t + \beta_1 y_{1,t-1} + \beta_2 y_{2,t-1}) + \begin{pmatrix} \varepsilon_{1t} \\ \varepsilon_{2t} \end{pmatrix}. \tag{12.69}$$

◇

3. Testing of Cointegration

Testing of cointegration should confirm that the tested VAR model contains just the given number r of cointegration relations, which is important information when constructing a suitable EC model. The cointegration is declared in the case when $r > 0$ (in particular, the case of stationary model VAR with $r = m$ can be also regarded as a special cointegrated state, in which each equation represents directly one of m cointegration relations).

Engle and Granger (1987) suggested a simple testing of cointegration among variables $y, x_2, \ldots, x_k$. Their *EG test* is based on a simple idea: if the given variables are cointegrated, then the OLS residuals $\widehat{\varepsilon}_t$ calculated by the least squares method in the model

$$y_t = \beta_1 + \beta_2 x_{t2} + \beta_3 x_{t3} + \cdots + \beta_k x_{tk} + \varepsilon_t \tag{12.70}$$

should be of the type I(0). Therefore, it is sufficient to modify DF test (Dickey and Fuller (1979); see (6.81)) and to test the null hypothesis

$$\mathrm{H}_0 : \Delta\widehat{\varepsilon}_t = \psi\widehat{\varepsilon}_{t-1} + u_t \quad \text{for } \psi = 0. \tag{12.71}$$

The only difference from the classical DF test is due to the fact that in (12.71) one applies the residuals estimated from a specific model so that the critical values of the classical DF test cannot be used. The relevant critical values which are more negative than the ones for DF test are tabulated by means of simulations in Engle and Granger (1987) and Engle and Yoo (1987). On the other hand, this approach has some drawbacks, namely:

- In the case of nonstationary variables, the OLS estimate of model (12.70) may not be reliable.
- In the case of more cointegration relations, one cannot decide which of them is in fact by means of (12.70) just estimated (what about receiving a "more intensive" cointegration relation after reordering given variables?).

Nowadays in practice the cointegration is mostly tested by means of *Johansen tests* (see Johansen (1991)). The method is based on ML estimate of so-called *canonical correlations* which measure the partial correlations among m-variate vectors $\Delta\mathbf{y}_t$ and $\mathbf{y}_{t-1}$ under fixed values of vectors $\Delta\mathbf{y}_{t-1}, \ldots, \Delta\mathbf{y}_{t-p+1}$ in the EC model (12.64). These canonical correlations are the square roots of eigenvalues $\lambda_1, \ldots, \lambda_m$ of a positive definite matrix that is closely related to the matrix $\mathbf{\Pi}$. The ML estimates $\widehat{\lambda}_1, \ldots, \widehat{\lambda}_m$ of these eigenvalues based on $\mathbf{y}_1, \ldots, \mathbf{y}_n$ fulfill

$$1 \geq \widehat{\lambda}_1 \geq \widehat{\lambda}_2 \geq \cdots \geq \widehat{\lambda}_m \geq 0. \tag{12.72}$$

In particular, the number of nonzero (positive) values $\lambda_1, \ldots, \lambda_r (\lambda_{r+1} = \lambda_{r+2} = \ldots = \lambda_m = 0)$ is equal to the rank r of matrix $\mathbf{\Pi}$, i.e., to the number of cointegration relations in the EC model (12.64).

Johansen tests are constructed to test that the eigenvalues λ are zero. As a matter of fact, these are LR tests (i.e., tests based on the concept of *likelihood ratio*), whose critical values are not generated by means of χ^2-distribution, but by means of simulations (see MacKinnon et al. (1999)). Moreover, two types of these tests are used in practice:

- Johansen test with statistics

$$\lambda_{trace}(r) = -n \sum_{i=r+1}^{m} \ln\left(1 - \widehat{\lambda}_i\right) \tag{12.73}$$

  is a compound test of the null hypothesis that the number of cointegration relations is at most r against the alternative that this number is higher than r. This test rejects the null hypothesis, if $\lambda_{trace}(r)$ is higher than the corresponding critical value (by intuition, the closer to the upper unit bound is the value $\widehat{\lambda}_i$, the higher is the value $-\ln\left(1 - \widehat{\lambda}_i\right)$, while for the lower bound $\widehat{\lambda}_i = 0$ one has $-\ln\left(1 - \widehat{\lambda}_i\right) = 0$). It is tested gradually for $r = 0, 1, \ldots, m - 1$.

- Johansen test with statistics

$$\lambda_{max}(r) = -n \ln\left(1 - \widehat{\lambda}_{r+1}\right) \tag{12.74}$$

  is a test of the null hypothesis that the number of cointegration relations is r against the alternative that it is $r + 1$. This test rejects the null hypothesis, if $\lambda_{max}(r)$ is higher than the corresponding critical value. It is tested gradually for $r = 0, 1, \ldots, m - 1$.

Example 12.9 Table 12.16 and Fig. 12.10 present the monthly yields to maturity for 3-month T-bills (*TB*3 in % p.a.) and for corporate bonds of the highest rating AAA by S&P (*AAA* in % p.a.) during 10-year period 1985–1994 in the USA. The first differences of these time series have been analyzed as the stationary model VAR(2) in Example 12.3. Here we shall test an eventual cointegration of these time series in their non-differenced (i.e., nonstationary) form, since cointegration is a typical relationship among important (log) returns on financial markets (Fig. 12.10 reflects similar courses of time series *TB*3 and *AAA* as well).

Table 12.17 presents the results of Johansen tests (12.73) and (12.74) by means of EViews if one models the time series *TB*3 and *AAA* as VAR(3) (i.e., the model (12.64) for the first differences *DTB*3 and *DAAA* has the order $p - 1 = 2$). Both the test with statistics $\lambda_{trace}(r)$ and the test with statistics $\lambda_{max}(r)$ indicate with significance level of 5% just one cointegration relation ($r = 1$).

◇

Table 12.16 Monthly data in Example 12.9 (the monthly yields to maturity for 3-month T-bills and corporate bonds AAA in the USA in % p.a.)

Obs	TB3	AAA	Obs	TB3	AAA	Obs	TB3	AAA
1985M01	7.76	12.08	1988M05	6.27	9.90	1991M09	5.25	8.61
1985M02	8.22	12.13	1988M06	6.50	9.86	1991M10	5.03	8.55
1985M03	8.57	12.56	1988M07	6.73	9.96	1991M11	4.60	8.48
1985M04	8.00	12.23	1988M08	7.02	10.11	1991M12	4.12	8.31
1985M05	7.56	11.72	1988M09	7.23	9.82	1992M01	3.84	8.20
1985M06	7.01	10.94	1988M10	7.34	9.51	1992M02	3.84	8.29
1985M07	7.05	10.97	1988M11	7.68	9.45	1992M03	4.05	8.35
1985M08	7.18	11.05	1988M12	8.09	9.57	1992M04	3.81	8.33
1985M09	7.08	11.07	1989M01	8.29	9.62	1992M05	3.66	8.28
1985M10	7.17	11.02	1989M02	8.48	9.63	1992M06	3.70	8.22
1985M11	7.20	10.55	1989M03	8.83	9.80	1992M07	3.28	8.07
1985M12	7.07	10.16	1989M04	8.70	9.79	1992M08	3.14	7.95
1986M01	7.04	10.05	1989M05	8.40	9.57	1992M09	2.97	7.92
1986M02	7.03	9.67	1989M06	8.22	9.10	1992M10	2.84	7.99
1986M03	6.59	9.00	1989M07	7.92	8.93	1992M11	3.14	8.10
1986M04	6.06	8.79	1989M08	7.91	8.96	1992M12	3.25	7.98
1986M05	6.12	9.09	1989M09	7.72	9.01	1993M01	3.06	7.91
1986M06	6.21	9.13	1989M10	7.59	8.92	1993M02	2.95	7.71
1986M07	5.84	8.88	1989M11	7.67	8.89	1993M03	2.97	7.58
1986M08	5.57	8.72	1989M12	7.64	8.86	1993M04	2.89	7.46
1986M09	5.19	8.89	1990M01	7.64	8.99	1993M05	2.96	7.43
1986M10	5.18	8.86	1990M02	7.76	9.22	1993M06	3.10	7.33
1986M11	5.35	8.68	1990M03	7.87	9.37	1993M07	3.05	7.17
1986M12	5.49	8.49	1990M04	7.78	9.46	1993M08	3.05	6.85
1987M01	5.45	8.36	1990M05	7.78	9.47	1993M09	2.96	6.66
1987M02	5.59	8.38	1990M06	7.74	9.26	1993M10	3.04	6.67
1987M03	5.56	8.36	1990M07	7.66	9.24	1993M11	3.12	6.93
1987M04	5.76	8.85	1990M08	7.44	9.41	1993M12	3.08	6.93
1987M05	5.75	9.33	1990M09	7.38	9.56	1994M01	3.02	6.92
1987M06	5.69	9.32	1990M10	7.19	9.53	1994M02	3.21	7.08
1987M07	5.78	9.42	1990M11	7.07	9.30	1994M03	3.52	7.48
1987M08	6.00	9.67	1990M12	6.81	9.05	1994M04	3.74	7.88
1987M09	6.32	10.18	1991M01	6.30	9.04	1994M05	4.19	7.99
1987M10	6.40	10.52	1991M02	5.95	8.83	1994M06	4.18	7.97
1987M11	5.81	10.01	1991M03	5.91	8.93	1994M07	4.39	8.11
1987M12	5.80	10.11	1991M04	5.67	8.86	1994M08	4.50	8.07
1988M01	5.90	9.88	1991M05	5.51	8.86	1994M09	4.64	8.34
1988M02	5.69	9.40	1991M06	5.60	9.01	1994M10	4.96	8.57
1988M03	5.69	9.39	1991M07	5.58	9.00	1994M11	5.25	8.68
1988M04	5.92	9.67	1991M08	5.39	8.75	1994M12	5.64	8.46

Source: FRED (Federal Reserve Bank of St. Louis)

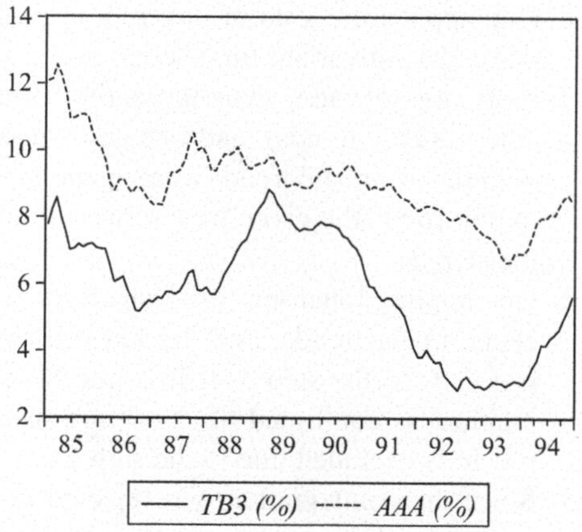

Fig. 12.10 Monthly data in Example 12.9 (the monthly yields to maturity for 3-month T-bills and corporate bonds AAA in the USA in % p.a.)

Table 12.17 Johansen tests of cointegration from Example 12.9 by means of EViews (*TB3 and AAA*)

Sample (adjusted): 1985M04 1994M12				
Included observations: 117 after adjustments				
Trend assumption: Linear deterministic trend				
Series: TB3 AAA				
Lags interval (in first differences): 1 to 2				
Unrestricted cointegration rank test (Trace)				
Hypothesized		Trace	0.05	
No. of CE(s)	Eigenvalue	Statistic	Critical value	Prob.**
None*	0.132054	18.61642	15.49471	0.0164
At most 1	0.017337	2.046241	3.841466	0.1526
Trace test indicates 1 cointegrating eqn(s) at the 0.05 level				

Unrestricted cointegration rank test (Maximum eigenvalue)				
Hypothesized		Max-Eigen	0.05	
No. of CE(s)	Eigenvalue	Statistic	Critical value	Prob.**
None*	0.132054	16.57018	14.26460	0.0212
At most 1	0.017337	2.046241	3.841466	0.1526

Max-eigenvalue test indicates 1 cointegrating eqn(s) at the 0.05 level
*Rejection of the hypothesis at the 0.05 level
**MacKinnon–Haug–Michelis (1999) *p*-values

4. Construction of EC Model

The construction of EC models can be described in the following steps (for simplicity, we assume that the m-variate time series $\mathbf{y}_1, \ldots, \mathbf{y}_n$ is either stationary or nonstationary of the type I(1), i.e., it can be stationarized by transferring it to the time series of first differences):

1. One applies the tests of unit root (e.g., DF and ADF tests from Sect. 6.4.1) for particular univariate time series $\{y_{1t}\}, \ldots, \{y_{mt}\}$. If the null hypotheses of unit roots are rejected, then these time series are stationary (except for possible deterministic trends), and one constructs for $\mathbf{y}_1, \ldots, \mathbf{y}_n$ a model VAR (possibly with deterministic trends as exogenous variables; see Sect. 12.2). Otherwise due to unit roots, the given time series contain stochastic trends, and one proceeds to the step 2.
2. One applies Johansen (or other) tests of cointegration (possibly including intercepts, linear trends, and the like). If the cointegration is rejected ($r = 0$), one proceeds to the step 3. If it is not the case and the existence of r cointegration relations is confirmed ($0 < r < m$), then one proceeds to the step 4 (the case of $r = m$ is excluded due to the step 1).
3. Since the cointegration was rejected in the previous step of the algorithm, one constructs the corresponding model VAR for the stationary time series $\Delta\mathbf{y}_1, \ldots, \Delta\mathbf{y}_n$.
4. Since there exist r cointegration relations ($0 < r < m$), the step 3 is ignored, and one constructs the corresponding EC model (12.66) for the original time series $\mathbf{y}_1, \ldots, \mathbf{y}_n$. The estimation procedure can combine LM method and OLS method (see Johansen (1995)). In particular, the maximal value of logarithmic likelihood is

$$c - \frac{n-p}{2} \sum_{i=1}^{r} \ln\left(1 - \widehat{\lambda}_i\right), \tag{12.75}$$

where c is a constant (independent of r) and $\widehat{\lambda}_1, \ldots, \widehat{\lambda}_r$ are positive numbers as in (12.72). This estimation procedure can be supplemented by a priori constraints for the parameters in matrices $\boldsymbol{\alpha}$ and $\boldsymbol{\beta}$ in (12.66).

Example 12.10 In Table 12.18, the time series $\{TB3_t\}$ and $\{AAA_t\}$ from Examples 12.1 and 12.3 are estimated by means of EViews (the first differences of these time series are denoted as $\{DTB3_t\}$ and $\{DAAA_t\}$). Here the order of the VAR model (12.66) is $p - 1 = 2$ (the order of the original model before differencing is $p = 3$), and the intercepts are explicitly estimated both in the model (12.66) and in its cointegration relation. The estimated EC model from Table 12.18 has the explicit form

$$\begin{pmatrix} DTB3_t \\ DAAA_t \end{pmatrix} = \begin{pmatrix} -0.01 \\ -0.03 \end{pmatrix} + \begin{pmatrix} 0.01 \\ 0.03 \end{pmatrix} (23.06 + TB3_{t-1} - 3.21 AAA_{t-1}) +$$
$$+ \begin{pmatrix} 0.44 & 0.11 \\ -0.10 & 0.53 \end{pmatrix} \begin{pmatrix} DTB3_{t-1} \\ DAAA_{t-1} \end{pmatrix} + \begin{pmatrix} 0.04 & -0.16 \\ 0.10 & -0.27 \end{pmatrix} \begin{pmatrix} DTB3_{t-2} \\ DAAA_{t-2} \end{pmatrix} + \begin{pmatrix} \varepsilon_{1t} \\ \varepsilon_{2t} \end{pmatrix}.$$

The cointegration vector $(1, -3.21)'$ is normed by means of an a priori constraint in order of its first component being unit.

◇

Table 12.18 Construction of EC model from Example 12.10 by means of EViews (*TB3 and AAA*)

Vector error correction estimates		
Sample (adjusted): 1985M04 1994M12		
Included observations: 117 after adjustments		
Cointegrating Eq	CointEq1	
TB3(−1)	1.000000	
AAA(−1)	−3.212049	
	(0.53460)	
C	23.06448	
Error correction:	D(TB3)	D(AAA)
CointEq1	0.014189	0.029656
	(0.00774)	(0.00725)
D(TB3(−1))	0.438843	−0.097404
	(0.11419)	(0.10687)
D(TB3(−2))	0.043494	0.103530
	(0.11166)	(0.10451)
D(AAA(−1))	0.110342	0.534480
	(0.11089)	(0.10378)
D(AAA(−2))	−0.161987	−0.267855
	(0.11037)	(0.10330)
C	−0.014468	−0.027313
	(0.01882)	(0.01761)
S.E. equation	0.200969	0.188087

12.6 Exercises

Exercise 12.1

Repeat Johansen tests of cointegration for time series $\{TB3_t\}$ and $\{AAA_t\}$ from Example 12.9, but only for the 5-year period 1990_1994 (*hint: both the test with* $\lambda_{trace}(r)$ *and* $\lambda_{max}(r)$ *indicate with significance level of 5% no cointegration relation* ($r = 1$)).

Chapter 13
Multivariate Volatility Modeling

The models of volatility in Chap. 8 are univariate, i.e., they model the volatility quite independently on other time series. It may be a drawback (particularly in finance) since

- The effect of volatility *spillover* among various financial markets or among various assets within the same financial market is a typical phenomenon in finance.
- Correlations among particular components play a key role when constructing and managing (diversified) investment portfolios.

For instance, let us consider so-called *dynamic hedging* applied frequently when reducing investment risk. The dynamic hedging is mostly realized in such a way that the investor simultaneously enters opposite positions on markets with a mutually inverse behavior, e.g., on spot and future markets (in this context, the position characterized as a purchase operation is denoted as *long*, and similarly, the position characterized as a sale operation is denoted as *short*). The pragmatic investors suppose that potential losses in one market may be balanced by profits from another market that behaves just inversely. Therefore, they follow so-called *hedging ratio* h that presents the number of units of future contracts per one unit of (spot) assets and should be optimal in terms of the risk reduction. For example, in the case of so-called short hedge of an investor, which means a long position in assets and simultaneously a short position in futures, the value of total investor's position (during hedging till the maturity of futures) changes by $\Delta S - h{\cdot}\Delta F$, where ΔS and ΔF are the corresponding change in spot and future price, respectively. One can easily show that the optimal hedging ratio minimizing var($\Delta S - h{\cdot}\Delta F$) is

$$h_t = \rho_t \frac{\sigma_{st}}{\sigma_{ft}},$$

where σ_{st} is the risk of spot market (i.e., the standard deviation of spot prices), σ_{ft} is the risk of future market (i.e., the standard deviation of future prices), and finally ρ_t is

T. Cipra, *Time Series in Economics and Finance*,
https://doi.org/10.1007/978-3-030-46347-2_13

the correlation coefficient between both time series of prices (the time index t emphasizes the fact that the described hedging is dynamic with corrections of hedging ratio realized in particular times). Obviously, in order to calculate $\{h_t\}$ one makes use of the time series $\{\rho_t\}$ that records the dynamics of correlation between both time series.

In general, the multivariate volatility modeling plays an important role for the risk control (e.g., for portfolio investment, but also for internal models in the framework of regulatory methodologies *Basel III* of capital adequacy in banks or *Solvency II* in insurance companies including commercial products of the type *RiskMetrics*, and the like).

13.1 Multivariate Models EWMA

The modeling of multivariate volatility may be approached in an analogous way as in the univariate case. However, in addition to the univariate modeling used in Chap. 8, now one must model also the correlatedness among time series denoted as *mutual volatility* (or also *covolatility*).

Commercially (e.g., in the commercial system RiskMetrics; see Sect. 8.3), one frequently exploits the *multivariate models EWMA* (i.e., the *exponentially weighted moving averages* from Sect. 8.3.1), where the covariances or "covolatilities" among particular time series $\{y_{it}\}$ and $\{y_{jt}\}$ are calculated recursively as

$$\begin{aligned} \widehat{\sigma}_{ij,t} &= (1-\lambda)\sum_{k=0}^{\infty}\lambda^k\left(y_{i,t-1-k}-\overline{y}_i\right)\left(y_{j,t-1-k}-\overline{y}_j\right) \\ &= (1-\lambda)\left(y_{i,t-1}-\overline{y}_i\right)\left(y_{j,t-1}-\overline{y}_j\right)+\lambda\widehat{\sigma}_{ij,t-1}. \end{aligned} \tag{13.1}$$

Here the estimated covariance $\widehat{\sigma}_{ij,t}$ presents the mutual (co)volatility prediction from time $t-1$, $\overline{y}_i$ and $\overline{y}_j$ are mean levels of given time series $\{y_{it}\}$ and $\{y_{jt}\}$, and λ $(0<\lambda<1)$ is a discount constant chosen in advance. In the case of time series with levels close to zero (which is mostly the case of log returns) one usually applies zero mean returns in (13.1) so that the problem of their suitable choice is removed. Some authors (e.g., Fleming et al. (2003)) summarize multivariate EWMA relations to the matrix form

$$\widehat{\boldsymbol{\Sigma}}_t = \alpha\exp(-\alpha)(\mathbf{y}_{t-1}-\overline{\mathbf{y}})(\mathbf{y}_{t-1}-\overline{\mathbf{y}})' + \exp(-\alpha)\widehat{\boldsymbol{\Sigma}}_{t-1} \tag{13.2}$$

using a decay rate α.

Remark 13.1 For the sake of completeness, the historical volatility approach (8.17) can be also generalized to the following multivariate form:

$$\widehat{\boldsymbol{\Sigma}}_t = \frac{\sum_{\tau= t-k}^{t-1} (\mathbf{y}_\tau - \overline{\mathbf{y}})(\mathbf{y}_\tau - \overline{\mathbf{y}})'}{k-1} \tag{13.3}$$

for a suitable length of sample period k. The values (13.3) are also denoted as *multivariate SMA* of length k (*simple moving average*; see, e.g., Chiriac and Voev (2011)).

◇

13.2 Implied Mutual Volatility

It is a direct analogy of the univariate implied volatility from Sect. 8.3.2. The mutual (i.e., multivariate) volatility may be implied, e.g., by means of currency option. For instance, if we deal with implied mutual volatility between currency rates USD/EUR and USD/CNY, then it can be calculated (in time) by means of relation

$$\widehat{\sigma}_{USD/EUR,USD/CNY} = \frac{\widehat{\sigma}^2_{USD/EUR} + \widehat{\sigma}^2_{USD/CNY} - \widehat{\sigma}^2_{EUR/CNY}}{2}, \tag{13.4}$$

where $\widehat{\sigma}^2_{\;USD/EUR}$ is the implied volatility of currency rate return USD/EUR and similarly for $\widehat{\sigma}^2_{\;USD/CNY}$ and $\widehat{\sigma}^2_{\;EUR/CNY}$ (these implied volatilities are constructed by means of quoted option premiums for returns of particular currency rates; see Sect. 8.3.2).

13.3 Multivariate GARCH Models

It is not surprising that one tries to construct multivariate generalization of univariate models GARCH extending the principle of univariate conditional heteroscedasticity from Sect. 8.3 to mutual volatility. For instance, in a very simple bivariate GARCH (1,1) model, one can exploit the following (vector) volatility equation:

$$\begin{pmatrix} \sigma_{11,t} \\ \sigma_{22,t} \end{pmatrix} = \begin{pmatrix} \alpha_{10} \\ \alpha_{20} \end{pmatrix} + \begin{pmatrix} \alpha_{11} & \alpha_{12} \\ \alpha_{21} & \alpha_{22} \end{pmatrix} \begin{pmatrix} e^2_{1,t-1} \\ e^2_{2,t-1} \end{pmatrix} + \begin{pmatrix} \beta_{11} & \beta_{12} \\ \beta_{21} & \beta_{22} \end{pmatrix} \begin{pmatrix} \sigma_{11,t-1} \\ \sigma_{22,t-1} \end{pmatrix}, \tag{13.5}$$

where, e.g., $\sigma_{11,t}$ denotes the volatility in the first component $\{y_{1t}\}$. Apparently, it is a direct generalization of the volatility equation from the univariate model (8.59)

$$\sigma_t^2 = \alpha_0 + \alpha_1 e_{t-1}^2 + \beta_1 \sigma_{t-1}^2. \tag{13.6}$$

Moreover, in practice one often combines the volatility equation with the mean equation (e.g., the mean equation in Example 13.1 is based on the VAR methodology).

Example 13.1 Tsay (2002) constructed for monthly log returns of stocks IBM (time series $\{r_{t1}\}$) and index S&P 500 (time series $\{r_{t2}\}$) during period 1926–1999 the bivariate model of the form

$$r_{1t} = 1.351 + 0.072 r_{1,t-1} + 0.055 r_{1,t-2} - 0.119 r_{2,t-2} + e_{1t},$$
$$r_{2t} = 0.703 + e_{2t},$$
$$\begin{pmatrix} \sigma_{11,t} \\ \sigma_{22,t} \end{pmatrix} = \begin{pmatrix} 2.98 \\ 2.09 \end{pmatrix} + \begin{pmatrix} 0.079 & \cdot \\ 0.042 & 0.045 \end{pmatrix} \begin{pmatrix} e_{1,t-1}^2 \\ e_{2,t-1}^2 \end{pmatrix} + \begin{pmatrix} 0.873 & -0.031 \\ -0.066 & 0.913 \end{pmatrix} \times \begin{pmatrix} \sigma_{11,t-1} \\ \sigma_{22,t-1} \end{pmatrix}.$$

This model obviously combines the bivariate model VAR(2) from Sect. 12.2 representing the mean equation and the bivariate GARCH(1,1) model (13.5) representing the volatility equation (insignificant parameters are omitted). One assumes the constant correlation between $\{e_{t1}\}$ and $\{e_{t2}\}$ estimated as 0.614.

◇

In general, one can extend the univariate principle of conditional heteroscedasticity in (8.16) to the *m*-variate case as

$$\mathbf{y}_t = \boldsymbol{\Sigma}_t^{1/2} \cdot \boldsymbol{\varepsilon}_t, \tag{13.7}$$

where $\boldsymbol{\varepsilon}_t$ are *iid* *m*-variate random vectors with zero mean vector and unit covariance matrix, i.e., $\{\boldsymbol{\varepsilon}_t\}$ is an *iid* multivariate white noise with

$$\mathrm{E}(\boldsymbol{\varepsilon}_t) = \mathbf{0}, \qquad \mathrm{var}(\boldsymbol{\varepsilon}_t) = \mathrm{E}\left(\boldsymbol{\varepsilon}_t \boldsymbol{\varepsilon}_t'\right) = \mathbf{I} \tag{13.8}$$

and $\boldsymbol{\Sigma}_t^{1/2}$ is the square root matrix of the conditional covariance matrix $\boldsymbol{\Sigma}_t$ expressed in time t as a suitable function of the information known till time $t-1$. The matrix $\boldsymbol{\Sigma}_t$ (or also $\mathbf{H}_t$ by some authors) may be looked upon as *(co)volatility matrix* since its diagonal elements are univariate volatilities of particular univariate components of $\{\mathbf{y}_t\}$ and the elements outside the main diagonal are mutual volatilities (or covolatilities) of these components (here the logical denotation is $\sigma_{11,t} = \sigma_{1t}^2$, and the like). Moreover, it is necessary to respect some practical aspects which are important for construction of such models using real data:

- One must guarantee that the matrix $\boldsymbol{\Sigma}_t$ produced by the model at time t is positive definite (and therefore also symmetric).

- The model parametrization must be parsimonious (otherwise for higher dimensions $m \geq 3$, there can appear serious problems to identify and estimate such models).

The models, which are based on the principle (13.7) (and in addition are acceptable from the practical point of view), can be roughly classified to (1) *models of conditional covariance matrix*, (2) *models of conditional variances and correlations*, and (3) *factor models*. We shall show important representatives of these classes now, even though the development of new approaches in this context is dramatic (see also Bauwens et al. (2006), Clements et al. (2012), Kroner and Ng (1998), McNeil et al. (2005), Silvennoinen and Teräsvirta (2009), Tsay (2002), and others). Moreover, the models can be completed by leverage effect arrangements, etc.

13.3.1 Models of Conditional Covariance Matrix

These models attempt to model the conditional covariance matrix (i.e., volatility matrix) directly using similar model instruments as the univariate GARCH models:

1. VEC Model

Vector model GARCH denoted simply as *VEC* (suggested by Bollerslev et al. (1988)) models the volatility matrix $\mathbf{\Sigma}_t$ in (13.7) as

$$\text{vech}(\mathbf{\Sigma}_t) = \boldsymbol{a}_0^* + \sum_{i=1}^{r} \mathbf{A}_i^* \text{vech}\left(\mathbf{y}_{t-i}\mathbf{y}_{t-i}'\right) + \sum_{j=1}^{s} \mathbf{B}_j^* \text{vech}\left(\mathbf{\Sigma}_{t-j}\right), \tag{13.9}$$

where $\mathbf{a}_0{}^*$ is a vector and $\mathbf{A}_i{}^*$ and $\mathbf{B}_j{}^*$ are square matrices of parameters. The symbol vech($\cdot$) is so-called *vector half operator* which stacks the lower triangular part of a $m \times m$ matrix as a $m(m + 1)/2 \times 1$ vector. For instance, the bivariate model GARCH (1,1) of this type has

$$\text{vech}(\mathbf{\Sigma}_t) = \begin{pmatrix} \sigma_{11,t} \\ \sigma_{21,t} \\ \sigma_{22,t} \end{pmatrix} = \begin{pmatrix} \sigma_{1t}^2 \\ \sigma_{21,t} \\ \sigma_{2t}^2 \end{pmatrix}, \quad \text{vech}\left(\mathbf{y}_{t-1}\mathbf{y}_{t-1}'\right) = \begin{pmatrix} y_{1,t-1}^2 \\ y_{1,t-1}y_{2,t-1} \\ y_{2,t-1}^2 \end{pmatrix} \tag{13.10}$$

(the matrices $\mathbf{A}_1{}^*$ and $\mathbf{B}_1{}^*$ must be of the type 3×3 so that the corresponding volatility equation contains 21 unknown parameters). In the general VEC model, each element of $\mathbf{\Sigma}_t$ is a linear function of the lagged squared and cross-product observations and lagged values of the elements of $\mathbf{\Sigma}_t$. The total number of unknown parameters $[1 + (r + s)m(m + 1)/2]m(m + 1)/2$ increases with growing dimensions to an intolerable level so that for higher m the model VEC cannot be recommended in routine practice.

2. DVEC Model

Diagonal vector model DVEC (see Bollerslev et al. (1988)) is a special case of VEC model (13.9) with diagonal parametric matrices $\mathbf{A}_i^*$ and $\mathbf{B}_j^*$. It reduces not only the number of parameters to be estimated but also enables to rewrite the model (13.9) to a more transparent form

$$\boldsymbol{\Sigma}_t = \mathrm{A}_0 + \sum_{i=1}^{r} \mathrm{A}_i \circ \left(\mathbf{y}_{t-i}\mathbf{y}'_{t-i}\right) + \sum_{j=1}^{s} \mathbf{B}_j \circ \boldsymbol{\Sigma}_{t-j}, \tag{13.11}$$

where the parametric matrix $\mathbf{A}_0$ is $m \times m$ symmetric with positive elements on the main diagonal, the parametric matrices $\mathbf{A}_i$ and $\mathbf{B}_j$ are $m \times m$ symmetric with nonnegative elements on the main diagonal, and the symbol "$\circ$" denotes *Hadamar* (i.e., elementwise) *product* of two matching matrices. Then, e.g., the volatility matrix (13.11) for the bivariate model GARCH(1,1) of type DVEC can be written by means of three scalar (co)volatility equations:

$$\begin{aligned} \sigma_{11,t} &= \alpha_{11,0} + \alpha_{11,1} y_{1,t-1}^2 + \beta_{11,1}\sigma_{11,t-1}, \\ \sigma_{12,t} &= \alpha_{12,0} + \alpha_{12,1} y_{1,t-1} y_{2,t-1} + \beta_{12,1}\sigma_{12,t-1}, \\ \sigma_{22,t} &= \alpha_{22,0} + \alpha_{22,1} y_{2,t-1}^2 + \beta_{22,1}\sigma_{22,t-1}. \end{aligned} \tag{13.12}$$

The volatilities of both components of $\{\mathbf{y}_t\}$ are modeled in the same way as in the univariate GARCH(1,1) models of Sect. 8.3.5. The (scalar) equation of mutual volatility has a similar structure, but now with the product of lagged values $y_{1,t-1} \cdot y_{2,t-1}$. Unfortunately, the natural property that the volatility of a component is impacted by higher absolute values of another component in past time is not guaranteed here.

The sufficient condition of positive definiteness of matrix $\boldsymbol{\Sigma}_t$ requires that the matrix $\mathbf{A}_0$ is positive definite and the matrices $\mathbf{A}_i$ and $\mathbf{B}_j$ positive semidefinite. It can be achieved by several alternative parametrizations:

(*i*) $\mathbf{A}_0 = \mathbf{A}_0^{1/2}(\mathbf{A}_0^{1/2})'$, $\mathbf{A}_i = \mathbf{A}_i^{1/2}(\mathbf{A}_i^{1/2})'$, $\mathbf{B}_j = \mathbf{B}_j^{1/2}(\mathbf{B}_j^{1/2})'$, where matrices $\mathbf{A}_0^{1/2}$, $\mathbf{A}_i^{1/2}$, and $\mathbf{B}_j^{1/2}$ are lower triangular matrices of Cholesky decomposition (see (12.16)).

(*ii*) $\mathbf{A}_0 = \mathbf{A}_0^{1/2}(\mathbf{A}_0^{1/2})'$, $\mathbf{A}_i = \mathbf{a}_i\, \mathbf{a}_i'$, $\mathbf{B}_j = \mathbf{b}_j\, \mathbf{b}_j'$, where $\mathbf{a}_i$ and $\mathbf{b}_j$ are m-variate parametric vectors.

(*iii*) $\mathbf{A}_0 = \mathbf{A}_0^{1/2}(\mathbf{A}_0^{1/2})'$, $\mathbf{A}_i = \alpha_i\, \mathbf{I}_m$, $\boldsymbol{B}_j = \beta_j\, \mathbf{I}_m$, where α_i and β_j are positive scalar parameters and $\mathbf{I}_m$ denotes the unit matrix of dimension m (the multivariate model EWMA (13.1) can be looked upon as a special case of it). In particular, this case can reduce the number of parameters substantially when the dimension m is higher.

3. BEKK Model

Model BEKK denoted by initials of its authors (Baba, Engle, Kraft, Kroner; see Engle and Kroner (1995)) guarantees (automatically in comparison with the previous models) the positive definiteness of volatility matrix $\boldsymbol{\Sigma}_t$. Namely, this matrix is modeled as

$$\boldsymbol{\Sigma}_t = \mathbf{A}_0 + \sum_{i=1}^{r} \mathbf{A}_i' \mathbf{y}_{t-i} \mathbf{y}_{t-i}' \mathbf{A}_i + \sum_{j=1}^{s} \mathbf{B}_j' \boldsymbol{\Sigma}_{t-j} \mathbf{B}_j, \tag{13.13}$$

where all parametric matrices are quite general of dimension $m \times m$ and only $\mathbf{A}_0$ is required to be symmetric and positive definite (it can be achieved by a suitable parametrization of $\mathbf{A}_0$ as in the case of model DVEC; see above). For instance, the volatility matrix (13.13) for the bivariate model GARCH(1,1) of type BEKK can be written by means of three scalar (co)volatility equations:

$$\sigma_{11,t} = \alpha_{11,0} + \alpha_{11,1}^2 y_{1,t-1}^2 + 2\alpha_{11,1}\alpha_{21,1} y_{1,t-1} y_{2,t-1} + \alpha_{21,1}^2 y_{2,t-1}^2 + \beta_{11,1}^2 \sigma_{11,t-1} + \\ +2\beta_{11,1}\beta_{21,1}\sigma_{12,t-1} + \beta_{21,1}^2 \sigma_{22,t-1}, \tag{13.14}$$

$$\sigma_{12,t} = \alpha_{12,0} + \alpha_{11,1}\alpha_{12,1} y_{1,t-1}^2 + (\alpha_{11,1}\alpha_{22,1} + \alpha_{12,1}\alpha_{21,1}) y_{1,t-1} y_{2,t-1} + \alpha_{21,1}\alpha_{22,1} y_{2,t-1}^2 + \\ +\beta_{11,1}\beta_{12,1}\sigma_{11,t-1} + (\beta_{11,1}\beta_{22,1} + \beta_{12,1}\beta_{21,1})\sigma_{12,t-1} + \beta_{21,1}\beta_{22,1}\sigma_{22,t-1}, \tag{13.15}$$

$$\sigma_{22,t} = \alpha_{22,0} + \alpha_{12,1}^2 y_{1,t-1}^2 + 2\alpha_{12,1}\alpha_{22,1} y_{1,t-1} y_{2,t-1} + \alpha_{22,1}^2 y_{2,t-1}^2 + \beta_{12,1}^2 \sigma_{11,t-1} + \\ +2\beta_{12,1}\beta_{22,1}\sigma_{12,t-1} + \beta_{22,1}^2 \sigma_{22,t-1}. \tag{13.16}$$

The model BEKK has finally the desirable property, namely that the volatility of any component may be impacted by higher absolute values of other component in past time (e.g., the higher absolute value $y_{2,\ t-1}$ in (13.14) raises volatility $\sigma_{11,\ t}$). If no interactions among volatilities occur, then it should be $\alpha_{21,1} = \beta_{21,1} = 0$ in Eq. (13.14) and $\alpha_{12,1} = \beta_{12,1} = 0$ in Eq. (13.16) so that only the parameters of the type $a_{ii,k}$ impact the mutual volatilities, namely the parameters $\alpha_{11,1}, \alpha_{22,1}, \beta_{11,1}, \beta_{22,1}$ in Eq. (13.15) (however, it does not mean that the model BEKK is transferred to the model DVEC in such a case).

Particular parametrizations reducing the total number of parameters are, e.g.:

(*i*) *Diagonal* BEKK models with diagonal matrices $\mathbf{A}_i$ and $\mathbf{B}_j$.

(*ii*) *Scalar* BEKK models with matrices $\mathbf{A}_i = \alpha_i\,\mathbf{I}_m$ and $\mathbf{B}_j = \beta_j\,\mathbf{I}_m$, where α_i and β_j are positive scalar parameters.

13.3.2 Models of Conditional Variances and Correlations

The models of this type primarily model the conditional correlation matrix, while the volatilities of particular scalar components are modeled by means of the univariate GARCH instruments as in Sects. 8.3.5 and 8.3.6.

1. CCC Model

Model CCC (*constant conditional correlations*; see Bollerslev (1990)) applies for (13.7) the conditional covariance matrix $\mathbf{\Sigma}_t$ of the form

$$\mathbf{\Sigma}_t = \mathbf{\Delta}_t \mathbf{R}\ \mathbf{\Delta}_t, \tag{13.17}$$

where $\mathbf{R}$ is the constant correlation matrix (particularly it must be positive definite) and $\mathbf{\Delta}_t = \text{diag}\{\sqrt{\sigma_{1t}}, \ldots, \sqrt{\sigma_{mt}}\}$ is the diagonal matrix with diagonal elements, which are the square roots of univariate volatilities and therefore can be modeled by means of univariate GARCH models as in (8.55) including the parameter constraints (8.56), i.e.:

$$\sigma_{kk,t} = \alpha_{0k} + \sum_{i=1}^{r_k} \alpha_{ki} y_{k,t-i}^2 + \sum_{j=1}^{s_k} \beta_{kj} \sigma_{kk,t-j}, \qquad k = 1, \ldots, m \tag{13.18}$$

(the modifications from Sect. 8.3.6 are also possible, e.g., EGARCH and others).

According to (13.7) and (13.17) the transformed process

$$\mathbf{z}_t = \mathbf{\Delta}_t^{-1} \mathbf{y}_t \tag{13.19}$$

is obviously a multivariate white noise with the constant correlation matrix $\mathbf{R}$. Therefore, the transformation (13.19) is sometimes denoted as *devolatilization* (or standardization) and $\mathbf{R}$ can be looked upon as the (unconditional) correlation matrix of the devolatilized values $z_{kt} = y_{kt}/\sqrt{\sigma_{kk,t}}$. To perform the devolatilization in practice, one can utilize the volatilities estimated by means of the univariate GARCH models (13.18) (see Sect. 8.3.5).

A simple special case of the model CCC is a model with the unit correlation matrix $\mathbf{R} = \mathbf{I}$. Then the process $\{\mathbf{y}_t\}$ is the multivariate white noise with mutually uncorrelated components modeled simply by means of univariate models GARCH.

On the other hand, the model CCC is not usually suitable for financial applications that require not only the dynamical modeling of conditional variances but also the dynamical modeling of conditional correlations. Moreover, the model CCC does not enable to model the situation common in the context of multivariate risk when high absolute returns in one component give rise to higher future volatility of other components. Therefore, the models DCC are preferred (see below).

2. DCC Model

Model DCC (*dynamic conditional correlations*; see, e.g., Engle (2002), Tse and Tsui (2002)) has the conditional covariance matrix $\mathbf{\Sigma}_t$ figuring in (13.7) of the form

$$\mathbf{\Sigma}_t = \mathbf{\Delta}_t \mathbf{R}_t \mathbf{\Delta}_t, \tag{13.20}$$

where $\mathbf{\Delta}_t = \text{diag}\{\sqrt{\sigma_{1t}}, \ldots, \sqrt{\sigma_{mt}}\}$ is similarly to (13.17) the diagonal matrix with diagonal elements, which are the square roots of univariate volatilities so that one can again utilize univariate GARCH models to estimate them. In DCC models, the matrix $\mathbf{R}_t$ can vary in time and one obtains it as

$$\mathbf{R}_t = \text{diag}\{\mathbf{Q}_t\}^{-1/2}\mathbf{Q}_t \ \text{diag}\{\mathbf{Q}_t\}^{-1/2}, \tag{13.21}$$

namely by rescaling a dynamic matrix process $\{\mathbf{Q}_t\}$. Engle (2002) suggested a possible alternative for this process as

$$\mathbf{Q}_t = \left(1 - \sum_{i=1}^{r} \alpha_i - \sum_{j=1}^{s} \beta_j\right)\mathbf{R} + \sum_{i=1}^{r} \alpha_i \mathbf{z}_{t-i}\mathbf{z}'_{t-i} + \sum_{j=1}^{s} \beta_j \mathbf{Q}_{t-j}, \tag{13.22}$$

where $\mathbf{R}$ is the unconditional (positive definite) correlation matrix of devolatilized values $\mathbf{z}_t = \mathbf{\Delta}_t^{-1}\mathbf{y}_t$ and the parameters α_i and β_j in (13.22) fulfill the following constraints:

$$\alpha_i \geq 0, \quad \beta_j \geq 0, \quad \alpha_1 + \ldots + \alpha_r + \beta_1 + \ldots + \beta_s < 1. \tag{13.23}$$

The model DCC is inspired by the univariate GARCH model (8.55) since such a model under the assumption of stationarity with a constant (unconditional) variance σ^2 can be rewritten in the form

$$\sigma_t^2 = \left(1 - \sum_{i=1}^{r} \alpha_i - \sum_{j=1}^{s} \beta_j\right)\sigma^2 + \sum_{i=1}^{r} \alpha_i y_{t-i}^2 + \sum_{j=1}^{s} \beta_j \sigma_{t-j}^2. \tag{13.24}$$

According to this analogy, the correlation matrix $\mathbf{R}$ in (13.22) can be looked upon as such a part of the volatility equation that models the systematic correlatedness. Apparently if all parameters α_i and β_j are zero, then we return to the CCC model (13.17). To estimate DCC models, one can use special methods which are again based on the devolatilized values estimated by means of univariate GARCH models. In particular, the matrix $\mathbf{R}$ for (13.22) can be set to its empirical counterpart (e.g., applying the rolling window sample estimation to devolatilized values) or it can be estimated as a parametric (positive definite) matrix in addition to remaining parameters of the type α and β. An alternative form of the dynamic matrix process $\{\mathbf{Q}_t\}$ in (13.22) was suggested by Tse and Tsui (2002).

13.3.3 Factor Models

Engle et al. (1990) suggested a parametrization of the conditional covariance matrix $\mathbf{\Sigma}_t$ using the idea that the comovements of asset returns (e.g., stock returns) are driven by a small number of common underlying variables called factors (see also

Lin (1992), Vrontos et al. (2003), and others). Moreover, these factors are frequently conditionally heteroscedastic and possess the GARCH structure.

The factor approach has an advantage that it reduces the dimensionality when the number of factors K relative to the dimension m of given multivariate time series $\{\mathbf{y}_t\}$ is small. Engle et al. (1990) defined their factor models as follows. They assumed that the conditional covariance matrix $\mathbf{\Sigma}_t$ is generated by K ($K < m$) factor volatilities $f_{kk,t}$ corresponding to K underlying (not necessarily uncorrelated) factors, i.e.:

$$\mathbf{\Sigma}_t = \mathbf{\Omega} + \mathbf{W}\mathbf{F}_t\mathbf{W}' = \mathbf{\Omega} + \sum_{k=1}^{K} \mathbf{w}_k \mathbf{w}_k' f_{kk,t}, \tag{13.25}$$

where $\mathbf{\Omega}$ is an $m \times m$ positive semidefinite matrix, $\mathbf{W}$ is $m \times K$ weight (or transformation) matrix with linearly independent $m \times 1$ columns $\mathbf{w}_k$, and $\mathbf{F}_t = \text{diag}\{f_{11,t}, \ldots, f_{KK,t}\}$ is a $K \times K$ diagonal matrix. It is assumed that particular factors have GARCH(1,1) structure with volatility equations of the form

$$f_{kk,t} = \omega_k + \alpha_k \left(\mathbf{v}_k' \mathbf{y}_{t-1}\right)^2 + \beta_k f_{kk,t-1}, \tag{13.26}$$

where ω_k, α_k, and β_k are scalar parameters, $\mathbf{v}_k$ are $m \times 1$ vectors of weights, and $f_{kk,t}$ are factor volatilities ($k = 1, \ldots, K$). In any case, the number of factors K is intended to be much smaller than the number of assets m which makes the model feasible even for a large number of assets.

In the previous model the factors are generally correlated. This may be undesirable when it turns out that several of the factors capture very similar characteristics of the data. On the other hand, if the factors were uncorrelated, they would represent really different components that drive the data. The uncorrelatedness of factors can be achieved by means of various orthogonal transformations in O-GARCH (*orthogonal* GARCH) models (see Alexander and Chibumba (1997)) and GO-GARCH (*generalized orthogonal* GARCH) models (see van der Weide (2002), Lanne and Saikkonen (2007) and others).

Another possibility how to reduce the number of factors in a parsimonious orthogonal way consists in application of the principal component analysis (PCA) which is based (similarly to orthogonal GARCH models) on the eigenvalues and eigenvectors of the conditional covariance matrix $\mathbf{\Sigma}_t$ of given multivariate time series (in more details, first few principal components that explain a high percentage of variability of the process are identified as common factors; see Example 13.2).

Example 13.2 For the same bivariate time series as in Example 13.1 (the component $\{r_{t1}\}$ of monthly log returns of stocks IBM and $\{r_{t2}\}$ of monthly log returns of index S&P 500 during period 1926–1999), Tsay (2002) constructed the bivariate GARCH model applying the factor approach. At first the single common factor $\{x_t\}$ as the first principal component was constructed explaining 82.5% of variability of $\{r_{t1}\}$ and $\{r_{t2}\}$:

$$x_t = 0.796r_{1t} + 0.605r_{2t}.$$

Then this factor $\{x_t\}$ was modeled by means of the following univariate GARCH model:

$$x_t = 1.317 + 0.096x_{t-1} + e_t, \qquad e_t = \sigma_t \varepsilon_t,$$
$$\sigma_t^2 = 3.834 + 0.110e_{t-1}^2 + 0.825\sigma_{t-1}^2.$$

Finally, the bivariate model exploiting $\{\sigma_t^2\}$ as the common volatility factor was constructed:

$$r_{1t} = 1.140 + 0.079r_{1,t-1} + 0.067r_{1,t-2} - 0.122r_{2,t-2} + e_{1t},$$
$$r_{2t} = 0.537 + e_{2t},$$
$$\begin{pmatrix} \sigma_{11,t} \\ \sigma_{22,t} \end{pmatrix} = \begin{pmatrix} 19.08 \\ -5.62 \end{pmatrix} + \begin{pmatrix} 0.098 & \cdot \\ \cdot & \cdot \end{pmatrix} \begin{pmatrix} e_{1,t-1}^2 \\ e_{2,t-1}^2 \end{pmatrix} + \begin{pmatrix} 0.333 \\ 0.596 \end{pmatrix} \sigma_t^2$$

(insignificant parameters are omitted). Obviously, the constructed volatility equation makes use of the lagged squared residuals only in a very limited scope. It is not surprising as the single common factor explains 82.5% of the total variability of data. ◇

13.3.4 *Estimation of Multivariate GARCH Models*

In the literature and various software systems, one can find various approaches how to estimate particular types of multivariate GARCH models. The most recommended methods consist in maximum likelihood approach similarly to that in the univariate case (see, e.g., (8.48)). This approach usually maximizes the log likelihood function of the form (up to a constant)

$$-\frac{1}{2}\sum_{t=1}^{n} \ln|\mathbf{\Sigma}_t| + \sum_{t=1}^{n} \ln g\left(\mathbf{\Sigma}_t^{-1/2}\mathbf{y}_t\right), \tag{13.27}$$

where $g(\cdot)$ is an unspecified probability density function of the standardized residuals $\boldsymbol{\varepsilon}_t = \mathbf{\Sigma}_t^{-1/2}\mathbf{y}_t$ (see (13.7)). In addition, the conditional mean value can be included in (13.27).

For normally distributed innovations $\boldsymbol{\varepsilon}_t$ one obtains the log likelihood of the form

$$-\frac{1}{2}\sum_{t=1}^{n} \ln|\mathbf{\Sigma}_t| - \frac{1}{2}\sum_{t=1}^{n} \ln \mathbf{y}_t'\mathbf{\Sigma}_t^{-1}\mathbf{y}_t. \tag{13.28}$$

One should again remind here that the normality of innovations is often rejected in financial applications (mainly with daily or weekly data): the kurtosis of most financial asset returns is larger than three and the tails are often fatter than what is implied by a conditional normal distribution. Fortunately, Bollerslev and Wooldridge (1992) have shown that a consistent estimator of unknown parameters (and under some assumptions even a strong consistent one; see Gourieroux (1997) or Jeantheau (1998)) can be obtained when maximizing (13.28) even if the distribution of generating process is not normal. Then one denotes it as (Gaussian) *quasi-maximum likelihood QML* or *pseudo-maximum likelihood PML* estimator. Example 13.3 demonstrates the application of this estimation in financial practice.

Example 13.3 Hendrych and Cipra (2016) analyzed the mutual currency risk. By means of the software system EViews for multivariate GARCH models, one estimated the mutual volatilities (covolatilities) of six European currencies in the period from January 5, 2007, to April 27, 2012 (i.e., 1362 observations for each currency), namely for the Czech crown (CZK), the British pound sterling (GBP), the Hungarian forint (HUF), the Polish zloty (PLN), the Romanian leu (RON), and the Swedish krona (SEK). In the EU27, 17 member countries used the Euro currency; other three states (Denmark, Latvia, and Lithuania) were members of the ERM II regime (the European Exchange Rate Mechanism II), i.e., the national currencies were allowed to fluctuate around their assigned value with respect to limiting bounds; the Bulgarian lev was pegged to the euro. Therefore, only six remaining currencies (see above) were not linked to a currency mechanism and were used in the case study.

More precisely, one modeled the daily log returns on the bilateral exchange rates of six currencies with euro as the denominator. Table 13.1 delivers the sample characteristics of the data collected from the European Central Bank in 2013, e.g., the maximum log return of the Czech crown versus euro in the given period was 3.17%.

The analysis of corresponding six-variate process must start by modeling its conditional mean value. For this purpose vector autoregression (VAR) appears suitable similarly to that in Examples 13.1 and 13.2, namely VAR(3) (see

Table 13.1 Sample characteristics of daily log returns on exchange rates for selected currencies versus euro in the period from January 5, 2007, to April 27, 2012 (1362 observations for each currency) from Example 13.3

	CZK	GBP	HUF	PLN	RON	SEK
Mean	−0.00007	0.00014	0.00010	0.00006	0.00019	−0.00001
Median	−0.00007	0.00012	−0.00024	−0.00014	0.00000	0.00004
Maximum	0.03165	0.03461	0.05069	0.04164	0.02740	0.02784
Minimum	−0.03274	−0.02657	−0.03389	−0.03680	−0.01992	−0.02260
Std. deviation	0.00478	0.00601	0.00763	0.00721	0.00462	0.00497
Skewness	0.20218	0.30655	0.42056	0.30802	0.54616	0.31526
Kurtosis	8.49754	6.49258	7.80556	8.05110	7.37830	6.05079

Source: Hendrych and Cipra (2016)

(12.25)). The analysis of correlation structure is then performed using only the deviations from the conditional mean (prediction errors) e_{it} that originate in particular components of the process applying alternatively the six-variate models GARCH(1,1) of the type CCC, DCC, or scalar BEKK (denoted as sBEKK):

1. In the case of model CCC one must construct at first particular models for univariate volatilities. Here the EGARCH(1,1) models (see (8.74)) seem to be acceptable; e.g., for the deviations from the conditional mean $\{e_{1t}\}$ in the case of log returns of the Czech crown versus euro one obtains

$$\ln \sigma_{11,t} = -0.280 + 0.154\left|\frac{e_{1,t-1}}{\sqrt{\sigma_{11,t-1}}}\right| + 0.985 \ln \sigma_{11,t-1} + 0.004\frac{e_{1,t-1}}{\sqrt{\sigma_{11,t-1}}}.$$

   Then it suffices to estimate the constant correlation matrix **R** (see Table 13.2) by means of the devolatilization (13.19). Figure 13.1 plots the constant conditional correlation 0.664 for the pair HUF/EUR and PLN/EUR only.

Table 13.2 CCC and DCC estimation of (constant) correlation matrix from Example 13.3 (daily log returns on exchange rates for six selected currencies versus euro in period from January 5, 2007, to April 27, 2012)

	CZK	GBP	HUF	PLN	RON	SEK
CZK	1.00000	−0.05632	0.39278	0.41581	0.18822	0.16915
GBP	−0.05632	1.00000	−0.04448	−0.02724	0.01421	0.07426
HUF	0.39278	−0.04448	1.00000	0.66421	0.44590	0.32100
PLN	0.41581	−0.02724	0.66421	1.00000	0.41214	0.36136
RON	0.18822	0.01421	0.44590	0.41214	1.00000	0.21656
SEK	0.16915	0.07426	0.32100	0.36136	0.21656	1.00000

Source: Hendrych and Cipra (2016)

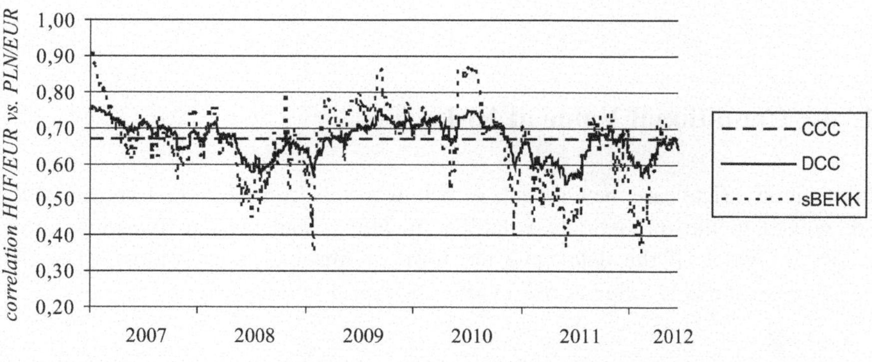

Fig. 13.1 Conditional correlations among daily log returns of exchange rates HUF/EUR and PLN/EUR estimated by means of models GARCH(1,1) of type CCC, DCC, and scalar BEKK from Example 13.3 (daily log returns on exchange rates for six selected currencies versus euro in the period from January 5, 2007, to April 27, 2012). Source: Hendrych and Cipra (2016)

2. In the case of model DCC one starts similarly to that in the previous case of CCC model with the univariate volatilities EGARCH(1,1) for the deviations from the conditional mean $\{e_{ti}\}$. The estimation of the dynamic volatility matrix $\mathbf{\Sigma}_t$ is obtained according to the formulas (13.20)–(13.22), where

$$\mathbf{Q}_t = (1 - 0.011 - 0.978)\mathbf{R} + 0.011\mathbf{z}_{t-1}\mathbf{z}'_{t-1} + 0.978\mathbf{Q}_{t-1}$$

with the same estimated correlation matrix $\mathbf{R}$ as in the previous case of the model CCC (see Table 13.2) and devolatilized process $\{\mathbf{z}_t\}$ according to (13.19). Figure 13.1 plots the dynamic conditional correlation again for the pair HUF/EUR and PLN/EUR only.

3. Finally in the case of scalar BEKK model (denoted as sBEKK), the estimated model (13.13) with $\mathbf{A}_1 = \alpha_1\, \mathbf{I}_6, \boldsymbol{B}_1 = \beta_1\, \mathbf{I}_6$ for the dynamic volatility matrix $\mathbf{\Sigma}_t$ has the form

$$\mathbf{\Sigma}_t = 0.043^2\mathbf{e}_{t-1}\mathbf{e}'_{t-1} + 0.951^2\mathbf{\Sigma}_{t-1},$$

where $\{\mathbf{e}_t\}$ is the corresponding multivariate process of the deviations from the conditional mean. Figure 13.1 again plots the dynamic conditional correlation for the pair HUF/EUR and PLN/EUR only.
From the pragmatic point of view, the estimated models help to do conclusions, e.g., on the average level of particular conditional correlations (see the estimated correlation matrix $\mathbf{R}$ in Table 13.2): as the currency risk is concerned, the British pound influences the behavior of remaining five currencies in a negligible scope; the Hungarian forint and the Polish zloty are rather strongly correlated both mutually (see also Fig. 13.1) and to the Czech crown, etc. The models DCC and sBEKK inject an important dynamic aspect to the analysis.

◇

13.4 Conditional Value at Risk

In the multivariate case, new aspects of risk measures from Sect. 11.1 appear which are related to the multivariate volatility modeling (and even to the multivariate GARCH models if the data have the form of financial asset returns). The risk measures of the type value at risk (VaR) are typical in this context.

1. CoVaR
Adrian and Brunnermeier (2008) showed empirically that the stress state of some financial institutions (mainly big banks, but also insurance companies, mortgage agencies, and others) can raise significantly the value at risk of the global financial system (even by 50%). Therefore, specific risk measures were introduced:

$CoVaR^{j|i}$(*conditional VaR* or *contagion VaR*) is the value at risk (11.1) of the *j*th subject (e.g., a bank) with a possible loss X_j under the condition that the *i*th subject (e.g., another bank) with a possible loss X_i finds oneself in a crisis situation or emergency ($i, j = 1, \ldots, N$), i.e.

$$P\left(X_j \leq CoVaR_\alpha^{j|i} | \, X_i = VaR_\alpha^i\right) = \alpha, \tag{13.29}$$

where VaR_α^i is the value at risk of the *i*th subject on the confidence level α (e.g., $\alpha = 0.99$).

$CoVaR^{|i}$ is another conditional value at risk which measures the "contagion" spread caused by the subject with loss X_i to the global system with loss X (e.g., the impact of a defaulting bank on the whole bank system), i.e.:

$$P\left(X \leq CoVaR_\alpha^{|i} | \, X_i = VaR_\alpha^i\right) = \alpha. \tag{13.30}$$

$\Delta CoVaR^{|i}$ (*delta conditional VaR*) is defined as the following difference:

$$\Delta CoVaR_\alpha^{|i} = CoVaR_\alpha^{|i} - VaR_\alpha, \tag{13.31}$$

where $CoVaR^{|i}$ is given by (13.30) and VaR_α is the value at risk corresponding to the loss X of the global system.

This methodology can be modified for the log returns $\{r_{it}\}$ of financial assets from a global system with the log return $\{r_{mt}\}$ (a global financial market or a security index; see Brownlees and Engle (2012)). As a special case, one could even consider a simple portfolio situation

$$r_{mt} = \sum_{i=1}^{N} w_{it} r_{it}, \tag{13.32}$$

where w_{it} is the relative market capitalization (i.e., the weight) of *i*th asset at time *t* (the scheme (13.32) corresponds, e.g., to the construction of stock indices). In any case, one can deal with suitable values at risk also in this modified situation:

$CoVaR_t^{m|r_{it}=VaR_{it}(\alpha)}$ is the conditional value at risk corresponding to the value at risk of the market return under the condition that the *i*th asset finds oneself in a crisis situation:

$$\mathrm{P}\left(r_{mt} \geq CoVaR_t^{m|r_{it}=VaR_{it}(\alpha)} | r_{it} = VaR_{it}(\alpha)\right) = \alpha, \tag{13.33}$$

so that (13.33) is the modification of (13.30) in this (portfolio) context. Note the inequality sign in (13.33) since the loss consists in drops of returns so that typical values at risk are negative return (losses have the negative sign in this context).

$CoVaR_{it}(\alpha)$ is defined as the difference between the value at risk of the global market conditionally on the ith asset being in financial distress and the value at risk of the global market conditionally on the asset i being in its median state:

$$CoVaR_{it}(\alpha) = CoVaR_t^{m|r_{it}=VaR_{it}(\alpha)} - CoVaR_t^{m|r_{it}=median(r_{it})}. \tag{13.34}$$

2. MES

Marginal expected shortfall (MES) is based on the concept of the expected shortfall ES (the ES at level α is the expected return in the worst $\alpha\%$ of the cases; see (11.14)). The expected shortfall is usually preferred among risk measures in today's financial practice (due to its coherence and other properties giving to it preferences, e.g., in comparison with the classical value at risk approach; see Artzner et al. (1999) or Yamai and Yoshiba (2005)):

$MES_{it}(C)$ is the conditional version of ES, in which the global returns exceed a given market drop C which is chosen as a suitable threshold value ($C < 0$, i.e., measures of the type MES similarly to CoVaR are again in the context of (log) returns typically negative):

$$MES_{it}(C) = \mathrm{E}_{t-1}(r_{it}|r_{mt} < C). \tag{13.35}$$

The symbol E_{t-1} means that one understands the symbols $MES_{it}(C)$ conditionally at time as $MES_{i,t|t-1}(C)$, i.e., computed at time t given the information available at time $t-1$; see also the commentary to (8.12) concerning the symbols of the type $\sigma^2_{t|t-1}$.

Such a concept seems to be productive in various applications, e.g., when dealing with so-called *systemic risk* and systemically important financial institutions (SIFI) whose distress or disorderly failure, because of their size, complexity, and systemic interconnectedness, would cause significant disruption to the wider financial system and economic activity. If the conditional ES of the system is formally defined as

$$ES_{mt}(C) = \mathrm{E}_{t-1}(r_{mt}|r_{mt} < C) = \sum_{i=1}^{N} w_{it}\mathrm{E}_{t-1}(r_{it}|r_{mt} < C) \tag{13.36}$$

then it holds

$$MES_{it}(C) = \frac{\partial ES_{mt}(C)}{\partial w_{it}}. \tag{13.37}$$

Hence MES measures the increase in the risk of the system (measured by the ES) induced by a marginal increase in the weight of ith subject of the system (the higher the subject's MES, the higher the individual contribution of this subject to the risk of the financial system).

Table 13.3 Constituents of Prague Stock Exchange index (PX index) from Example 13.4

Stock name (abbrev.)	Stock name	Obs. from	Obs. to
PX	Prague Stock Exchange Index	Jan 6, 2000	May 9, 2016
AAA	AAA Auto	Sep 25, 2007	Jul 3, 2013
VIG	Vienna Insurance Group	Feb 6, 2008	May 9, 2016
CEZ	CEZ	Jan 6, 2000	May 9, 2016
CETV	Central European Media Enterprises	Jun 28, 2005	May 9, 2016
ECM	ECM Real Estate Investments	Dec 8, 2006	Jul 20, 2011
ERSTE	Erste Group Bank	Oct 2, 2002	May 9, 2016
KB	Komercni banka	Jan 6, 2000	May 9, 2016
NWR	New World Resources PLC	May 7, 2008	May 9, 2016
O2	O2 CR	Jan 6, 2000	May 9, 2016
ORCO	Orco Property Group SA	Feb 2, 2005	Sep 19, 2014
PEGAS	Pegas Nonwovens SA	Dec 19, 2006	May 9, 2016
PHILMOR	Philip Morris CR	Oct 9, 2000	May 9, 2016
UNIPETROL	Unipetrol	Jan 6, 2000	May 9, 2016
ZENTIVA	Zentiva	Jun 29, 2004	Apr 27, 2009

Source: Cipra and Hendrych (2017)

Example 13.4 The case study by Cipra and Hendrych (2017) examines the systemic risk for the Prague Stock Exchange index (PX index) constituents (see Table 13.3). In order to calculate MES for each involved firm i ($i = 1, \ldots, N$), one implemented GARCH modeling schemes for each bivariate process of the daily firm and market log returns r_{it} and r_{mt} (see also Brownlees and Engle (2012)):

$$\begin{aligned} r_{it} &= \sigma_{it}\varepsilon_{it} = \sigma_{it}\rho_{im,t}\varepsilon_{mt} + \sigma_{it}\sqrt{1-\rho_{im,t}^2}\;\zeta_{it}, \\ r_{mt} &= \sigma_{mt}\varepsilon_{mt}, \end{aligned} \tag{13.38}$$

where the shocks (ζ_{it}, ε_{mt}) are independent and identically distributed in time with zero mean, unit variance, and zero covariance. A mutual independence of these shocks is not assumed: on the contrary, there are reasons to believe that extreme values of ε_{mt} and ζ_{it} interact (when the market is in its tail, the firm disturbances may be even further in the tail if there is serious risk of default). Obviously, the modeling scheme (13.38) guarantees that the conditional variances and correlation of r_{it} and r_{mt} are $\sigma_{it}^2(= \sigma_{ii,t})$, $\sigma_{mt}^2(= \sigma_{mm,t})$ and $\rho_{im,t}$, respectively.

The specification is completed by description of conditional (co)moments. The volatilities σ_{it}^2 and σ_{mt}^2 were modeled as the univariate GJR GARCH(1,1) models (see (8.72))

$$\begin{aligned}\sigma_{it}^2 &= \omega_i + \alpha_i r_{i,t-1}^2 + \beta_i \sigma_{i,t-1}^2 + \gamma_i r_{i,t-1}^2 I_{i,t-1}^-, \\ \sigma_{mt}^2 &= \omega_m + \alpha_m r_{m,t-1}^2 + \beta_m \sigma_{m,t-1}^2 + \gamma_m r_{m,t-1}^2 I_{m,t-1}^-\end{aligned} \tag{13.39}$$

with $I_{i,t}^- = 1$ for $r_{it} < 0$ and 0 otherwise, $I_{m,t}^- = 1$ for $r_{mt} < 0$ and 0 otherwise (this threshold GARCH modification covers the leverage effect, i.e., the tendency of volatility to increase more with bad news (negative log returns) rather than with good ones (positive log returns). The time-varying correlations are captured by using GARCH(1,1) of the type DCC (also with an asymmetric modification; see Engle (2009)). For example, the conditional covariance matrix (13.20) has the form

$$\boldsymbol{\Sigma}_t = \boldsymbol{\Delta}_t \mathbf{R}_t \boldsymbol{\Delta}_t = \begin{pmatrix} \sigma_{it} & 0 \\ 0 & \sigma_{mt} \end{pmatrix} \begin{pmatrix} 1 & \rho_{im,t} \\ \rho_{im,t} & 1 \end{pmatrix} \begin{pmatrix} \sigma_{it} & 0 \\ 0 & \sigma_{mt} \end{pmatrix}. \tag{13.40}$$

The previous modeling scheme was applied to 14 firms, which have been included into the PX index basis according to their market capitalization as of the end of June 2008. One extracted the daily log returns from January 6, 2000, to May 9, 2016 (these data are unbalanced in that sense that not all companies have been continuously traded during the sample period; see Table 13.3). Selected sample characteristics of the studied log returns are presented in Table 13.4. One made use of the estimation methodology for multivariate GARCH models from Sect. 13.3.4.

Figure 13.2 displays conditional volatilities of all investigated firms jointly with the PX index (market) conditional volatility. Apparently, all graphs are significantly influenced by the explosion in variability during the financial crisis 2008. Furthermore, one identifies the similar trend over many charts that is in line with the market volatility trend. On the contrary, several log return time series are dominated by other effects, which are not common for the whole market or for other returns. For instance, one can mention the volatility of O2 log returns, which was increased due to the split of the company in 2015.

Figure 13.3 shows the estimated time-varying correlations $\rho_{im,t}$ between returns of the ith company and the PX index (market). It is evident that the financial returns of involved firms are significantly positively correlated with the market financial returns. However, one can identify different behavior of correlations displayed in particular plots. Some correlations are relatively stable when comparing with others; see, e.g., CEZ, KB, PHILMOR, or VIG; others demonstrate trends varying in time, e.g., O2, PEGAS, or UNIPETROL.

Table 13.5 reports the examined stocks listed in ascending order regarding the one-step ahead MES predicted for October 20, 2008, which was a very critical date from the point of view of the financial crisis 2008. The threshold C (see (13.35)) was set as the unconditional VaR of the PX index log returns with the confidence level 99%. Under the distress condition of the global market, the short-run prediction produced by the model, e.g., for ERSTE indicates a deep drop over 25%.

Finally, Table 13.6 contains the estimated multi-period ahead MES predictions ($h = 125$, i.e., the half-year ahead) starting from May 9, 2016 (the end of the

Table 13.4 Sample characteristics of the log returns from Example 13.4 (systemic risk analysis for constituents of PX index)

Stock	# obs	Mean	Std. dev	Median	Min	Max	Skew	Kurt
PX	4101	0.00014	0.01429	0.00051	−0.16185	0.12364	−0.45054	12.21528
AAA	1450	−0.00057	0.02885	0.00000	−0.23107	0.34179	1.14937	23.07040
VIG	2071	−0.00049	0.02215	0.00000	−0.17920	0.13539	−0.51478	8.42960
CEZ	4099	0.00039	0.01932	0.00090	−0.19834	0.19517	−0.38745	9.95323
CETV	2705	−0.00114	0.03921	0.00000	−0.70628	0.47994	−1.98131	56.46028
ECM	1155	−0.00356	0.03748	−0.00253	−0.41313	0.24481	−1.02567	20.24787
ERSTE	3127	−0.00005	0.02668	0.00000	−0.26834	0.19382	−0.37664	11.05523
KB	4099	0.00048	0.02138	0.00000	−0.19392	0.10410	−0.43406	6.76452
NWR	2008	−0.00436	0.05186	0.00000	−0.49590	0.51083	−0.86226	16.88669
O2	4101	−0.00022	0.02509	0.00000	−0.94253	0.13056	−13.02806	487.11037
ORCO	2385	−0.00181	0.03614	−0.00168	−0.26397	0.29480	0.14712	11.66799
PEGAS	2353	0.00000	0.01784	0.00000	−0.23370	0.13509	−0.93750	21.07422
PHILMOR	3882	0.00020	0.01756	0.00000	−0.13709	0.14842	−0.63086	8.43172
UNIPETROL	4097	0.00028	0.02233	0.00000	−0.21770	0.26472	−0.12378	14.04297
ZENTIVA	1214	0.00061	0.01929	0.00000	−0.17839	0.09733	−1.31104	13.24937

Source: Cipra and Hendrych (2017)

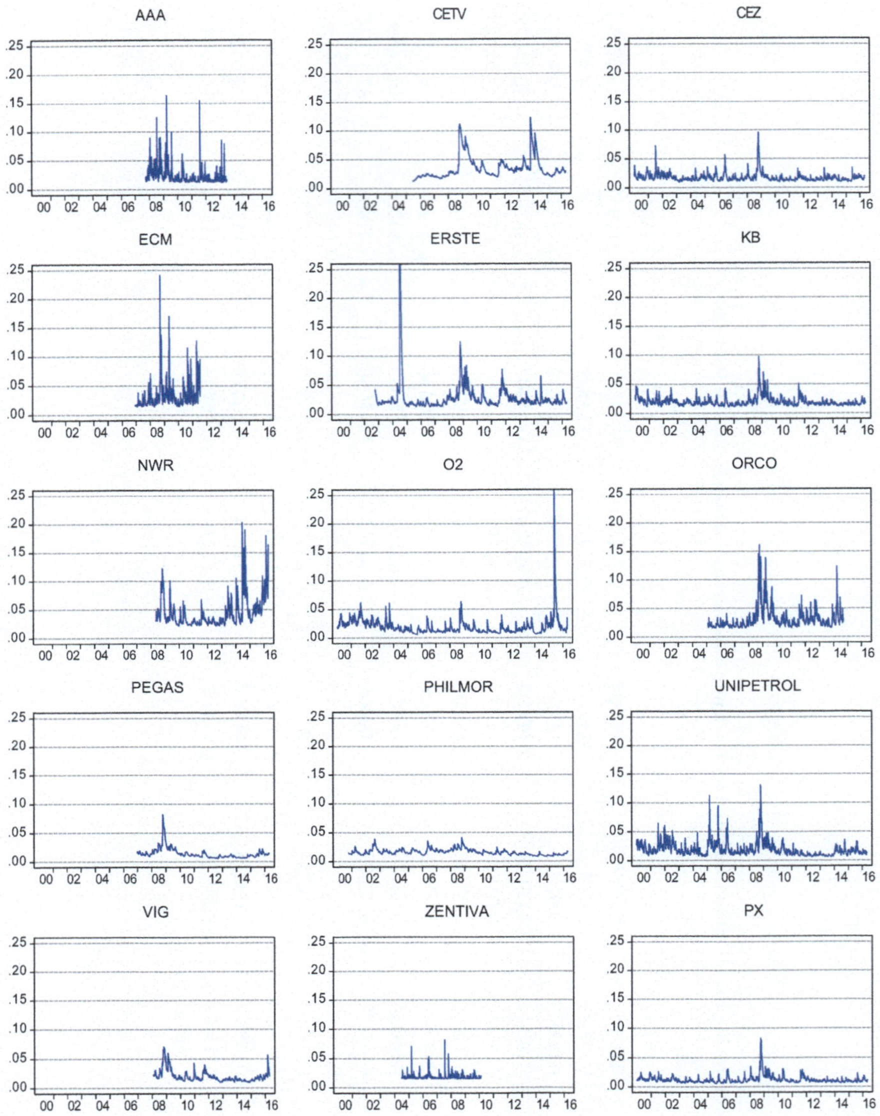

Fig. 13.2 Conditional volatilities of PX index constituents and PX index itself from Example 13.4 (systemic risk analysis for constituents of PX index). Source: Cipra and Hendrych (2017)

examined data set). Here the threshold C was set as minus 5% and minus 20%, respectively. To be more precise, an investor can identify and anticipate potential capital shortfall under the condition that a systemic event occurs half a year after the investment (i.e., when the market global return is less than the threshold C at that time moment). Consequently, the stocks NWR were identified as the most risky

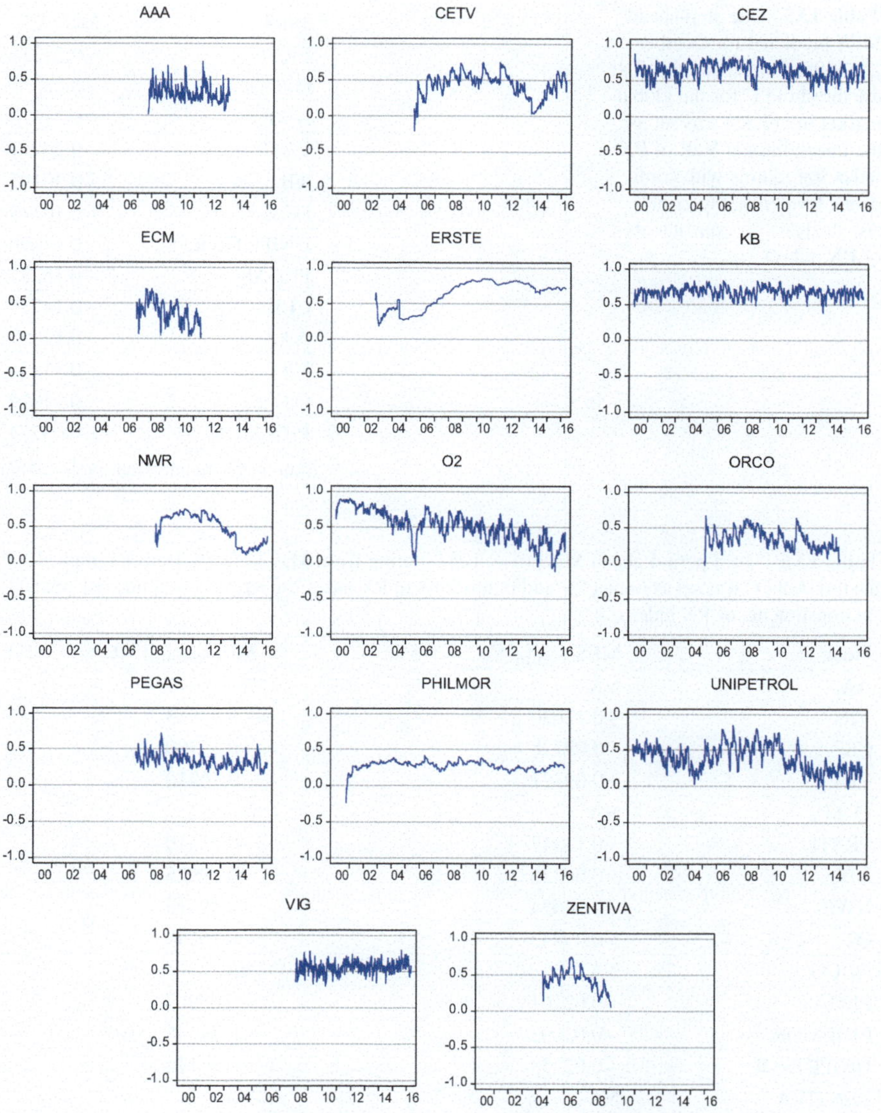

Fig. 13.3 Conditional correlations among PX index constituents and PX index from Example 13.4 (systemic risk analysis for constituents of PX index). Source: Cipra and Hendrych (2017)

assets assuming $C = -0.05$ (the multi-period ahead MES of -14.7%) and the stocks ERSTE as the most risky assets assuming $C = -0.20$ (the multi-period ahead MES of -25.5%).

◊

Table 13.5 One-step-ahead MES predicted for October 20, 2008, from Example 13.4; the threshold C for the global market in (13.35) was set as the unconditional VaR of PX index log returns with confidence level 99% (systemic risk analysis for constituents of PX index)

Stock	MES
ERSTE	−0.25211
ORCO	−0.23861
NWR	−0.21770
CETV	−0.20791
KB	−0.20197
ECM	−0.18318
UNIPETROL	−0.17690
PEGAS	−0.15227
CEZ	−0.14778
AAA	−0.12997
VIG	−0.11844
O2	−0.11019
PHILMOR	−0.05774

Source: Cipra and Hendrych (2017)

Table 13.6 Multi-period-ahead MES ($h = 125$) starting from May 9, 2016, from Example 13.4; the threshold C was set as minus 5% and minus 20% of PX index log returns (systemic risk analysis for constituents of PX index)

Stock	$MES_{9/5/2016}^{125}(C = -5\%)$	$MES_{9/5/2016}^{125}(C = -20\%)$
AAA	NA	NA
VIG	−0.12630	−0.21913
CEZ	−0.09328	−0.19380
CETV	−0.09321	−0.16910
ECM	NA	NA
ERSTE	−0.13347	−0.25510
KB	−0.09235	−0.19925
NWR	−0.14654	−0.19410
O2	−0.09833	−0.16540
ORCO	NA	NA
PEGAS	−0.02161	−0.08087
PHILMOR	−0.03092	−0.07247
UNIPETROL	−0.02253	−0.09114
ZENTIVA	NA	NA

Source: Cipra and Hendrych (2017)

13.5 Exercises

Exercise 13.1

Apply the multivariate EWMA methodology for time series $\{DTB3_t\}$ and $\{DAAA_t\}$ from Table 12.1 (the first differences of monthly yields to maturity for three-month T-bills and corporate bonds AAA in the USA in % p.a.).

Chapter 14
State Space Models of Time Series

14.1 Kalman Filter

Kalman filter presents a theoretical background for various recursive methods in (linear) systems, particularly in (multivariate) time series models. In general, one speaks on so-called *Kalman* (or *Kalman–Bucy*) *recursions* for *filtering*, *predicting*, and *smoothing* in the framework of so-called *state space model*; see, e.g., Brockwell and Davis (1993, 1996), Durbin and Koopman (2012), Hamilton (1994), Harvey (1989), and others.

Originally, state space modeling was suggested for technical disciplines (e.g., for the fire control of missiles and in the telecommunications); later it has shown to be useful also for (Bayesian) statistics and econometrics. This methodological approach is based on the principle that the state of given dynamic system is determined in time by *state vectors* (*state variables*), which are unobservable in time, but one can draw conclusions on their behavior by means of their observations in the form of *observation vectors* (in practice, it is usually a multivariate time series).

The formal base of state space modeling is the *dynamic linear model DLM*, which can be formulated under various levels of complexity (this model can be even nonlinear). We confine ourselves to the simplest form

$$\mathbf{x}_{t+1} = \mathbf{F}_t\mathbf{x}_t + \mathbf{v}_t, \qquad t = 1, 2, \ldots, \tag{14.1}$$

$$\mathbf{y}_t = \mathbf{G}_t\mathbf{x}_t + \mathbf{w}_t, \qquad t = 1, 2, \ldots, \tag{14.2}$$

where (14.1) is the (vector) *state equation* describing the development of state vector in time and (14.2) is the (vector) *observation equation* describing the relationship between observation vectors and state vectors. The meaning of particular symbols is the following:

$\mathbf{x}_t$ state vector of dimension $d \times 1$ (at time t);
$\mathbf{y}_t$ observation vector of dimension $m \times 1$ (at time t);

T. Cipra, *Time Series in Economics and Finance*,
https://doi.org/10.1007/978-3-030-46347-2_14

$\mathbf{F}_t$ parameter matrix of state equation of dimension $d \times d$ (at time t);
$\mathbf{G}_t$ parameter matrix of observation equation of dimension $m \times d$ (at time t);
$\mathbf{v}_t$ (vector) random residual of state equation of dimension $d \times 1$ (at time t);
$\mathbf{w}_t$ (vector) random residual of observation equation of dimension $m \times 1$ (at time t);
$\mathbf{V}_t$ covariance matrix of random residual $\mathbf{v}_t$ of dimension $d \times d$ (at time t);
$\mathbf{W}_t$ covariance matrix of random residual $\mathbf{w}_t$ of dimension $m \times m$ (at time t).

Moreover, one usually assumes that

$$\begin{aligned}&\mathbf{v}_t \sim iid\ N(\mathbf{0}, \mathbf{V}_t), \quad \mathbf{w}_t \sim iid\ N(\mathbf{0}, \mathbf{W}_t), \quad \operatorname{cov}(\mathbf{v}_s, \mathbf{w}_t) = \mathrm{E}(\mathbf{v}_s \cdot \mathbf{w}_t') \\ &= \mathbf{0}, \quad s, t = 1, 2, \ldots\end{aligned} \tag{14.3}$$

(sometimes one admits more generally the contemporaneous correlatedness between the random residuals $\mathbf{v}_t$ and $\mathbf{w}_t$, their non-normality, and other modifications).

Example 14.1 Let us consider the univariate process AR(p) (see also (6.31))

$$y_t = \varphi_1 y_{t-1} + \cdots + \varphi_p y_{t-p} + \varepsilon_t, \quad \varepsilon_t \sim N(0,\ \sigma^2). \tag{14.4}$$

1. On one hand, by means of the state space modeling one can solve the problem of recursive estimation of parameters of this process, even by an adaptive way in real time (i.e., the parameters can change in time). In such a case, the state vector is the parameter vector

$$\mathbf{x}_t = (\varphi_{1t}, \ldots, \varphi_{pt})' \tag{14.5}$$

and the dynamic linear model (14.1) and (14.2) will be formulated as

$$\mathbf{x}_{t+1} = \mathbf{x}_t, \tag{14.6}$$

$$y_t = (y_{t-1}, \ldots, y_{t-p})\mathbf{x}_t + \varepsilon_t \tag{14.7}$$

(i.e., particularly, $\mathbf{y}_t = y_t$, $\mathbf{F}_t = \mathbf{I}$, $\mathbf{G}_t = (y_{t-1}, \ldots, y_{t-p})$, $\mathbf{v}_t = \mathbf{0}$, $\mathbf{w}_t = \varepsilon_t$, $\mathbf{V}_t = \mathbf{0}$, $\mathbf{W}_t = \sigma^2$).

2. On the other hand, by means of the state space modeling one can also predict or filter the given process AR(p). In such a case, the state vector is the following vector of values:

$$\mathbf{x}_t = (y_{t-p+1}, y_{t-p+2}, \ldots, y_t)' \tag{14.8}$$

and the dynamic linear model (14.1) and (14.2) will have the form

$$\mathbf{x}_{t+1} = \begin{pmatrix} 0 & 1 & 0 & \cdots & 0 \\ 0 & 0 & 1 & \cdots & 0 \\ \vdots & \vdots & \vdots & \cdots & \vdots \\ 0 & 0 & 0 & \cdots & 1 \\ \varphi_p & \varphi_{p-1} & \varphi_{p-2} & \cdots & \varphi_1 \end{pmatrix} \mathbf{x}_t + \begin{pmatrix} 0 \\ 0 \\ \vdots \\ 0 \\ 1 \end{pmatrix} \varepsilon_{t+1}, \tag{14.9}$$

$$y_t = (0, \ 0, \ldots, 1) \ \mathbf{x}_t \tag{14.10}$$

(i.e., particularly, $\mathbf{y}_t = y_t$, $\mathbf{G}_t = (0, 0, \ldots, 1)$, $\mathbf{v}_t = (0, 0, \ldots, \varepsilon_{t+1})'$, $\mathbf{w}_t = 0$, $\mathbf{W}_t = 0$).

◇

According to the previous commentaries, the state space representation enables us to solve recursively (namely in an effective way by means of Kalman recursive formulas) the problem of filtering, smoothing, and predicting in a given DLM. The key role in this context plays the conditional distribution of the state vector $\mathbf{x}_t$ conditioned by information contained in the observations $\mathbf{y}_s$, $\mathbf{y}_{s\text{-}1}$, $\mathbf{y}_{s\text{-}2}$, . . . till time s. Due to practical purposes, we confine ourselves only to the first two moments of this distribution and denote

$$\widehat{\mathbf{x}}_{t|s} = \mathrm{E}_s(\mathbf{x}_t), \qquad \mathbf{P}_{t|s} = \mathrm{E}_s\left(\left(\mathbf{x}_t - \widehat{\mathbf{x}}_{t|s}\right) \left(\mathbf{x}_t - \widehat{\mathbf{x}}_{t|s}\right)' \right), \tag{14.11}$$

where the index s in the symbol $\mathrm{E}_s(\cdot)$ means that the mean value is conditioned by information till time s. Important values in this context are the following ones:

- The prediction of state vector $\mathbf{x}_t$ from time $t - 1$ by one-step-ahead and the corresponding error matrix:

$$\widehat{\mathbf{x}}_{t|t-1} = \mathrm{E}_{t-1}(\mathbf{x}_t), \qquad \mathbf{P}_{t|t-1} = \mathrm{E}_{t-1}\left(\left(\mathbf{x}_t - \widehat{\mathbf{x}}_{t|t-1}\right) \left(\mathbf{x}_t - \widehat{\mathbf{x}}_{t|t-1}\right)' \right). \tag{14.12}$$

- The estimated (filtered) value of state vector $\mathbf{x}_t$ at time t and the corresponding error matrix:

$$\widehat{\mathbf{x}}_{t|t} = \mathrm{E}_t(\mathbf{x}_t), \qquad \mathbf{P}_{t|t} = \mathrm{E}_t\left(\left(\mathbf{x}_t - \widehat{\mathbf{x}}_{t|t}\right) \left(\mathbf{x}_t - \widehat{\mathbf{x}}_{t|t}\right)' \right). \tag{14.13}$$

These predictions and estimations are the best ones according to the criterion *MSE* (i.e., in the sense of mean squared error; see (2.11)). Moreover, under the given assumptions (i.e., in the described DML under the assumption of normality), they even have the form of linear functions, the argument of which is always the corresponding conditioning information (i.e., corresponding observation vectors). Simultaneously one can also obtain the prediction of vector $\mathbf{y}_t$ from time $t - 1$ (i.e., by one-step-ahead) and the corresponding error matrix as

$$\hat{\mathbf{y}}_{t|t-1} = \mathrm{E}_{t-1}(\mathbf{y}_t) = \mathbf{G}_t\hat{\mathbf{x}}_{t|t-1}, \quad \mathrm{E}_{t-1}\left(\left(\mathbf{y}_t - \hat{\mathbf{y}}_{t|t-1}\right)\left(\mathbf{y}_t - \hat{\mathbf{y}}_{t|t-1}\right)'\right)$$
$$= \mathbf{G}_t\mathbf{P}_{t|t-1}\mathbf{G}_t' + \mathbf{W}_t. \tag{14.14}$$

Remark 14.1 Sometimes the matrices $\mathbf{F}_t$, $\mathbf{G}_t$, $\mathbf{V}_t$, $\mathbf{W}_t$ (eventually others) in DML contain unknown parameters which must be estimated. In practice, one applies usually so-called *EM algorithm* (*expectation-maximization*; see, e.g., Brockwell and Davis (1996), Dempster et al. (1977), Wu (1983)) which combines the maximum likelihood method with optimization algorithms and can be used in the situations with incomplete information where some data are missing.

◇

1. Filtering in State Space Model

Filtering in a given state space model consists in the (recursive) estimation of state vector $\mathbf{x}_t$ exploiting information contained in $\mathbf{y}_t$, $\mathbf{y}_{t-1}$, $\mathbf{y}_{t-2}$, ... The corresponding Kalman recursive formulas, which are called *Kalman* (or *Kalman-Bucy*) *filter* in such a case, have the form

$$\begin{aligned}\hat{\mathbf{x}}_{t|t} &= \hat{\mathbf{x}}_{t|t-1} + \mathbf{P}_{t|t-1}\mathbf{G}_t'\left(\mathbf{G}_t\mathbf{P}_{t|t-1}\mathbf{G}_t' + \mathbf{W}_t\right)^{-1}\left(\mathbf{y}_t - \mathbf{G}_t\hat{\mathbf{x}}_{t|t-1}\right),\\ \mathbf{P}_{t|t} &= \mathbf{P}_{t|t-1} - \mathbf{P}_{t|t-1}\mathbf{G}_t'\left(\mathbf{G}_t\mathbf{P}_{t|t-1}\mathbf{G}_t' + \mathbf{W}_t\right)^{-1}\mathbf{G}_t\mathbf{P}_{t|t-1},\end{aligned} \tag{14.15}$$

where

$$\begin{aligned}\hat{\mathbf{x}}_{t|t-1} &= \mathbf{F}_{t-1}\hat{\mathbf{x}}_{t-1|t-1},\\ \mathbf{P}_{t|t-1} &= \mathbf{F}_{t-1}\mathbf{P}_{t-1|t-1}\mathbf{F}_{t-1}' + \mathbf{V}_{t-1}.\end{aligned} \tag{14.16}$$

Example 14.2 Kalman filter can be used to construct the recursive OLS estimate in the classical model of linear regression which is rewritten in the form of dynamic linear model (14.1) and (14.2) with the state vector $\boldsymbol{\beta}_t$, i.e.,

$$\beta_{t+1} = \beta_t, \tag{14.17}$$

$$y_t = (1,\ x_{t2},\ \ldots,\ x_{tk})\boldsymbol{\beta}_t + \varepsilon_t = \mathbf{x}_{t\cdot}\boldsymbol{\beta}_t + \varepsilon_t, \tag{14.18}$$

where $\mathbf{x}_{t\cdot}$ is the tth row of regression matrix $\mathbf{X}$ and $\varepsilon_t \sim iid\ N(0, \sigma_t^2)$. After putting into (14.15) and (14.16), one obtains the following recursive formulas for OLS estimate (using a simpler denotation, namely $\mathbf{b}_t$ instead of $\mathbf{b}_{t|t}$ and $\mathbf{P}_t$ instead of $\mathbf{P}_{t|t}/\sigma_t^2$ since, e.g., $\mathbf{b}_{t|t-1} = \mathbf{b}_{t|t}$):

$$\mathbf{b}_t = \mathbf{b}_{t-1} + \frac{\mathbf{P}_{t-1}\mathbf{x}'_{t.}}{\mathbf{x}_{t.}\mathbf{P}_{t-1}\mathbf{x}'_{t.} + 1}(\mathbf{y}_t - \mathbf{x}_{t.}\mathbf{b}_{t-1}),$$
$$\mathbf{P}_t = \mathbf{P}_{t-1} - \frac{\mathbf{P}_{t-1}\mathbf{x}'_{t.}\mathbf{x}_{t.}\mathbf{P}_{t-1}}{\mathbf{x}_{t.}\mathbf{P}_{t-1}\mathbf{x}'_{t.} + 1}. \tag{14.19}$$

In contrast to the classical non-recursive OLS estimate, the algorithm (14.19) is applicable online and it does not demand to invert any matrix during calculations. However, the formulas (14.19) must be completed by a simple recursive estimate of the white noise variance σ_t^2.

◇

Example 14.3 In this example, we will show the application of Kalman filter for recursive estimation of linear time series models, namely autoregressive models. For this purpose, the model AR(p) can be rewritten in the form of dynamic linear model (14.1) and (14.2) with the state vector $\boldsymbol{\varphi}_t$ as

$$\boldsymbol{\varphi}_{t+1} = \boldsymbol{\varphi}_t, \tag{14.20}$$

$$y_t = (y_{t-1},\ y_{t-2}, \ldots, y_{t-p})\boldsymbol{\varphi}_t + \varepsilon_t = \mathbf{y}_{t.}\boldsymbol{\varphi}_t + \varepsilon_t, \tag{14.21}$$

where $\mathbf{y}_{t.} = (y_{t-1}, y_{t-2}, \ldots, y_{t-p})$ and $\varepsilon_t \sim iid\ N(0, \sigma_t^2)$. Again after putting into (14.15) and (14.16), we receive the following recursive formulas for estimating parameters of this heteroscedastic model AR(p) (using a simpler denotation again, namely $\widehat{\boldsymbol{\varphi}}_t$ instead of $\widehat{\boldsymbol{\varphi}}_{t|t}$ and $\mathbf{P}_t$ instead of $\mathbf{P}_{t|t}/\sigma_t^2$):

$$\widehat{\boldsymbol{\varphi}}_t = \widehat{\boldsymbol{\varphi}}_{t-1} + \frac{\mathbf{P}_{t-1}\mathbf{y}'_{t.}}{\mathbf{y}_{t.}\mathbf{P}_{t-1}\mathbf{y}'_{t.} + 1}(y_t - \mathbf{y}_{t.}\widehat{\boldsymbol{\varphi}}_{t-1})\ ,$$
$$\mathbf{P}_t = \mathbf{P}_{t-1} - \frac{\mathbf{P}_{t-1}\mathbf{y}'_{t.}\mathbf{y}_{t.}\mathbf{P}_{t-1}}{\mathbf{y}_{t.}\mathbf{P}_{t-1}\mathbf{y}'_{t.} + 1}\ ,$$
$$\widehat{\sigma}_t^2 = \frac{1}{t-p}\left((t-p-1)\widehat{\sigma}_{t-1}^2 + \frac{(y_t - \mathbf{y}_{t.}\widehat{\boldsymbol{\varphi}}_{t-1})^2}{\mathbf{y}_{t.}\mathbf{P}_{t-1}\mathbf{y}'_{t.} + 1}\right)\ . \tag{14.22}$$

Particularly for the process AR(1), i.e., $y_t = \varphi\ y_{t-1} + \varepsilon_t$, these recursive formulas are simplified to the form

$$\widehat{\varphi}_t = \widehat{\varphi}_{t-1} + P_t y_{t-1}(y_t - \widehat{\varphi}_{t-1}y_{t-1})\ ,$$
$$P_t = \frac{P_{t-1}}{P_{t-1}y_{t-1}^2 + 1}\ ,$$
$$\widehat{\sigma}_t^2 = \frac{1}{t-1}\left((t-2)\widehat{\sigma}_{t-1}^2 + \frac{(y_t - \widehat{\varphi}_{t-1}y_{t-1})^2}{P_{t-1}y_{t-1}^2 + 1}\right)\ . \tag{14.23}$$

Table 14.1 Recursive estimation based on Kalman filter in simulated process $y_t =$ $= 0.6\, y_{t-1} + \varepsilon_t$, $\varepsilon_t \sim iid\ N(0, 100)$, $t = 1, \ldots, 100$ in Example 14.4

t	$\widehat{\varphi}_t$	P_t	$\widehat{\sigma}_t^2$
10	0.552	0.0024	26.54
20	0.518	0.0034	110.69
30	0.508	0.0003	94.24
40	0.448	0.0002	81.74
50	0.613	0.0001	104.39
60	0.640	0.0001	114.00
70	0.608	0.0001	113.42
80	0.630	0.0001	123.06
90	0.641	0.0001	114.14
100	0.616	0.0001	115.25

The initial values in the recursive formulas (14.22) can be chosen as $\widehat{\boldsymbol{\varphi}}_0 = \mathbf{0}$, $\mathbf{P}_0 = c\mathbf{I}$ for a high positive constant c; the initial value of variance of white noise can be taken as the sample variance of $\{y_t\}$ from an initial segment of this time series (or $\widehat{\sigma}_0^2 = 0$).

◇

Example 14.4 Table 14.1 presents selected values obtained by means of recursive estimation (14.23) based on a simulated trajectory of the process AR(1) modeled as

$$y_t = 0.6y_{t-1} + \varepsilon_t, \quad \varepsilon_t \sim iid\ N(0,\ 100), \quad t = 1,\ 2, \ldots, 100$$

(see Kalman filter in Example 14.3). The initial values were chosen as $\widehat{\varphi}_0 = 0$, $P_0 = 1$ and $\widehat{\sigma}_0^2 = 0$.

◇

2. Predicting in State Space Model

Predicting in state space model (also *predictor*) consists in (recursive) estimation of the state vector $\mathbf{x}_{t+h}$ for particular t using information contained in $\mathbf{y}_t, \mathbf{y}_{t\text{-}1}, \mathbf{y}_{t\text{-}2}, \ldots$ (h is fixed). One constructs recursively the predictions of the type

$$\widehat{\mathbf{x}}_{t+h|t} = \mathrm{E}_t(\mathbf{x}_{t+h}), \qquad \mathbf{P}_{t+h|t} = \mathrm{E}_t\Big(\big(\mathbf{x}_{t+h} - \widehat{\mathbf{x}}_{t+h|t}\big)\big(\mathbf{x}_{t+h} - \widehat{\mathbf{x}}_{t+h|t}\big)'\Big). \tag{14.24}$$

The corresponding recursive formulas for one-step-ahead prediction are

$$\begin{aligned}\widehat{\mathbf{x}}_{t+1|t} &= \mathbf{F}_t\widehat{\mathbf{x}}_{t|t-1} + \mathbf{F}_t\mathbf{P}_{t|t-1}\mathbf{G}_t'\big(\mathbf{G}_t\mathbf{P}_{t|t-1}\mathbf{G}_t' + \mathbf{W}_t\big)^{-1}\big(\mathbf{y}_t - \mathbf{G}_t\widehat{\mathbf{x}}_{t|t-1}\big),\\ \mathbf{P}_{t+1|t} &= \mathbf{F}_t\mathbf{P}_{t|t-1}\mathbf{F}_t' + \mathbf{V}_t - \mathbf{F}_t\mathbf{P}_{t|t-1}\mathbf{G}_t'\big(\mathbf{G}_t\mathbf{P}_{t|t-1}\mathbf{G}_t' + \mathbf{W}_t\big)^{-1}\mathbf{G}_t\mathbf{P}_{t|t-1}\mathbf{F}_t'\end{aligned} \tag{14.25}$$

and for more steps h ahead ($h \geq 2$)

$$\begin{aligned}\widehat{\mathbf{x}}_{t+h|t} &= \mathbf{F}_{t+h-1}\mathbf{F}_{t+h-2}\ldots\ \mathbf{F}_{t+1}\widehat{\mathbf{x}}_{t+1|t},\\ \mathbf{P}_{t+h|t} &= \mathbf{F}_{t+h-1}\mathbf{P}_{t+h-1|t}\mathbf{F}'_{t+h-1} + \mathbf{V}_{t+h-1}.\end{aligned} \tag{14.26}$$

Hence one can also calculate the prediction of $\mathbf{y}_{t+h}$ and the corresponding error matrix as

$$\begin{aligned}\widehat{\mathbf{y}}_{t+h|t} &= \mathrm{E}_t(\mathbf{y}_{t+h}) = \mathbf{G}_{t+h}\widehat{\mathbf{x}}_{t+h|t},\ \mathrm{E}_t\left(\left(\mathbf{y}_{t+h} - \widehat{\mathbf{y}}_{t+h|t}\right)\left(\mathbf{y}_{t+h} - \widehat{\mathbf{y}}_{t+h|t}\right)'\right)\\ &= \mathbf{G}_{t+h}\mathbf{P}_{t+h|t}\mathbf{G}'_{t+h} + \mathbf{W}_{t+h}.\end{aligned} \tag{14.27}$$

3. Smoothing in State Space Model

Smoothing in state space model (also *smoother* or *Kalman fixed point smoothing*) consists in the estimation of state vector $\mathbf{x}_t$ for a fixed time t. The procedure is recursive for increasing n using gradually information contained in samples $\mathbf{y}_n$, $\mathbf{y}_{n-1}$, $\mathbf{y}_{n-2}$, ... In other words, one constructs recursively smoothed values of state vector $\mathbf{x}_t$ and corresponding error matrix:

$$\widehat{\mathbf{x}}_{t|n} = \mathrm{E}_n(\mathbf{x}_t), \qquad \mathbf{P}_{t|n} = \mathrm{E}_n\left(\left(\mathbf{x}_t - \widehat{\mathbf{x}}_{t|n}\right)\left(\mathbf{x}_t - \widehat{\mathbf{x}}_{t|n}\right)'\right). \tag{14.28}$$

The corresponding recursive formulas are

$$\begin{aligned}\widehat{\mathbf{x}}_{t|n} &= \widehat{\mathbf{x}}_{t|n-1} + \boldsymbol{\Omega}_{t,n}\mathbf{G}'_n\left(\mathbf{G}_n\mathbf{P}_{n|n-1}\mathbf{G}'_n + \mathbf{W}_n\right)^{-1}\left(\mathbf{y}_n - \mathbf{G}_n\widehat{\mathbf{x}}_{n|n-1}\right),\\ \mathbf{P}_{t|n} &= \mathbf{P}_{t|n-1} - \boldsymbol{\Omega}_{t,n}\mathbf{G}'_n\left(\mathbf{G}_n\mathbf{P}_{n|n-1}\mathbf{G}'_n + \mathbf{W}_n\right)^{-1}\mathbf{G}_n\boldsymbol{\Omega}'_{t,n},\\ \boldsymbol{\Omega}_{t,n+1} &= \boldsymbol{\Omega}_{t,n}\left[\mathbf{F}_n - \mathbf{F}_n\mathbf{P}_{n|n-1}\mathbf{G}'_n\left(\mathbf{G}_n\mathbf{P}_{n|n-1}\mathbf{G}'_n + \mathbf{W}_n\right)^{-1}\mathbf{G}_n\right]'.\end{aligned} \tag{14.29}$$

Hence one can also smooth the observed time series $\{\mathbf{y}_t\}$ as

$$\widehat{\mathbf{y}}_{t|n} = \mathbf{G}_t\widehat{\mathbf{x}}_{t|n}, \quad \mathrm{E}_n\left(\left(\mathbf{y}_t - \widehat{\mathbf{y}}_{t|n}\right)\left(\mathbf{y}_t - \widehat{\mathbf{y}}_{t|n}\right)'\right) = \mathbf{G}_t\mathbf{P}_{t|n}\mathbf{G}'_t + \mathbf{W}_t. \tag{14.30}$$

It is necessary to stress once more that the state space methodology is the theoretical concept for construction of various recursive procedures in time series analysis. Section 14.1.1 shows a possible application for recursive estimation of (multivariate) GARCH models of financial time series.

14.1.1 Recursive Estimation of Multivariate GARCH Models

The usual estimation of GARCH models is based on the maximum likelihood principle (see, e.g., Fan and Yao (2005)). On the other hand, GARCH models for high-frequency data (HFD) in finance necessitate an application of recursive (i.e., online) approaches which mostly consist in the state space modeling.

Hendrych and Cipra (2018) modified recursive algorithms suggested originally for system identification in engineering (see Ljung (1999), Ljung and Söderström (1983), Söderström and Stoica (1989)) to be applicable also for online estimation of (multivariate) GARCH models in finance. The method combines so-called *recursive pseudo-linear regression* with the ML estimation:

One uses the usual model framework for multivariate GARCH processes $\{\mathbf{r}_t\}$, which are formed mostly by m-variate vectors of log returns of financial assets from a given portfolio (see Sect. 13.3):

$$\mathbf{r}_t = \mathbf{H}_t^{1/2} \cdot \boldsymbol{\varepsilon}_t, \tag{14.31}$$

where $\{\boldsymbol{\varepsilon}_t\}$ is an *iid* multivariate white noise with normal distribution

$$\boldsymbol{\varepsilon}_t \sim N(\mathbf{0},\ \mathbf{I}) \tag{14.32}$$

and $\mathbf{H}_t^{1/2}$ is the square root matrix of conditional covariance matrix $\mathbf{H}_t$ expressed in time t as a suitable function of the information Ω_{t-1} known till time $t - 1$. In particular, $\mathbf{H}_t$ is a positive definite Ω_{t-1}-measurable matrix and $\mathbf{H}_t = \mathbf{H}_t^{1/2}\left(\mathbf{H}_t^{1/2}\right)'$.

As the corresponding conditional moments are

$$\mathrm{E}(\mathbf{r}_t|\Im_{t-1}) = \mathbf{0}, \qquad \mathrm{var}(\mathbf{r}_t|\Im_{t-1}) = \mathbf{H}_t, \tag{14.33}$$

the conditional probability density is obviously

$$f(\mathbf{r}_t|\Im_{t-1}) = |2\pi\ \mathbf{H}_t|^{-1/2} \exp\left\{-\frac{1}{2}\mathbf{r}_t'\ \mathbf{H}_t^{-1}\mathbf{r}_t\right\}. \tag{14.34}$$

Hence the conditional ML estimator of the true parameter vector $\boldsymbol{\theta}$ for modeling $\mathbf{H}_t(\boldsymbol{\theta})$ can be found by minimizing

$$\min_{\boldsymbol{\theta}} \sum_{t=1}^{T} \left[\ln|\mathbf{H}_t(\boldsymbol{\theta})| + r_t'\ \mathbf{H}_t(\boldsymbol{\theta})^{-1} r_t\right]. \tag{14.35}$$

In this phase, we apply the general scheme of recursive pseudo-linear regression (see, e.g., Hendrych and Cipra (2018), Ljung (1999)):

$$\widehat{\boldsymbol{\theta}}_t = \widehat{\boldsymbol{\theta}}_{t-1} - \eta_t \mathbf{R}_t^{-1}\mathbf{F}_t'\left(\widehat{\boldsymbol{\theta}}_{t-1}\right), \tag{14.36}$$

$$\mathbf{R}_t = \mathbf{R}_{t-1} + \eta_t \left[\widetilde{\mathbf{F}}''_t \left(\widehat{\boldsymbol{\theta}}_{t-1} \right) - \mathbf{R}_{t-1} \right], \tag{14.37}$$

$$\eta_t = \frac{1}{1 + \xi_t/\eta_{t-1}} \quad \text{for a forgetting factor } \xi_t, \tag{14.38}$$

where

$$F_t(\boldsymbol{\theta}) = \ln |\mathbf{H}_t(\boldsymbol{\theta})| + \boldsymbol{r}_t^{\mathrm{T}} \mathbf{H}_t(\boldsymbol{\theta})^{-1} \boldsymbol{r}_t. \tag{14.39}$$

Here $\mathbf{F}'_t(\boldsymbol{\theta})$ denotes the gradient of $F_t(\boldsymbol{\theta})$ and $\widetilde{\mathbf{F}}''_t(\boldsymbol{\theta})$ is an approximation of the Hessian matrix $\mathbf{F}''_t(\boldsymbol{\theta})$ such that

$$\mathbf{E}\left(\widetilde{\mathbf{F}}''_t \left(\widehat{\boldsymbol{\theta}}_{t-1} \right) - \mathbf{F}''_t \left(\widehat{\boldsymbol{\theta}}_{t-1} \right) | \Im_{t-1} \right) = \mathbf{0} \tag{14.40}$$

(the approximation based on the conditional mean value in (14.40) makes simpler the calculation of Hessian matrix). Finally, the *forgetting factor* $\{\xi_t\}$ in (14.38) substantially improves convergence and statistical properties of the given recursive estimation. The usual choice in practice is either a *constant* forgetting factor ξ (e.g., $\xi = 0.95$) or an *increasing* forgetting factor, e.g.,

$$\xi_t = \widetilde{\xi} \cdot \xi_{t-1} + \left(1 - \widetilde{\xi} \right), \quad \xi_0, \ \widetilde{\xi} \ \in \ (0, 1). \tag{14.41}$$

Besides the choice of forgetting factor, further technicalities must by solved before applying the estimation in practice, e.g., the initialization of the estimation algorithm. The special case of recursive estimation of univariate GARCH models is shown in Hendrych and Cipra (2018).

Example 14.5 Let us consider the recursive estimation of the (single) parameter λ in the multivariate EWMA (or MEWMA or *scalar* VEC-IGARCH(1,1)) model (13.1) which can be rewritten as

$$\mathbf{H}_t(\lambda) = (1 - \lambda) \ \mathbf{r}_t \mathbf{r}'_t + \lambda \mathbf{H}_{t-1}(\lambda), \quad \lambda \in (0, 1), \tag{14.42}$$

where the discount constant λ ($0 < \lambda < 1$) is the only parameter in this very simple multivariate GARCH model. Then after troublesome (matrix) arrangements one can rewrite the recursive pseudo-linear regression (14.36)–(14.38) to the form

$$\widehat{\lambda}_t = \widehat{\lambda}_{t-1} - \eta_t R_t^{-1} \left[\operatorname{tr} \left(\mathbf{H}_t^{-1} \left(\widehat{\lambda}_{t-1} \right) \frac{\partial \mathbf{H}_t \left(\widehat{\lambda}_{t-1} \right)}{\partial \lambda} \right) - \mathbf{r}'_t \, \mathbf{H}_t^{-1} \left(\widehat{\lambda}_{t-1} \right) \frac{\partial \mathbf{H}_t \left(\widehat{\lambda}_{t-1} \right)}{\partial \lambda} \mathbf{H}_t^{-1} \left(\widehat{\lambda}_{t-1} \right) \mathbf{r}_t \right], \tag{14.43}$$

$$R_t = R_{t-1} + \eta_t \left[\text{tr}\left(\mathbf{H}_t^{-1}\left(\widehat{\lambda}_{t-1}\right) \frac{\partial \mathbf{H}_t\left(\widehat{\lambda}_{t-1}\right)}{\partial \lambda} \mathbf{H}_t^{-1}\left(\widehat{\lambda}_{t-1}\right) \frac{\partial \mathbf{H}_t\left(\widehat{\lambda}_{t-1}\right)}{\partial \lambda} \right) - R_{t-1} \right], \tag{14.44}$$

$$\mathbf{H}_{t+1}\left(\widehat{\lambda}_t\right) = \left(1 - \widehat{\lambda}_t\right)\mathbf{r}_t\mathbf{r}_t' + \widehat{\lambda}_t\mathbf{H}_t\left(\widehat{\lambda}_{t-1}\right), \tag{14.45}$$

$$\frac{\partial \mathbf{H}_{t+1}\left(\widehat{\lambda}_t\right)}{\partial \lambda} = -\mathbf{r}_t\mathbf{r}_t' + \mathbf{H}_t\left(\widehat{\lambda}_{t-1}\right) + \widehat{\lambda}_t \frac{\partial \mathbf{H}_t\left(\widehat{\lambda}_{t-1}\right)}{\partial \lambda}, \tag{14.46}$$

$$\eta_t = \frac{1}{1 + \xi_t/\eta_{t-1}} \quad \text{for a forgetting factor } \{\xi_t\} \tag{14.47}$$

(the symbol tr(**A**) denotes the trace of matrix **A**). Note that all calculations (including the calculations of matrix derivatives in (14.46)) are recursive.

Figure 14.1 shows the simulation results in the bivariate case, where the process $\{\mathbf{r}_t\}$ in (14.42) was generated as a *iid* normal white noise with zero mean values, unit variances, and correlation coefficient 0.8. Four alternatives with true values of the parameter λ (namely $\lambda = 0.91, 0.94, 0.97, 0.99$) were considered (one realized 1000 simulations for each of them).

This recursive estimate was applied for 647 couples of daily log returns of 40 currency rates versus EUR (i.e., (EUR, CURR1) and (EUR, CURR2)) from January 1999 to December 2017 according to the European Central Bank. For estimating the parameter λ, three approaches were used (see Cipra and Hendrych (2019)):

- Fixed $\widehat{\lambda}_t = 0.94$.
- Recursive MEWMA method with fixed forgetting factor $\xi_t = 0.995$.
- Recursive MEWMA method with increasing forgetting factor $\xi_t = \widetilde{\xi} \cdot \xi_{t-1} + \left(1 - \widetilde{\xi}\right)$, where $\xi_0 = 0.95$, $\widetilde{\xi} = 0.99$.

For example, Fig. 14.2 presents the parameter estimators for the couple EUR/USD and EUR/JPY. Moreover, the corresponding estimated conditional correlation and volatilities are shown using the results of recursive MEWMA method with increasing forgetting factor.

◇

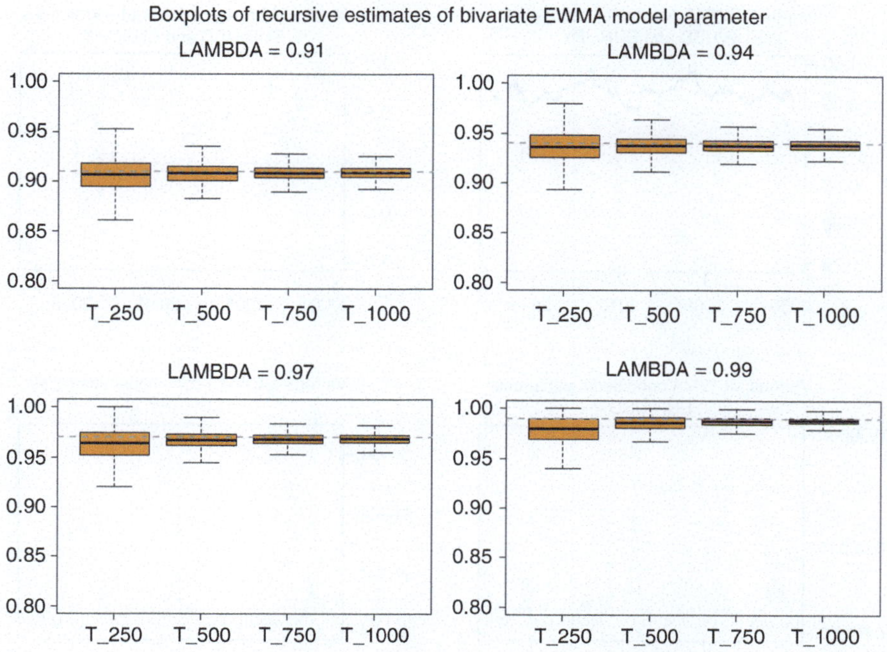

Fig. 14.1 Recursive estimation of parameter λ in bivariate EWMA model (14.42) (boxplots are based on 1000 simulations for four alternatives with true values $\lambda = 0.91, 0.94, 0.97, 0.99$). Source: Cipra and Hendrych (2019)

14.2 State Space Model Approach to Exponential Smoothing

Exponential smoothing from Sects. 3.3 and 4.1.3 including Holt's and Holt–Winters' method can be formulated as filtering and predicting based on state space modeling (see the monograph by Hyndman et al. (2008)). One can even systematically classify particular models according to the type of trend, seasonal, and residual (or error) components (see Sect. 2.2.2) and the type of decomposition of time series (additive or multiplicative).

For instance, let us consider the following DLM (14.1)–(14.2) for a (univariate) time series $\{y_t\}$:

$$\mathbf{x}_t = \begin{pmatrix} 1 & 1 \\ 0 & 1 \end{pmatrix} \mathbf{x}_{t-1} + \begin{pmatrix} \alpha \\ \gamma^* \end{pmatrix} \varepsilon_t, \tag{14.48}$$

$$y_t = (1 \;\; 1) \; \mathbf{x}_{t-1} + \varepsilon_t \tag{14.49}$$

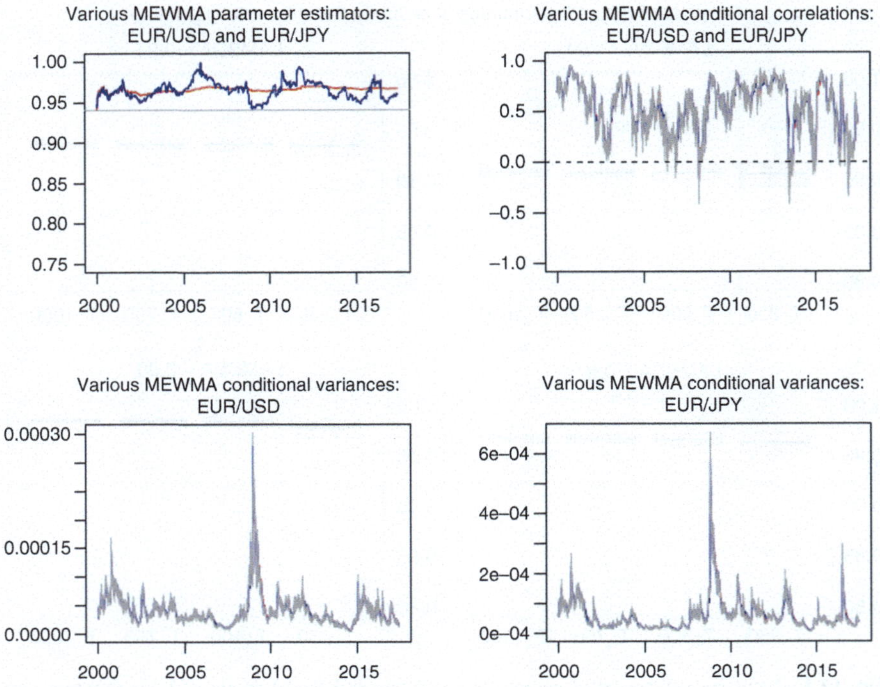

Fig. 14.2 *MEWMA method for log returns of currency rates for the couple EUR/USD and EUR/JPY* from January 1999 to December 2017 in Example 14.5: the parameter estimators (the smooth non-constant line plots the recursive MEWMA estimate with increasing forgetting factor) and the corresponding model estimates of conditional correlation coefficient and volatilities (for recursively estimated λ with increasing forgetting factor). Source: Cipra and Hendrych (2019)

with the state vector $\mathbf{x}_t = (L_t, T_t)'$, where the symbols L_t and T_t denote the level and slope of the given time series (see Sect. 3.1.1), respectively, and $\{\varepsilon_t\}$ is a white noise (note that the residuals in state and observation equations in time t are mutually correlated). Then one obtains gradually

$$\begin{aligned} L_t &= L_{t-1} + T_{t-1} + \alpha\varepsilon_{t-1} = L_{t-1} + T_{t-1} + \alpha(y_t - L_{t-1} - T_{t-1}) \\ &= \alpha y_t + (1-\alpha)(L_{t-1} + T_{t-1}), \end{aligned}$$

$$\begin{aligned} T_t &= T_{t-1} + \gamma^*\varepsilon_{t-1} = T_{t-1} + \alpha\gamma\varepsilon_{t-1} = T_{t-1} + \gamma(L_t - L_{t-1} - T_{t-1}) \\ &= \gamma(L_t - L_{t-1}) + (1-\gamma)T_{t-1}, \end{aligned}$$

$$\widehat{y}_{t+\tau}(t) = \mathrm{E}\big(y_{t+\tau} \big| \, \mathbf{x}_t\big) = L_t + T_t \cdot \tau \quad (\tau \geq 0),$$

which is equivalent to *Holt's method* (3.106)–(3.109) (we put $\gamma^* = \alpha\gamma$).

One can proceed analogously in the case of *additive Holt–Winters' method*. The corresponding DML can be chosen as

$$\mathbf{x}_t = \begin{pmatrix} 1 & 1 & 0 & 0 & \dots & 0 & 0 \\ 0 & 1 & 0 & 0 & \dots & 0 & 0 \\ 0 & 0 & 0 & 0 & \dots & 0 & 1 \\ 0 & 0 & 1 & 0 & \dots & 0 & 0 \\ 0 & 0 & 0 & 1 & \dots & 0 & 0 \\ \vdots & \vdots & \vdots & \vdots & \ddots & \vdots & \vdots \\ 0 & 0 & 0 & 0 & \cdot^{\cdot^{\cdot}} & 1 & 0 \end{pmatrix} \mathbf{x}_{t-1} + \begin{pmatrix} \alpha \\ \gamma^* \\ \delta^* \\ 0 \\ 0 \\ \vdots \\ 0 \end{pmatrix} \varepsilon_t, \tag{14.50}$$

$$y_t = (1\ 1\ 0\ 0\ \dots\ 0\ 1)\ \mathbf{x}_{t-1} + \varepsilon_t \tag{14.51}$$

with the state vector $\mathbf{x}_t = (L_t, T_t, I_t, I_{t-1}, \dots, I_{t-s+1})'$, where the symbols L_t, T_t, and I_t denote the level, slope, and seasonal index of the given time series in time t, respectively, and $\{\varepsilon_t\}$ is again a white noise. Hence it follows gradually

$$L_t = \alpha(y_t - I_{t-s}) + (1 - \alpha)(L_{t-1} + T_{t-1}), \tag{14.52}$$

$$T_t = \gamma(L_t - L_{t-1}) + (1 - \gamma)T_{t-1}, \tag{14.53}$$

$$I_t = \delta(y_t - L_t) + (1 - \delta)I_{t-s}, \tag{14.54}$$

$$\hat{y}_t = L_t + I_t, \tag{14.55}$$

$$\begin{aligned} \hat{y}_{t+\tau}(t) &= L_t + T_t \cdot \tau + I_{t+\tau-s} && \text{for} \quad \tau = 1, \dots, s, \\ &= L_t + T_t \cdot \tau + I_{t+\tau-2s} && \text{for} \quad \tau = s + 1, \dots, 2s, \\ &\vdots \end{aligned} \tag{14.56}$$

which is equivalent to the additive Holt–Winters' method (4.1.16)–(4.1.20) with seasonality s (we put $\gamma^* = \alpha\gamma$ and $\delta^* = (1 - \alpha)\,\delta$). Let us derive, e.g., the recursive formula (14.54). It holds

$$\begin{aligned} I_t &= I_{t-s} + \delta^*\varepsilon_t = I_{t-s} + (1-\alpha)\delta\varepsilon_t = I_{t-s} + \delta(y_t - L_t - I_{t-s}) \\ &= \delta(y_t - L_t) + (1-\delta)I_{t-s}. \end{aligned}$$

In this context, a broad class of state space models can be considered providing various types of exponential smoothing alternatives.

Table 14.2 Trend types used in state space approach to exponential smoothing (φ denotes a damping parameter $(0 < \varphi < 1)$)

Trend	
None (N)	L
Additive (A)	$L + T \cdot \tau$
Additive damped (A_d)	$L + (\phi + \phi^2 + \cdots + \phi^\tau) \cdot \tau$
Multiplicative (M)	$L \cdot T^\tau$
Multiplicative damped (M_d)	$L \cdot T^{\phi+\phi^2+\cdots+\phi^\tau}$

1. Classification of Exponential Smoothing Models
State space approach to exponential smoothing by Hyndman et al. (2008) starts with the classification of trend components. The five trend types (or growth patterns) are presented in Table 14.2.

The classification of state space models in the context of exponential smoothing is based on the additive character (A) or multiplicative character (M) of particular decomposition components. It is a triplet ETS(· , · , ·) for *E*rror component (A or M), *T*rend component (N, A, A_d, M, or M_d; see Table 14.2), and *S*easonal component (N, A, or M). According to this classification, e.g., the Holt's method is ETS(A, A, N) (see (14.48)–(14.49)), the additive Holt–Winters' method is ETS(A, A, A) (see (14.50)–(14.51)), and similarly the multiplicative Holt–Winters' method is ETS(A, A, M).

Tables 14.3, 14.4, and 14.5 present the recursive relations of the type (14.52)–(14.56) for models ETS(A, · , N), ETS(A, · , A), ETS(A, · , M), respectively, i.e., for additive errors only. In particular, Table 14.3 contains the Holt's method, Table 14.4 contains the additive Holt–Winters' method, and Table 14.5 contains the

Table 14.3 Recursive relations of exponential smoothing for state space models ETS(A, · , N) $(\varphi_\tau = \varphi + \varphi^2 + \cdots + \varphi^\tau)$

Trend	Recursive relations for ETS(A, · , N)
None (N)	$L_t = \alpha y_t + (1 - \alpha)L_{t-1}$ $\widehat{y}_{t+\tau}(t) = L_t$
Additive (A)	$L_t = \alpha y_t + (1 - \alpha)(L_{t-1} + T_{t-1})$ $T_t = \gamma(L_t - L_{t-1}) + (1 - \gamma)T_{t-1}$ $\widehat{y}_{t+\tau}(t) = L_t + T_t \cdot \tau$
Additive damped (A_d)	$L_t = \alpha y_t + (1 - \alpha)(L_{t-1} + \phi T_{t-1})$ $T_t = \gamma(L_t - L_{t-1}) + (1 - \gamma)T_{t-1} \cdot \phi$ $\widehat{y}_{t+\tau}(t) = L_t + T_t \cdot \phi_\tau$
Multiplicative (M)	$L_t = \alpha y_t + (1 - \alpha)L_{t-1}T_{t-1}$ $T_t = \gamma(L_t/L_{t-1}) + (1 - \gamma)T_{t-1}$ $\widehat{y}_{t+\tau}(t) = L_t T_t^\tau$
Multiplicative damped (M_d)	$L_t = \alpha y_t + (1 - \alpha)T_{t-1}^\phi$ $T_t = \gamma(L_t/L_{t-1}) + (1 - \gamma)T_{t-1}^\phi$ $\widehat{y}_{t+\tau}(t) = L_t T_t^{\phi_\tau}$

Table 14.4 Recursive relations of exponential smoothing for state space models ETS(A, · , A) ($\varphi_\tau = \varphi + \varphi^2 + \cdots + \varphi^\tau$, $\tau_s^+ = [(\tau - 1) \bmod s] + 1$)

Trend	Recursive relations for ETS(A, · , A)
None (N)	$L_t = \alpha(y_t - 1I_{t-s}) + (1-\alpha)L_{t-1}$ $I_t = \delta(y_t - L_{t-1}) + (1-\delta)I_{t-s}$ $\widehat{y}_{t+\tau}(t) = L_t + I_{t-s+\tau_s^+}$
Additive (A)	$L_t = \alpha(y_t - I_{t-s}) + (1-\alpha)(L_{t-1} + T_{t-1})$ $T_t = \gamma(L_t - L_{t-1}) + (1-\gamma)T_{t-1}$ $I_t = \delta(y_t - L_{t-1} - T_{t-1}) + (1-\delta)I_{t-s}$ $\widehat{y}_{t+\tau}(t) = L_t + T_t \cdot \tau + I_{t-s+\tau_s^+}$
Additive damped (A_d)	$L_t = \alpha(y_t - I_{t-s}) + (1-\alpha)(L_{t-1} + T_{t-1} \cdot \phi)$ $T_t = \gamma(L_t - L_{t-1}) + (1-\gamma)T_{t-1} \cdot \phi$ $I_t = \delta(y_t - L_{t-1} - T_{t-1} \cdot \phi) + (1-\delta)I_{t-s}$ $\widehat{y}_{t+\tau}(t) = L_t + T_t \cdot \phi_\tau + I_{t-s+\tau_s^+}$
Multiplicative (M)	$L_t = \alpha(y_t - I_{t-s}) + (1-\alpha)L_{t-1}T_{t-1}$ $T_t = \gamma(L_t/L_{t-1}) + (1-\gamma)T_{t-1}$ $I_t = \delta(y_t - L_{t-1}T_{t-1}) + (1-\delta)I_{t-s}$ $\widehat{y}_{t+\tau}(t) = L_t T_t^\tau + I_{t-s+\tau_s^+}$
Multiplicative damped (M_d)	$L_t = \alpha(y_t - I_{t-s}) + (1-\alpha)L_{t-1}T_{t-1}^\phi$ $T_t = \gamma(L_t/L_{t-1}) + (1-\gamma)T_{t-1}^\phi$ $I_t = \delta\left(y_t - L_{t-1}T_{t-1}^\phi\right) + (1-\delta)I_{t-s}$ $\widehat{y}_{t+\tau}(t) = L_t T_t^{\phi_\tau} + I_{t-s+\tau_s^+}$

Table 14.5 Recursive relations of exponential smoothing for state space models ETS(A, · , M) ($\varphi_\tau = \varphi + \varphi^2 + \ldots + \varphi^\tau$, $\tau_s^+ = [(\tau - 1) \bmod s] + 1$)

Trend	Recursive relations for ETS(A, · , M)
None (N)	$L_t = \alpha(y_t/I_{t-s}) + (1-\alpha)L_{t-1}$ $I_t = \delta(y_t/L_{t-1}) + (1-\delta)I_{t-s}$ $\widehat{y}_{t+\tau}(t) = L_t I_{t-s+\tau_s^+}$
Additive (A)	$L_t = \alpha(y_t/I_{t-s}) + (1-\alpha)(L_{t-1} + T_{t-1})$ $T_t = \gamma(L_t - L_{t-1}) + (1-\gamma)T_{t-1}$ $I_t = \delta(y_t/(L_{t-1} + T_{t-1})) + (1-\delta)I_{t-s}$ $\widehat{y}_{t+\tau}(t) = (L_t + T_t \cdot \tau)I_{t-s+\tau_s^+}$
Additive damped (A_d)	$L_t = \alpha(y_t/I_{t-s}) + (1-\alpha)(L_{t-1} + T_{t-1} \cdot \phi)$ $T_t = \gamma(L_t - L_{t-1}) + (1-\gamma)T_{t-1} \cdot \phi$ $I_t = \delta(y_t/(L_{t-1} + T_{t-1} \cdot \phi) + (1-\delta)I_{t-s}$ $\widehat{y}_{t+\tau}(t) = (L_t + T_t \cdot \phi_\tau)I_{t-s+\tau_s^+}$
Multiplicative (M)	$L_t = \alpha(y_t/I_{t-s}) + (1-\alpha)L_{t-1}T_{t-1}$ $T_t = \gamma(L_t/L_{t-1}) + (1-\gamma)T_{t-1}$ $I_t = \delta(y_t/(L_{t-1}T_{t-1})) + (1-\delta)I_{t-s}$ $\widehat{y}_{t+\tau}(t) = L_t T_t^\tau I_{t-s+\tau_s^+}$
Multiplicative damped (M_d)	$L_t = \alpha(y_t/I_{t-s}) + (1-\alpha)L_{t-1}T_{t-1}^\phi$ $T_t = \gamma(L_t/L_{t-1}) + (1-\gamma)T_{t-1}^\phi$ $I_t = \delta\left(y_t\left(L_{t-1}T_{t-1}^\phi\right)\right) + (1-\delta)I_{t-s}$ $\widehat{y}_{t+\tau}(t) = L_t T_t^{\phi_\tau} I_{t-s+\tau_s^+}$

multiplicative Holt–Winters' method. One uses a simplifying denotation in these Tables, namely $\varphi_\tau = \varphi + \varphi^2 + \cdots + \varphi^\tau$ for a damping parameter φ $(0 < \varphi < 1)$ and $\tau_s^+ = [(\tau - 1) \bmod s] + 1$ to simplify prediction relations of the type (14.56) (the *modulo* operation finds the remainder after division by s). Moreover, the smoothing relations of the type (14.55) are ignored since they can be obtained from the corresponding prediction relations if we put $\tau = 0$.

Remark 14.2 An example of state space models with multiplicative error is the model ETS(M, A, N) with DML of the form

$$\mathbf{x}_t = \begin{pmatrix} 1 & 1 \\ 0 & 1 \end{pmatrix} \mathbf{x}_{t-1} + (\, 1 \quad 1 \,) \mathbf{x}_{t-1} \begin{pmatrix} \alpha \\ \gamma^* \end{pmatrix} \varepsilon_t, \tag{14.57}$$

$$y_t = (\, 1 \quad 1 \,) \mathbf{x}_{t-1} \cdot (1 + \varepsilon_t) \tag{14.58}$$

which can be rewritten as

$$L_t = (L_{t-1} + T_{t-1})(1 + \alpha \varepsilon_t), \tag{14.59}$$

$$T_t = T_{t-1} + \gamma^* (L_{t-1} + T_{t-1}) \cdot \varepsilon_t, \tag{14.60}$$

$$\widehat{y}_t = (L_{t-1} + T_{t-1})(1 + \varepsilon_t). \tag{14.61}$$

Here the recursive relations are not presented due to their complexity. To derive them one should express at first the relative error as

$$\varepsilon_t = \frac{y_t - \mathrm{E}(y_t |\ \mathbf{x}_{t-1})}{\mathrm{E}(y_t |\ \mathbf{x}_{t-1})} \tag{14.62}$$

(from the observation relation $y_t = \mathrm{E}(y_t \mid \mathbf{x}_{t-1}) \cdot (1 + \varepsilon_t)$ of the corresponding DML).

◇

2. Construction of Exponential Smoothing Models

There is a lot of technicalities to be solved when constructing the exponential smoothing models described in the previous text (selection, estimation, initialization, assessing forecast accuracy; see Hyndman et al. (2008) and also Sects. 3.3 and 4.1.3). Here we deal briefly with the problem of model estimation only.

For this purpose, we apply the following general form of the corresponding DML:

$$\mathbf{x}_t = \mathbf{f}(\mathbf{x}_{t-1}) + \mathbf{g}(\mathbf{x}_{t-1}) \cdot \varepsilon_t, \tag{14.63}$$

$$y_t = w(\mathbf{x}_{t-1}) + r(\mathbf{x}_{t-1}) \cdot \varepsilon_t \tag{14.64}$$

with the state vector $\mathbf{x}_t = (L_t, T_t, I_t, I_{t-1}, \ldots, I_{t-s+1})'$ serving as an argument of scalar and vector (linear) functions. Further one assumes that $\{\varepsilon_t\}$ is a normal white noise with variance σ^2. The models with additive errors have $r(\mathbf{x}_{t-1}) = 1$ (so that $y_t = \mathrm{E}(y_t \mid \mathbf{x}_{t-1}) + \varepsilon_t$), while the models with multiplicative errors have $r(\mathbf{x}_{t-1}) = \mathrm{E}(y_t \mid \mathbf{x}_{t-1})$ (so that $y_t = \mathrm{E}(y_t \mid \mathbf{x}_{t-1})\cdot(1 + \varepsilon_t)$).

If $\boldsymbol{\theta} = (\alpha, \gamma, \delta, \varphi)'$ is the vector of unknown model parameters and $\mathbf{x}_0$ contains given initial state values, then the corresponding (normal) log likelihood function can be written as

$$L(\boldsymbol{\theta}, \sigma^2 | \mathbf{y}, \mathbf{x}_0) = -\frac{n}{2} \ln\left(2\pi\sigma^2\right) - \sum_{t=1}^{n} \ln|r(\mathbf{x}_{t-1})| - \frac{1}{2} \sum_{t=1}^{n} \varepsilon_t^2 / \sigma^2 \tag{14.65}$$

for observations y_t from the vector $\mathbf{y} = (y_1, \ldots, y_n)'$. If taking the partial derivative with respect to σ^2 and setting it to zero one obtains the maximum likelihood estimate of the innovation variance σ^2 as

$$\widehat{\sigma^2} = \frac{1}{n} \sum_{t=1}^{n} \varepsilon_t^2. \tag{14.66}$$

After putting (14.66) to (14.65) one obtains the concentrated log likelihood. Hence, maximum likelihood estimates of parameters $\boldsymbol{\theta} = (\alpha, \gamma, \delta, \varphi)'$ can be obtained by minimizing (twice) the negative log likelihood function, i.e.,

$$\min_{\boldsymbol{\theta}} \left\{ n \ln\left(\sum_{t=1}^{n} \varepsilon_t^2\right) + 2 \sum_{t=1}^{n} \ln|r(\mathbf{x}_{t-1})| \right\}, \tag{14.67}$$

where $\{\mathbf{x}_{t-1}\}$ and $\{\varepsilon_t\}$ are calculated recursively using initial state values $\mathbf{x}_0$ and observations y_t from the vector $\mathbf{y} = (y_1, \ldots, y_n)'$. This estimation method can be completed by information criteria (e.g., AIC; see Sect. 6.3.1) to identify (select) correct state space models.

Remark 14.3 One can generalize the given approach also for nonlinear state space models, e.g., for time series with conditional heteroscedasticity using the model

$$L_t = \mu + L_{t-1} + \alpha\varepsilon_t, \tag{14.68}$$

$$\ln h_{t+1} = \nu_0 + \nu_1 \ln h_t + \nu_2 \ln|\varepsilon_t|, \tag{14.69}$$

$$\ln y_t = L_{t-1} + \varepsilon_t, \tag{14.70}$$

where $\varepsilon_t \sim N(0, h_t)$ (sometimes the last term in (14.69) is supplemented by further positive parameter v_3 to the form $v_2\ln(|\varepsilon_t| + v_3)$ to reduce the problem of small residuals as arguments of logarithmic function).

◇

Example 14.6 Hyndman et al. (2008) estimated the model (14.68)–(14.70) for monthly closing prices of the Dow Jones Index (DJI) over the period January 1990–March 2007 as

$$L_t = 0.0074 + L_{t-1} + 0.960\varepsilon_t,$$

$$\ln h_{t+1} = 0.043 + 0.932 \ln h_t + 0.125 \ln |\varepsilon_t|,$$

$$\ln y_t = L_{t-1} + \varepsilon_t.$$

◇

14.3 Exercises

Exercise 14.1 Derive the recursive relations of exponential smoothing for particular state space models in Tables 14.3, 14.4, and 14.5 (*hint:* e.g., *for ETS(A,N,N) in the first row of* Table 14.3 *using model* $L_t = L_{t-1} + \alpha\, \varepsilon_t$ *and* $y_t = L_{t-1} + \varepsilon_t$ *one gets* $L_t = L_{t-1} + \alpha(y_t - L_{t-1}) = \alpha y_t + (1 - \alpha)L_{t-1}$).

Exercise 14.2 Derive in detail the minimized expression in (14.67) when constructing the maximum likelihood parameter estimates of state space models of exponential smoothing.

References

Abraham, B., Ledolter, J.: Statistical Methods for Forecasting. Wiley, New York (1983)

Acerbi, C.: Spectral measures of risk: a coherent representation of subjective risk aversion. J. Bank. Financ. **26**, 1505–1518 (2002)

Adrian, T., Brunnermeier, M.K.: CoVaR. Federal Reserve Bank of New York, Staff Report no. 348, September 2008

Ait-Sahalia, Y.: Testing continuous-time models for the spot interest rate. Rev. Financ. Stud. **9**, 385–426 (1996)

Ait-Sahalia, Y., Jacod, J.: High-Frequency Financial Econometrics. Princeton University Press, Princeton (2014)

Alexander, C.O., Chibumba, A.M.: Multivariate Orthogonal Factor GARCH. University of Sussex Discussion Papers in Mathematics (1997)

Almon, S.: The distributed lag between capital appropriations and expenditures. Econometrica. **33**, 178–196 (1965)

Al-Osh, M.A., Alzaid, A.A.: First-order integer-valued autoregressive (INAR(1)) process. J. Time Ser. Anal. **8**, 261–275 (1987)

Artzner, P., Delbaen, F., Eber, J.-M., Heath, D.: Coherent measures of risk. Math. Financ. **9**, 203–228 (1999)

Bauwens, L., Laurent, S., Rombouts, J.: Multivariate GARCH models: a survey. J. Appl. Econ. **21**, 79–109 (2006)

Baxter, M., Rennie, A.: Financial Calculus. An Introduction to Derivative Pricing. Cambridge University Press, Cambridge (1996)

Bollerslev, T.: Generalized autoregressive conditional heteroscedasticity. J. Econ. **31**, 307–327 (1986)

Bollerslev, T.: Modeling the coherence in short-run nominal exchange rates: a multivariate generalized ARCH model. Rev. Econ. Stat. **72**, 498–505 (1990)

Bollerslev, T., Wooldridge, J.M.: Quasi-maximum likelihood estimation and inference in dynamic models with time varying covariances. Econ. Rev. **11**, 143–172 (1992)

Bollerslev, T., Engle, R.F., Wooldridge, J.M.: A capital-asset pricing model with time-varying covariances. J. Polit. Econ. **96**, 116–131 (1988)

Bølviken, E.: New tests of significance in periodogram analysis. Scand. J. Statist. **10**, 1–10 (1983)

Bowerman, B.L., O'Connell, R.T.: Time Series Forecasting. Duxbury Press, Boston (1987)

Box, G.E.P., Jenkins, G.M.: Time Series Analysis, Forecasting and Control. Holden-Day, San Francisco (1970)

Box, G.E.P., Tiao, G.C.: Intervention analysis with applications to economic and environ-mental problems. J. Am. Stat. Assoc. **70**, 70–79 (1975)

T. Cipra, *Time Series in Economics and Finance*,
https://doi.org/10.1007/978-3-030-46347-2

Brock, W.A., Dechert, D., Scheinkman, H., LeBaron, B.: A test for independence based on the correlation dimension. Econ. Rev. **15**, 197–235 (1996)

Brockwell, P.J.: Lévy-driven continuous-time ARMA processes. In: Andersen, T.G., et al. (eds.) Handbook of Financial Time Series, Part III: Topics in Continuous Time Processes. Springer, Berlin (2009)

Brockwell, P.J., Davis, R.A.: Time Series: Theory and Methods. Springer, New York (1993)

Brockwell, P.J., Davis, R.A.: Introduction to Time Series and Forecasting. Springer, New York (1996)

Brownlees, C.T., Engle, R.: Volatility, correlation and tails for systemic risk measurement. Stern Center for Research Computing, New York University, New York (2012)

Campbell, J.Y., Lo, A.W., MacKinlay, A.C.: The Econometrics of Financial Markets. Princeton University Press, Princeton (1997)

Chan, K.S., Tong, H.: On estimating thresholds in autoregressive models. J. Time Ser. Anal. **7**, 179–190 (1986)

Chappel, D., Padmore, J., Mistry, P., Ellis, C.: A threshold model for French franc/Deutsch mark exchange rate. J. Forecast. **15**, 155–164 (1996)

Chen, R., Tsay, R.S.: Functional-coefficients autoregressive models. J. Am. Stat. Assoc. **88**, 298–308 (1993)

Chiriac, R., Voev, V.: Modeling and forecasting multivariate realized volatility. J. Appl. Econ. **26**, 922–947 (2011)

Cipra, T.: Financial and Insurance Formulas. Springer, New York (2010)

Cipra, T., Hendrych, R.: Systemic risk in financial risk regulation. Czech J. Econ. Financ. **67**, 15–38 (2017)

Cipra, T., Hendrych, R.: Modeling of currency covolatilities. Statistika. **99**(3), 259–271 (2019)

Clements, M., Harvey, D.: Forecast combination and encompassing. In: Mills, T., Patterson, K. (eds.) The Palgrave Handbook of Econometrics, Applied Econometrics, vol. 2. Palgrave, Oxford (2011)

Clements, A., Scott, A., Silvennoinen, A.: Forecasting Multivariate Volatility in Larger Dimensions: Some Practical Issues. Working Paper #80, NCER Working Paper Series (2012)

Conley, T.G., Hansen, L.P., Luttmer, E.G.J., Scheinkman, J.A.: Short-term interest rates as subordinate diffusions. Rev. Financ. Stud. **10**, 525–577 (1997)

Dagum, E.B., Bianconcini, S.: Seasonal Adjustment Methods and Real Time Trend-Cycle Estimation. Springer, New York (2016)

Davidson, J.: Econometric Theory. Blackwell, Oxford (2000)

Dempster, A.P., Laird, N.M., Rubin, D.B.: Maximum likelihood from incomplete data via the EM algorithm. J. R. Stat. Soc. **39**, 1–38 (1977)

Dickey, D.A., Fuller, W.A.: Distribution of estimators for time series regressions with a unit root. J. Am. Stat. Assoc. **74**, 427–431 (1979)

Dickey, D.A., Fuller, W.A.: Likelihood ratio statistics for autoregressive time series with a unit root. Econometrica. **49**, 1057–1072 (1981)

Ding, Z., Granger, C.W.J., Engle, R.F.: A long memory property of stock market returns and a new model. J. Empir. Financ. **1**, 83–106 (1993)

Duffie, D.: Security Markets: Stochastic Models. Academic, New York (1988)

Dupačová, J., Hurt, J., Štěpán, J.: Stochastic Modeling in Economics and Finance. Kluwer, Boston (2002)

Durbin, J., Koopman, S.J.: Time Series Analysis by State Space Methods. Oxford University Press, Oxford (2012)

Eilers, P.H.C., Marx, B.D.: Splines, knots and penalties. Wiley Interdiscip. Rev.: Comput. Stat. **2** (6), 637–653 (2010)

Elerian, O., Chib, S., Shephard, N.: Likelihood inference for discretely observed non-linear diffusions. Econometrica. **69**, 959–993 (2001)

Elliot, R.J., Kopp, P.E.: Mathematics of Financial Markets. Springer, New York (2004)

Embrechts, P., Kuppelberg, C., Mikosch, T.: Modelling Extremal Events. Springer, Berlin (1997)

Enders, W.: Applied Econometric Time Series. Wiley, New York (1995)
Engle, R.F.: Autoregressive conditional heteroscedasticity with the estimates of the variance of United Kingdom inflations. Econometrica. **50**, 987–1007 (1982)
Engle, R.F.: Dynamic conditional correlation—a simple class of multivariate GARCH models. J. Bus. Econ. Stat. **20**, 339–350 (2002)
Engle, R.F.: Anticipating Correlation. A New Paradigm for Risk Management. Theory and Practice. Princeton University Press, Princeton (2009)
Engle, R.F., Granger, C.W.J.: Co-integration, and error correction: representation, estimation and testing. Econometrica. **55**, 251–276 (1987)
Engle, R.F., Kroner, K.F.: Multivariate simultaneous generalized GARCH. Econ. Theory. **11**, 122–150 (1995)
Engle, R.F., Ng, V.K.: Measuring and testing the impact of news on volatility. J. Financ. **48**, 1749–1778 (1993)
Engle, R.F., Russell, R.J.: Autoregressive conditional duration: a new model for irregularly spaced transaction data. Econometrica. **66**, 1127–1162 (1998)
Engle, R.F., Yoo, B.S.: Forecasting and testing in cointegrated systems. J. Econ. **35**, 143–159 (1987)
Engle, R.F., Lilien, D.M., Robins, R.P.: Estimating time varying risk premia in the term structure: the ARCH-M model. Econometrica. **55**, 391–407 (1987)
Engle, R.F., Ng, V.K., Rothschild, M.: Asset pricing with a factor ARCH covariance structure: empirical estimates for treasury bills. J. Econ. **45**, 213–238 (1990)
Eraker, B.: MCMC analysis of diffusion models with applications to finance. J. Bus. Econ. Stat. **19**, 177–191 (2001)
EViews 10. IHS Global Inc., Englewood (2018)
Fan, J., Yao, Q.: Nonlinear Time Series: Nonparametric and Parametric Methods. Springer, New York (2005)
Fleming, J., Kirby, C., Ostdiek, B.: The economic value of volatility timing using "realized" volatility. J. Financ. Econ. **67**, 473–509 (2003)
Francq, C., Zakoian, J.-M.: GARCH Models. Wiley, Chichester (2010)
Franke, J., Härdle, W., Hafner, C.M.: Statistics of Financial Markets. Springer, New York (2004)
Franses, P.H., van Dijk, D.: Non-Linear Time Series Models in Empirical Finance. Cambridge University Press, Cambridge (2000)
Fuller, W.A.: Introduction to Statistical Time Series. Wiley, New York (1976)
Gallant, A.R., Long, J.R.: Estimating stochastic diffusion equations efficiently by minimum chi-squared. Biometrika. **84**, 125–141 (1997)
Glosten, L.R., Jagannathan, R., Runkle, D.E.: On the relation between the expected value and the volatility of the nominal excess return on stocks. J. Financ. **48**, 1779–1801 (1993)
Gómez, V.: Multivariate Time Series with Linear State Space Structure. Springer, New York (2016)
Gourieroux, C.: ARCH Models and Financial Applications. Springer, New York (1997)
Gourieroux, C., Jasiak, J.: Financial Econometrics: Problems, Models, and Methods. Princeton University Press, Princeton (2001)
Granger, C.W.J.: Investigating causal relations by econometric models and cross-spectral methods. Econometrica. **37**, 424–438 (1969)
Granger, C.W.J.: Long memory relationships and the aggregation of dynamic models. J. Econ. **14**, 227–238 (1980)
Granger, C.W.J., Andersen, A.P.: An Introduction to Bilinear Time Series Models. Vandenhoek and Ruprecht, Gottingen (1978)
Granger, C.W.J., Newbold, P.: Forecasting Economic Time Series. Academic, San Diego (1986)
Greene, W.H.: Econometric Analysis. Prentice Hall, New York (2012)
Haggan, V., Ozaki, T.: Modelling nonlinear vibrations using an amplitude-dependent auto-regressive time series models. Biometrika. **68**, 189–196 (1981)
Hamilton, J.D.: A new approach to the economic analysis of nonstationary time series and business cycle. Econometrica. **57**, 357–384 (1989)

Hamilton, J.D.: Time Series Analysis. Princeton University Press, Princeton (1994)
Härdle, W.: Applied Nonparametric Regression. Cambridge University Press, New York (1990)
Harvey, A.C.: Forecasting, Structural Time Series Models and the Kalman Filter. Cambridge University Press, Cambridge (1989)
Hatanaka, M.: Time-Series-Based Econometrics. Oxford University Press, Oxford (1996)
Hautsch, N.: Econometrics of Financial High-Frequency Data. Springer, Heidelberg (2012)
Heij, C., de Boer, P., Franses, P.H., Kloek, T., van Dijk, H.K.: Econometric Methods with Applications in Business and Economics. Oxford University Press, Oxford (2004)
Hendry, D.F.: Dynamic Econometrics. Oxford University Press, Oxford (1995)
Hendrych, R., Cipra, T.: On conditional covariance modelling: an approach using state space models. Comput. Stat. Data Anal. **100**, 304–317 (2016)
Hendrych, R., Cipra, T.: Systemic risk in financial risk regulation. Czech J. Econ. Financ. **67**, 15–38 (2017)
Hendrych, R., Cipra, T.: Self-weighted recursive estimation of GARCH models. Commun. Stat. Simul. Comput. **47**, 315–328 (2018)
Hull, J.: Options, Futures, and Other Derivative Securities. Prentice Hall, Englewood Cliffs (1993)
Hurst, H.: Long term storage capacity of reservoirs. Trans. Am. Soc. Civil Eng. **116**, 770–799 (1951)
Hyndman, R.J., Koehler, A.B., Ord, J.K., Snyder, R.D.: Forecasting with Exponential Smoothing. Springer, Berlin (2008)
Jacobs, P., Lewis, P.: Stationary discrete autoregressive-moving average time series generated by mixtures. J. Time Ser. Anal. **4**, 19–36 (1983)
Jeantheau, T.: Strong consistency of estimators for multivariate ARCH models. Econ. Theory. **14**, 70–86 (1998)
Johansen, S.: Estimation and hypothesis testing of cointegration vectors in Gaussian vector autoregressive models. Econometrica. **59**, 1551–1580 (1991)
Johansen, S.: Likelihood-Based Inference in Cointegrated Vector Autoregressive Models. Oxford University Press, Oxford (1995)
Karatzas, I., Shreve, S.E.: Brownian Motion and Stochastic Calculus. Springer, New York (1988)
Kedem, B.: Binary Time Series. Marcel Dekker, New York (1980)
Kedem, B., Fokianos, K.: Regression Models for Time Series Analysis. Wiley, Hoboken (2002)
Kendall, M.: Time-Series. Griffin, London (1976)
Kessler, M.: Estimation of an ergodic diffusion from discrete observations. Scand. J. Stat. **24**, 1–19 (1997)
Koopmans, L.H.: The Spectral Analysis of Time Series. Academic, San Diego (1995)
Kroner, K.F., Ng, V.K.: Modelling asymmetric co-movements of asset returns. Rev. Financ. Stud. **11**, 817–844 (1998)
Kwaitkovski, D., Phillips, P.C.B., Schmidt, P., Shin, Y.: Testing the null hypothesis of stationarity against the alternative of a unit root. J. Econ. **54**, 159–178 (1992)
Kwok, Y.-K.: Mathematical Models of Financial Derivatives. Springer, New York (1998)
Lanne, M., Saikkonen, P.: A multivariate generalized orthogonal factor GARCH model. J. Bus. Econ. Stat. **25**, 61–75 (2007)
Lim, K.G.: Financial Valuation and Econometrics. World Scientific, Singapore (2011)
Lin, W.L.: Alternative estimators for factor GARCH models – a Monte Carlo comparison. J. Appl. Econ. **7**, 259–279 (1992)
Ljung, L.: System Identification: Theory for the User. Prentice Hall PTR, Upper Saddle River (1999)
Ljung, L., Söderström, T.: Theory and Practice of Recursive Identification. MIT Press, Cambridge (1983)
Lo, A.W.: Maximum likelihood estimation of generalized Ito's processes with discretely sampled data. Econ. Theory. **4**, 231–247 (1988)
Lütkepohl, H.: New Introduction to Multiple Time Series Analysis. Springer, Berlin (2005)

MacDonald, I., Zucchini, W.: Hidden Markov and Other Models for Discrete-Valued Time Series. Chapman and Hall, London (1997)

MacKinnon, J.G.: Numerical distribution functions for unit root and cointegration tests. J. Appl. Econ. **11**, 601–618 (1996)

MacKinnon, J.G., Haugh, A.A., Michelis, L.: Numerical distribution functions of likelihood ratio tests for cointegration. J. Appl. Econ. **14**, 563–577 (1999)

Makridakis, S.: Accuracy measures: theoretical and practical concerns. Int. J. Forecast. **9**, 527–529 (1993)

Malliaris, A.G., Brock, W.A.: Stochastic Methods in Economics and Finance. North-Holland, Amsterdam (1982)

McKenzie, E.: Some ARMA models for dependent sequences of Poisson counts. Adv. Appl. Probab. **20**, 822–835 (1988)

McNeil, A.J., Frey, R., Embrechts, P.: Quantitative Risk Management. Princeton University Press, Princeton (2005)

Merigó, J.M., Yager, R.R.: Generalized moving averages, distance measures and OWA operators. Int. J. Uncertainty Fuzziness Knowledge-Based Syst. **21**, 533–559 (2013)

Mills, T.C.: The Econometric Modelling of Financial Time Series. Cambridge University Press, Cambridge (1993)

Montgomery, D.C., Johnson, L.A.: Forecasting and Time Series Analysis. McGraw-Hill, New York (1976)

Musiela, M., Rutkowski, M.: Martingale Methods in Financial Modelling. Springer, New York (2004)

Neftci, S.N.: Mathematics od Financial Derivatives. Academic, New York (2000)

Nelson, D.B.: Conditional heteroskedasticity in asset returns: a new approach. Econometrica. **59**, 347–370 (1991)

Newey, W.K., West, K.D.: A simple, positive semi-definite, heteroskedasticity and auto-correlation consistent covariance matrix. Econometrica. **55**, 703–708 (1987)

Nicholls, D.F., Quinn, B.G.: Random Coefficient Autoregressive Models: An Introduction, Lecture Notes in Statistics 11. Springer, New York (1982)

Parkinson, M.: The extreme value method for estimating the variance of the rate of return. J. Bus. **53**, 61–65 (1980)

Phillips, P.C.B., Perron, P.: Testing for a unit root in time series regression. Biometrika. **75**, 335–346 (1988)

Poon, S.-H.: A Practical Guide to Forecasting Financial Market Volatility. Wiley, Chichester (2005)

Priestley, M.B.: Non-Linear and Non-Stationary Time Series Analysis. Academic, London (1988)

Priestley, M.B.: Spectral Analysis and Time Series (Volume 1 and 2). Academic, London (2001)

Rachev, S.T., Mittnik, S., Fabozzi, F.J., Focardi, S.M., Jašić, T.: Financial Econometrics: from Basics to Advanced Modeling Techniques. Wiley, Chichester (2007)

Ramsey, J.B.: Tests for specification errors in classical linear least-squares regression analysis. J. R. Stat. Soc. B. **31**, 350–371 (1969)

Ripley, B.D.: Statistical aspects of neural network. In: Barndorff-Nielsen, O.B., et al. (eds.) Networks and Chaos-Statistical and Probability Aspects. Chapman and Hall, London (1993)

Risk Metrics-Technical Document. J. P. Morgan/Reuters, New York. www.riskmet-rics.com (1996)

Ruppert, D.: Statistics and Finance. Springer, New York (2004)

Schlicht, E.: A seasonal adjustment principle and a seasonal adjustment method derived from this principle. J. Am. Statist. Assoc. **76**, 374–378 (1982)

Sentana, E.: Quadratic ARCH models. Rev. Econ. Stud. **62**, 639–661 (1995)

Siegel, A.F.: Testing for periodicity in a time series. J. Am. Stat. Assoc. **75**, 345–348 (1980)

Silvennoinen, A., Teräsvirta, T.: Multivariate GARCH models. In: Andersen, T.G., Davis, R.A., Kreiss, J.-P., Mikosch, T. (eds.) Handbook of Financial Time Series. Springer, New York (2009)

Sims, C.A.: Money, income, and causality. Am. Econ. Rev. **62**, 540–552 (1972)

Söderström, T., Stoica, P.: System Identification. Prentice Hall, New York (1989)
Steele, J.M.: Stochastic Calculus and Financial Applications. Springer, New York (2001)
Taylor, S.: Modelling Financial Time Series. Wiley, New York (1986)
Taylor, S.J.: Modelling stochastic volatility. Math. Financ. **4**, 183–204 (1994)
Theil, H.: Applied Economic Forecasting. North-Holland, Amsterdam (1966)
Timmermann, A.: Forecast combinations. In: Handbook of Economic Forecasting, vol. 1, pp. 135–196. Elsevier (2006)
Tjøstheim, D.: Some doubly stochastic time series models. J. Time Ser. Anal. **7**, 51–72 (1986)
Tong, H.: Threshold Models in Non-Linear Time Series Analysis. Springer, New York (1983)
Tong, H.: Non-Linear Time Series: A Dynamical Systems Approach. Oxford University Press, Oxford (1990)
Tsay, R.S.: Analysis of Financial Time Series. Wiley, New York (2002)
Tse, Y.K., Tsui, A.K.C.: A multivariate GARCH model with time-varying correlations. J. Bus. Econ. Stat. **20**, 351–362 (2002)
van der Weide, R.: GO-GARCH: a multivariate generalized orthogonal GARCH model. J. Appl. Econ. **17**, 549–564 (2002)
Vasicek, O.: An equilibrium characterization of the term structure. J. Financ. Econ. **5**, 177–188 (1977)
Vrontos, I.D., Dellaportas, P., Politis, D.N.: A full-factor multivariate GARCH model. Econ. J. **6**, 311–333 (2003)
Wang, S.S.: A class of distortion operators for pricing financial and insurance risks. J. Risk Insur. **67**, 15–36 (2000)
Wang, P.: Financial Econometrics: Methods and Models. Routledge, London (2003)
Wecker, W.E.: Asymmetric time series. J. Am. Stat. Assoc. **76**, 16–21 (1981)
Wei, W.W.S.: Time Series Analysis: Univariate and Multivariate Methods. Addison-Wesley, Boston (1994)
Weiss, C.H.: An Introduction to Discrete-Valued Time Series. Wiley, Chichester (2018)
Wilmott, P.: Quantitative Finance. Wiley, Chichester (2000)
Wu, C.F.J.: On the convergence of the EM algorithm. Ann. Stat. **11**, 95–103 (1983)
Wu, W.B.: Nonlinear system theory: another look at dependence. Proc. Natl. Acad. Sci. **102**(40), 14150–14156 (2005)
Yamai, Y., Yoshiba, T.: Value-at-risk versus expected shortfall: a practical perspective. J. Bank. Financ. **29**, 997–1015 (2005)
Zakoian, J.M.: Threshold heteroskedastic models. J. Econ. Dyn. Control. **18**, 931–934 (1994)

Index

T. Cipra, *Time Series in Economics and Finance*,
https://doi.org/10.1007/978-3-030-46347-2

D

E

L

M

Q

R

The manufacturer's authorised representative in the EU is Springer Nature Customer Service Centre GmbH, Europaplatz 3, 69115 Heidelberg, Germany. If you have any concerns regarding our products, please contact ProductSafety@springernature.com

Printed and bound by CPI Group (UK) Ltd, Croydon, CR0 4YY
30/09/2026
02225697-0013

Time Series in Economics and Finance